2nd Edition
Revised for the 17th Edition IEE Wiring Regulations

Electrical Installations

NVQ and Technical Certificate Book 1

David Allan
John Blaus
on behalf of JTL

www.jtltraining.com

www.heinemann.co.uk
✓ Free online support
✓ Useful weblinks
✓ 24 hour online ordering

01865 888118

Heinemann is an imprint of Pearson Education Limited, a company incorporated in England and Wales, having its registered office at Edinburgh Gate, Harlow, Essex, CM20 2JE. Registered company number: 872828

www.heinemann.co.uk

Heinemann is a registered trademark of Pearson Education Limited

First published 2008

12 11 10 09 08
10 9 8 7 6 5 4 3 2 1

British Library Cataloguing in Publication Data
A catalogue record for this book is available from the British Library

ISBN 978 0 435 46704 3

Typeset by HL Studios
Layout by Ken Vail Graphic Design
Original illustrations © Pearson Education, 2008
Cover design by G D Associates
Cover photo/illustration © Pearson Education/Gareth Boden
Printed in the UK by Scotprint

Acknowledgements
Every effort has been made to contact copyright holders of material reproduced in this book. Any omissions will be rectified in subsequent printings if notice is given to the publishers.

Websites
The websites used in this book were correct and up-to-date at the time of publication. It is essential for tutors to preview each website before using it in class so as to ensure that the URL is still accurate, relevant and appropriate. We suggest that tutors bookmark useful websites and consider enabling students to access them through the school/college intranet.

Contents

Acknowledgements

JTL would like to express its appreciation to all those members of staff who contributed to the development of this book, ensuring that the professional standards expected were delivered and generally overseeing the high quality of the final product. Without their commitment and support – as much to each other as to the project, especially during some of the more fraught days – this project would not have been completed successfully. Particular thanks go to Dave Allan, Brian Tucker, Mike Crossley, Nigel Buckhurst, Phil Cunningham and John Langley. Our thanks also to Paul and Rita Hurt for their patience and assistance during the photoshoot and to John Blaus for his help and advice throughout the project.

July 2005

The authors and publishers would like to thank the following for assistance in the production of this book and permission to reproduce copyright material:

Danfoss Randall
The IEE
C.K. Tools (Carl Hammerling International Ltd.)
Paul and Rita Hurt
JTL
Gareth Boden
Michael Hawkins
Ginny Stroud-Lewis
Unite

Picture acknowledgements
The authors and publishers would like to thank the following for permission to reproduce photographs:

All pictures **Pearson Education Ltd/Gareth Boden** apart from the following:

Alamy Images pages 1, 93, 98 (top), 144; **Arco** pages 57 (left and centre), 58 (bottom), 59, 60, 61 (bottom left, right and centre); **Art Directors and Trip** pages 57 (right), 58 (top); **Construction Photography** pages 55, 97, 98 (bottom); **Corbis/Monatiuk** page 130; **Digital Vision** page 12; **Getty Images** page 89, 260, 319; **Getty Images/Stone** page 13; **Pearson Education/ Jules Selmes** page 72; **Pearson Education Ltd/Ginny Stroud-Lewis** page 61 (top three); **JTL/Dave Allan** 92 (bottom); **Photodisc** page 36; **Science Photo Library** page 147; **Shutterstock/Christophe Tesli** page 36 (bottom); **Shutterstock/Elena Elisseeva** page 134

Introduction

What is this book?

This is the first of two books designed with you in mind. They have a dual purpose:

- To lead you through the City & Guilds 2330 Technical Certificate In Electrotechnical Technology (Buildings & Structures)
- To provide a future reference book that you will find useful to dip into, long after you have gained your qualification.

As we said above, the book has been specially produced to help you achieve the City & Guilds 2330 and can therefore contribute towards the underpinning knowledge requirement of the relevant NVQ qualification. The scheme is available throughout England and Wales and is designed to set a quality standard for learning and training in the electrotechnical sector, whether you are a new entrant to the industry or an experienced worker wishing to update your qualifications.

The books' contents underpin the various topics on which you will be examined as part of the City & Guilds 2330 Technical Certificate. Each chapter concludes with knowledge tests that will allow you to measure your knowledge and understanding of the various topics.

Qualifications

Generally speaking there are two qualifications that should be considered at this point:

1. *National Vocational Qualifications (NVQs)*: The central feature of any NVQ is the National Occupational Standards (NOS) on which they are based. NOS are statements of performance that describe what competent people in a particular occupation are expected to be able to do … think of it as being a bit like a job description. They cover all the main aspects of an occupation, including current best practice, the ability to adapt to future requirements and the knowledge and understanding that underpins competent performance. Therefore, an NVQ is a qualification that is awarded when an individual can demonstrate that they are actually competent to do the job within the workplace. Consequently, the most valid form of 'evidence of ability' is by someone watching you do the job at your workplace. However, you will still be expected to demonstrate

your knowledge and understanding of these tasks and this is normally measured using external written examinations (such as the Technical Certificate) and oral questioning by an assessor. There are actually five levels of qualification within the NVQ system, with level five being the highest. For the electrical contracting industry, it is currently taken that the NVQ Level 3 is the minimum standard that must be attained for the award of electrician status via the Joint Industry Board (JIB).

2. *Technical Certificate*: For this industry, this is the primary means of providing the underpinning job knowledge and understanding that supports the NVQ competencies. Please note that, for apprentices, success is required in both the Level 2 and Level 3 component of the C&G 2330.

As a rough guide, within the electrotechnical industry, meeting the requirements of points 1 and 2 above allow an operative to be able to meet the current industry level for recognition with the JIB as a qualified electrician. However, specific requirements for recognition should always be checked with the JIB.

Note:
This publication is designed to complement studies towards the Level 2 and 3 certification for Electrical Installations. It should be noted that its use and interpretation is for the purpose of training and should not be regarded as being relevant to an actual installation. Specific reference should always be made to the British Standards or manufacturer's data when designing electrical installations.

How this book can help you

There are other key features of this book which are designed to help you make progress and reinforce the learning that has taken place, such features are:

- **Photographs:** easy to follow sequences of key operations.
- **Illustrations:** clear drawings, many in colour, showing essential information about complex components and procedures.
- **Margin notes:** short helpful hints to aid you to good practice.
- **Tables, bullet points and flowcharts:** easy to follow features giving information at a glance.
- **On the job scenarios:** typical things that happen on the job, what would you do?
- **Did you know?:** useful information about things you always wondered about.
- **End of section knowledge checks:** test yourself to see if you have absorbed all the information, are you ready for the real test?
- **Glossary:** definitions and explanations of strange words and phrases. Glossary and definition box terms are identified in bold the first time they appear in the text.

Why use this book?

Because it is structured to give you all the basic information required to help you gain the current industry Technical Certificate qualification and set you on course to an exciting long term career inside a challenging and varied industry ... Well done for choosing such a good start!

Please note that there is also another book like this to provide the balance of information.

chapter 1

Industry and communication

Overview

When we leave full-time education (nearly) all of us hope to move into the world of work. It is important to understand the legislation – our legal rights and responsibilities – that will govern our working lives. In this chapter we will see what employers and the people who work for them do, where they do it, and what rules and systems they have to follow.

We'll discover how important it is to deal successfully with other people, whether in conversation or in writing, and go on to look at the different types of information we are likely to come across in the course of our work.

The chapter also outlines current thinking regarding career and employment development within the electrotechnical industry. It will cover:

- **The world of work**
- **The construction industry**
- **The electrotechnical industry**
- **Project roles and responsibilities**
- **Sources of technical information**
- **Site documentation**
- **Storing and retrieving information**
- **Working with documentation, drawings and specifications**
- **Relationships and teamworking**

The world of work

Employment legislation

Within England and Wales, the law regarding employment both protects and imposes obligations on employees during their employment and after it ends. The law sets certain minimum rights and an employer cannot give you less than what the law stipulates. The principal rights and obligations imposed on employers and employees arise from three sources:

- common law, which governs any contract of employment between employer and employee and includes the body of law created by historical practice and decisions
- UK legislation
- European legislation and judgements from the European Court of Justice (ECJ).

UK employment law has been heavily influenced by European law, particularly in the areas of equal pay and equal treatment. Consequently, many of our statutory minimum rights began their life in European legislation.

The Employment Rights Act 1996

Subject to certain qualifications, employees have a number of statutory minimum rights (e.g. the right to a minimum wage) and the main vehicle for employment legislation is the Employment Rights Act 1996 – Chapter 18. If you did not agree certain matters at the time of commencing employment, your legal rights will apply automatically.

The Employment Rights Act 1996 deals with many matters such as: right to statement of employment; right to pay statement; minimum pay; minimum holidays; maximum working hours; and right to maternity/paternity leave.

The Sex Discrimination Act 1975

The Sex Discrimination Act 1975 makes discrimination unlawful on the grounds of sex and marital status and, to a certain degree, gender reassignment. The Act originated out of the Equal Treatment Directive, which made provisions for equality between men and women in terms of access to employment, vocational training, promotion and other terms and conditions of work.

The Equal Opportunities Commission (EOC) has since published a Code of Practice. While this is not a legally binding document, it does gives guidance on best practice in the promotion of equality of opportunity in employment and failure to follow it may be taken into account by the courts.

Disability Discrimination Act 1995

The Disability Discrimination Act tackles the discrimination faced by many disabled people. This Act gives disabled people rights in the areas of employment, access to goods, facilities and services and buying or renting land or property.

The employment rights and first rights of access came into force in December 1996; further rights of access came into force on 1 October 1999; and the final rights of access came into force in October 2004.

In addition, this Act allows the Government to set minimum standards so that disabled people can use public transport easily.

The Human Rights Act 1998

The Human Rights Act 1998 covers many different types of discrimination – including some which are not covered by other discrimination laws. However, it can be used only when one of the other 'articles' (the specific principles) of the Act applies, such as the right to 'respect for private and family life'.

Also, rights under the Act can only be used against a public authority (for example, the police, a local council or Jobcentre) and not a private company. However, court decisions on discrimination will generally have to take into account what the Human Rights Act says. The main articles within this Act are:

- right to life
- prohibition of torture
- prohibition of slavery and forced labour
- right to liberty and security
- right to a fair trial
- no punishment without law
- right to respect for private and family life
- freedom of thought, conscience and religion
- freedom of expression
- freedom of assembly and association
- right to marry
- prohibition of discrimination
- restrictions on political activity of aliens
- prohibition of abuse of rights
- limitation on use of restrictions on rights.

The Race Relations Act 1976 and Amendment Act 2000

When originally passed, the Race Relations Act 1976 made it unlawful to discriminate on racial grounds in relation to employment, training and education, the provision of goods, facilities and services and certain other specified activities. The 1976 Act applied to race discrimination by public authorities in these areas but not all functions of public authorities were covered.

The 1976 Act also made employers vicariously (explicitly) liable for acts of race discrimination committed by their employees in the course of their employment, subject to a defence that the employer took all reasonable steps to prevent the employee discriminating. However, police officers are office-holders, not employees. Chief officers of police were, therefore, not vicariously liable under the 1976 Act for acts of race discrimination by police officers. The Commission for Racial Equality (CRE) therefore proposed that the Act should be extended to all public services and that vicarious liability should be extended to the police. The main purposes of the 2000 Act were to:

- extend further the 1976 Act in relation to public authorities, thus outlawing race discrimination in functions not previously covered
- place a duty on specified public authorities to work towards the elimination of unlawful discrimination and promote equality of opportunity and good relations between persons of different racial groups
- make chief officers of police vicariously liable for acts of race discrimination by police officers
- amend the exemption under the 1976 Act for acts done for the purposes of safeguarding national security.

The Employment Act 2002

The main areas covered by the Act are paternity and adoption leave and pay, maternity leave and pay, flexible working, employment tribunal reform and resolving disputes between employers and employees. In many cases this Act amends current legislation. More specifically, it amends the Employment Rights Act 1996 to make provision for statutory rights to paternity and adoption leave and amends the law relating to statutory maternity leave.

Types of discrimination and victimisation

Currently, there are two main forms of discrimination – direct and indirect.

Direct discrimination

Did you know?

An employer cannot argue that it was not their intention to discriminate; the law only considers the end affect

Direct discrimination occurs when someone is treated less favourably because of their sex, race or disability. With regard to employment, this could happen if an employer treats a job applicant or existing employee less favourably on the grounds of their sex, race or disability.

The applicable test in law is to apply a comparison with someone of the opposite sex or alternative racial group. The question is then whether or not the applicant would have been treated differently and more favourably had it not been for their sex etc. It is expected that this definition will continue under foreseeable legislation.

An example of direct discrimination could be a woman of superior qualifications and experience being denied promotion in favour of a less experienced and less qualified man.

Indirect discrimination

Indirect discrimination occurs where the effect of certain requirements, conditions or practices imposed by an employer have an adverse impact disproportionately on one group or other. Courts tend to consider three factors:

- the number of people from a racial group or of one sex that can meet the job criteria is considerably smaller than the rest of the population
- the criteria cannot actually be justified by the employer as being a real requirement of the job. So an applicant who could not meet the criteria could still do the job as well as anyone else
- because the person cannot comply with these criteria they have actually suffered in some way (this may seem obvious, but a person cannot complain unless they have lost out in some way).

With cases of indirect discrimination, employers may argue that there may be discrimination, but that it is actually required for the job. As an example, one individual claimed indirect discrimination on religious grounds against his employer as he was requested to shave off his beard. The court agreed that discrimination had been applied, but as the employer was a factory involved in food preparation, the particular case was rejected on the grounds of hygiene.

Definition

Disability – a person has a disability if they have a physical or mental impairment that has a substantial and long-term affect on their ability to carry out normal everyday activities

Disability discrimination

Disability discrimination relies on the same basic principles, but the complainant must be treated less favourably due to their disability.

Victimisation

This is where an employee is singled out for using their workplace complaints procedures or exercising their legal rights – for example, making a complaint of discrimination or giving evidence and information on behalf of another employee who has brought proceedings for discrimination.

Positive discrimination and positive action

Positive discrimination occurs when someone is selected to do a job purely on the basis of their gender or race, not on their ability to do the job. This is illegal under the Sex Discrimination Act and the Race Relations Act and is generally unlawful other than for **Genuine Occupational Requirements**.

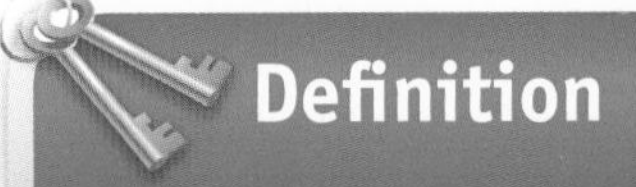

Genuine Occupational Requirements – where an employer can demonstrate that there is a genuine identified need for someone of specific race, gender etc., to the exclusion of other races, genders etc.

Positive action is activity to increase the numbers of men, women or minority ethnic groups in a workforce where they have been shown through monitoring to be under-represented. This may be in proportion to the total employed by the employer or in relation to the profile of the local population.

Examples of positive action might be carefully targeted advertising and courses to develop the careers of those from under-represented groups who are already employed by an organisation. Positive action is legal and is designed to help employers achieve a more balanced workforce. For example, in the UK women are typically under-represented at senior levels in organisations – in spite of being experienced and well qualified. Action to remedy this by providing encouragement and training would be legal.

Genuine Occupational Requirements can be controversial, but some typical examples of these could be:

- a film company is auditioning for *A Passage to India* and requires actors of Asian ethnic origin
- a modelling agency requires a woman to model female clothes
- a theatre company requires black actors for street scenes in a multiracial neighbourhood.

The Data Protection Act 1998

The Information Commissioner enforces and oversees the Data Protection Act 1998 and the Freedom of Information Act 2000. The Commissioner is a UK independent supervisory authority reporting directly to the UK Parliament and has an international role as well as a national one.

There are eight principles put in place by the Data Protection Act 1998 to make sure that information is handled properly. They say that data must be:

- fairly and lawfully processed
- processed for limited purposes

- adequate, relevant and not excessive
- accurate
- not kept for longer than is necessary
- processed in line with your rights
- secure
- not transferred to other countries without adequate protection. By law 'data controllers' have to keep to these principles.

Quality systems

Although the legislation we've just looked at places a legal requirement on both employer and employee, none of it reflects the quality of an employer. There are two commonly used standards that help to signify the quality of a company: Investors In People and ISO 9001.

Investors in People

Investors in People (IiP) is a national quality standard which sets a level of good practice for improving the performance of an organisation through its people. Developed in 1990, the standard sets out a level of good practice for training and development of people to achieve business goals.

The IiP standard provides a national framework for improving business performance and competitiveness through a planned approach to setting and communicating business objectives and then developing people to meet these objectives. The aim is to create an environment where what people can do and are motivated to do matches what the organisation needs them to do. Because the award of IiP status is time restricted, the process should bring about a culture of continuous improvement.

The Investors in People standard is based on four key principles:

- commitment to investing in people to achieve business goals
- planning how skills, individuals and teams are to be developed to achieve these goals
- taking action to develop and use necessary skills in a well defined and continuous programme directly tied to business objectives
- evaluating the outcomes of training and development for an individual's progress towards goals, the value achieved and future needs.

ISO 9001

Customers are becoming better informed and their expectations are growing. For any business, the only way to keep up is to offer a commitment to quality. In fact, any organisation, whatever their size or industry sector, can introduce a quality management system such as ISO 9000.

A quality management system in accordance with ISO 9001:2000 will provide an organisation with a set of processes that ensure a common-sense approach to the management of the organisation.

The system should ensure consistency and improvement of working practices, which in turn should provide products and services that meet customers' requirements. ISO 9000 is the most commonly used international standard that provides a framework for an effective quality management system.

The benefits of implementing a quality management system include:

- policies and objectives set by 'top management'
- understanding customers' requirements with a view to achieving customer satisfaction
- improved internal and external communications
- greater understanding of the organisation's processes
- understanding how statutory and regulatory requirements impact on the organisation and your customers
- clear responsibilities and authorities agreed for all staff
- improved use of time and resources
- reduced wastage
- greater consistency of products and services
- improved staff morale and motivation.

Registration of a quality management system is to the requirements set out in ISO 9001. All the other standards in the family then exist to help an organisation implement an effective quality management system and ultimately help gain approval.

Revised in December 2000, the existing three standards within ISO 9000 (ISO 9001, ISO 9002 and ISO 9003) were merged into a single standard, ISO 9001. Consequently, the ISO 9000 suite of standards was restructured to comprise four core standards:

- ISO 9000 Concepts and Terminology
- ISO 9001 Requirements for Quality Assurance
- ISO 9004 Guidelines for Quality Management of Organisations
- ISO 19011 Guidelines for Auditing Quality Management Systems (formerly ISO 10011).

ISO 9001:2000, the requirement standard, now includes the following main sections:

- Quality Management System
- Management Responsibility
- Resource Management
- Product Realisation
- Measurement Analysis and Improvement.

Even with the best legislation in place, circumstances can change. Some of the things that can affect an individual or an organisation are as follows.

Remember

If poorly handled, extra demand can be very stressful and can cause damage to an organisation's reputation.

A decrease in demand can prove equally stressful as organisations try to ensure their existence

Did you know?

Change in legislation can bring about work as installations are brought up to new required levels and standards

Changing demand

Demand can normally do one of two things: increase or decrease. An increase in demand may bring changes to a workplace as a company tries to handle this extra demand. If the extra demand is caused by winning more contracts, some companies handle this easily with recruitment of extra staff and training of both new and existing staff. Extra demand can also sometimes allow individuals within an organisation to progress their career and perhaps move from craft to supervisory duties.

A drop in workload, irrespective of the cause, invariably means that companies look to save money and, sadly, this can result in cutting back services and staffing levels.

The reasons for changes in demand are many and varied. For example, an increase in the amount of electrical contractors in an area will almost certainly result in an increased level of competition as they all try to stay in business, and it is not unusual for employers to quote very low prices, with almost no profit margins, to try to make sure that they get the work.

As competition increases, so does the marketing of that competition. Some organisations carry out little or no marketing, but exist happily on a reputation built upon their ability to regularly carry out high quality work. Other companies create work by 'aggressive' marketing with sales staff, leaflets, letters and advertising. Market forces will often play a role, as some clients are happy to pay for a high quality product and service, where other clients seem happy 'as the long as the job works'.

Technological and legal changes can also affect demand. There are now sections of industry that make their living by installing the specialist cabling needed for computers and computerised systems.

The effects of change

Changes in demand can bring about great change for individuals. Most apprentice electricians will undertake a structured apprenticeship, regarded as broad-based training, that gives a range of experience in most aspects of electrical installation work and consequently most will become qualified to the national standards. However, depending upon the employing company, everyone's experiences will be different.

If you move between companies, you'll find that you may need additional training in certain aspects of the job. Or you may find, even if you stay with the same company, that the job changes and you need to learn new skills. Some people see any form of additional training as being a burden they should not have to bear. Others, perhaps more rightly, see this as being the acquisition of new skills that ensure their usefulness to the employer, as well as giving them new skills that can help them remain a viable force in the workplace for many years.

These new skills also give individuals the opportunity of transferring their skills to other employers. As an example, if you have been trained to cut trunking correctly, then the amount of time it should take to teach you to cut conduit should be small as you already know how to use a hacksaw and measure. In other words, experience of measuring and cutting are the **transferable skills** you bring to the task of cutting conduit.

On the job: Leisure opportunities

Sanjay runs a successful small electrical maintenance business. He employs four permanent electricians and Matt, an apprentice. His clients are mainly other small businesses but he is delighted when his company wins the maintenance contract for a new private leisure centre. However, neither he nor his staff have ever worked on saunas or gym equipment before, and he realises he needs to train up one of his staff. Sanjay is keen that Matt should be trained, but Matt says he doesn't want to do it. Matt's apprenticeship is almost up and Sanjay has to decide whether to give Matt a job at the end of it.

1. Why do you think Matt was not keen to do the extra training? What would your reaction be if you were asked to train for some specialist work?
2. How might Sanjay persuade Matt to take on the training?
3. If you were Sanjay, what would you do now?

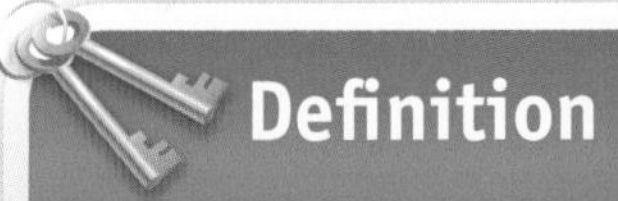

Definition

Transferable skills – skills that you gain which can be applied to different working areas and thus make you an attractive proposition to employers

Maintenance work can often create a more technical job content. For those that embrace this change and undergo additional training, it is likely that an increase in responsibility and change of working activities may result in an increase in salary or status within the company.

Maintenance work can also bring about a need for a flexible approach to working hours. After all, no one can predict when the sauna in a leisure centre is going to stop working, but we do know that the client will want it repaired as soon as possible.

Flexible working hours and increased demand mean that an organisation must be communicating and operating well together as a cohesive unit. If one part of the team isn't performing, then the whole company could suffer.

Hopefully, the example above demonstrates that we should expect to encounter changes in our working situation and that some of these may result in a change to our original career. It should also show that there will always be a need for continued retraining and updating of individual skills. Each revised edition of the IEE Wiring Regulations BS 7671 is another typical example that will create a retraining need.

Changing career

What of the future?

Some people are happy to remain in their original job role until they retire. Others have ambitions to become engineers, supervisors or even manage their own company. If you do have ambitions, then it is definitely an advantage if you can recognise them as early as possible, so that you can plan any relevant education as soon as possible. It is always easier to learn while you are still in education rather than having to relearn a subject in later years.

Remember

A happy worker is a good worker

Career development

We spend a large part of our lives in work. It therefore makes sense to be as happy at work as possible.

For many people this will mean pursuing ambitions and developing their career. For an electrician, there are a variety of options available. The list below is only meant as a guide to typical arrangements; it isn't intended to be definitive or to be followed in any particular order.

Occupation	Qualifications needed
JIB recognised Electrician	NVQ Level 3 + Technical Certificate
JIB recognised Approved Electrician	As above + C & G 2391 + two years' experience
Electrical Supervisor	Approved Electrician + additional experience
Contract/Site Engineer	Supervised + BTEC HNC in Building Services or Foundation degree
Contract Manager	Contract Engineer + BTEC HNC in Building Services or Foundation degree
Consultant	Contract Manager + Degree + Experience

Table 1.1 Industry progression route

Foundation degrees

Traditional degree qualifications are often seen as mainly 'theory' courses. Upon completion the successful students are often lacking the ability to apply the theoretical knowledge they have gained within real workplace situations.

Did you know?

Once you have completed the foundation degree, you can progress to a full honours degree with just a further 15 months of study

Foundation degrees are new work-related higher education qualifications, designed to equip young people with the higher-level skills that employers need. The flexibility of this new qualification opens up opportunities for students already in the workplace who thought higher education wasn't for them. Higher education institutions have developed this new qualification directly with employers, to ensure the skills they need are met. It is awarded by both colleges and universities.

The course can be studied over two years or on a pro-rata basis part-time. There are more than 70 foundation degrees on offer throughout the country, and courses include aircraft engineering, classroom assistance, construction, e-commerce, hospitality, multimedia design, police studies and textiles.

The construction industry

The construction industry is one of the biggest industries in the UK, employing well over one million people. The electrical contracting industry is just one sector within the construction industry. Organisations in the construction industry range from 'sole traders' (for example, a jobbing builder), to large multinational companies employing thousands of workers.

The work done by these companies is very varied, but we can broadly think of it under three headings:

- Building and structural engineering
- Civil engineering
- Maintenance.

Building and structural engineering covers the construction and installation of services for buildings such as factories, offices, shops, hospitals, schools, leisure centres and, of course, houses.

Civil engineering involves the construction and installation of services for large structures such as bridges, roads, motorways, docks, harbours and mines.

Maintenance covers the repair, refurbishment and restoration of existing buildings and structures.

Larger companies will be able to undertake work in all these areas, but smaller companies may specialise in one area.

Nearly every project, large or small, will involve a variety of different trades, such as bricklaying, plastering, plumbing and joining, as well as electrical work. The ability to work well with others and establish good professional relationships with colleagues and people in other trades is very important for the successful completion of a project.

Did you know?

Only about half the people working in the construction industry are skilled craft operatives (such as electricians), the other half are management, technical and clerical staff

The electrotechnical industry

Most people are unaware of the vast range of activities and occupations that make up the electrical industry. Their understanding is usually limited to a person who installs lights and sockets in their house. In reality, during his or her career an electrician could be involved with the installation, maintenance and repair of electrical services (both inside and outside) associated with buildings and structures such as houses, hospitals, schools, factories, car parks, leisure centres and shops. There are also specialist areas that call for additional knowledge and training.

Electricians may be employed within the electrical contracting industry, but they may also be employed directly by organisations needing their skills.

The rail industry is one major sector that employs electricians

These include:

- factories and manufacturing plant
- processing plant (e.g. chemical works, refineries)
- local councils
- commercial buildings
- leisure centres
- shopping complexes
- universities
- panel builders
- railways
- hospitals
- the armed forces.

Building services

Some of the electrical services that can be found in and around our buildings and structures are:

- lighting and power installations
- alarm, security and emergency installations
- building management and control installations
- communication, data and computer installations
- specialist areas.

Lighting and power installations

This covers all the varied types of lighting we see in houses and offices, from normal overhead lights through to the spotlights in car parks and floodlights in leisure centres and stadiums. Power installations cover everything from the installation of sockets in a house through to the installation of supplies for machines in factories.

Alarm, security and emergency installations

Many buildings have access control and security systems to prevent loss of, or damage to, valuable assets and information. If a fire or other emergency occurs, there must be a way of quickly alerting people to the danger. Most buildings contain some sort of fire alarm system, from a basic smoke detector in a house to complex systems in offices and shops (these are often linked directly to the emergency services).

Did you know?

Emergency lighting systems, capable of operating when the mains power fails, can help to guide people safely out of a building

Building management and control installations

There is an increasing use of complex electronic systems to control building services and installations, such as:

- 'intelligent' lighting (which switches on only when someone is present)
- air conditioning, heating and climate control systems
- energy management
- alarms and access control
- machine and motor control systems
- computer systems.

These developments place further demands on the skills and responsibilities of an electrician. More complex installations may require specialist contractors.

Communication, data and computer installations

Communications today, even within individual buildings, go well beyond simple speech and the telephone. The widespread use of computers enables people to communicate via email, to access data directly from other computers and the Internet, and to manage many aspects of an organisation from their desks. This interconnection of computers is called **networking**, and it uses a structured cable installation. There are several different types of cable in use, such as Cat 5, Cat 5e, Cat 6 and fibre optic. In many cases, voice, data, video and control signals for building services are carried on a single cable.

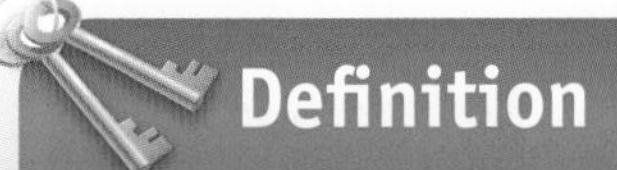

Networking – connecting computers together so that they all have access to the same common information

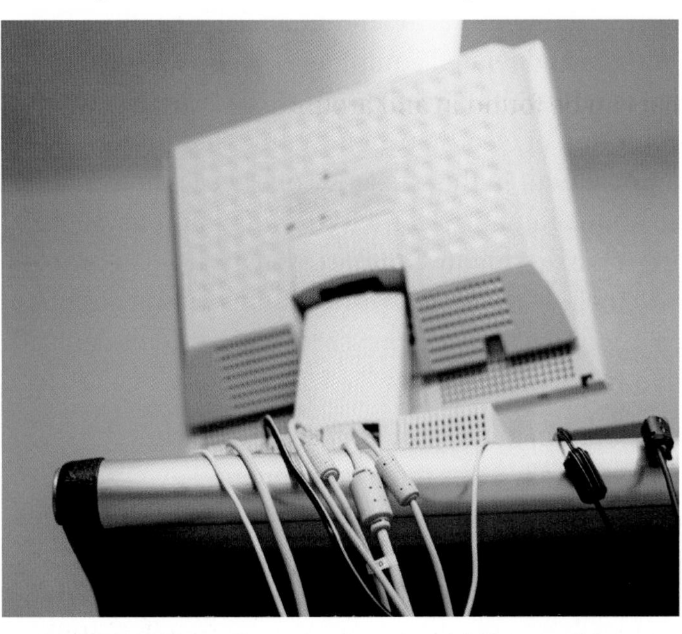

Networking computers can involve several different cable types

Specialist areas

The installation electrician could deal with all of the previous systems. However, the following areas within the electrotechnical sector are more likely to be undertaken by specialist contractors:

Cable jointing

This work involves connecting main power cables (high or low voltage, live or disconnected) to buildings, or within power stations and substations. It includes:

- making the joint
- placing insulating material around it
- testing the performance of the cables and insulation.

It must often be done outdoors. Sometimes non-specialist electricians can make small low voltage joints using cast resin kits.

Variable speed drive

Did you know?

Motor system efficiency drops quickly when motors operate below 40% of full-rated load

Highway electrical systems

Electricians undertake some public lighting work, but this category of work also includes more complex areas like street lighting and traffic control systems such as motorway signs and traffic lights. These tend to be catered for either by specialist sections within a local council or by electricians within specialist contractors who have received additional training.

Electrical machine drive installations

A growing number of countries are keen to encourage their industries to use electricity more efficiently. One way of doing this is to offer financial incentives for installing more energy-efficient motors and related equipment, such as **electronic variable speed drives** (also known as adjustable speed drives). This is an important measure, because with a high proportion of electricity being converted by motors into mechanical work, a small improvement in the efficiency of electric motors and motor-driven systems can lead to quite dramatic electricity savings. Motor systems can be simple (e.g. a motor driving a fan), or highly complex systems used in process plants such as refineries. These will include motors, control systems, power supplies and motor-driven machines such as pumps and compressors. Such systems can consume around 40 per cent of industrial drive energy.

Improving the overall design and operation of the process, rather than simply improving the drive motor efficiency, usually gives the biggest energy savings. Performance improvement therefore focuses first on reducing process inefficiencies, before selecting new drive components and control strategies to match the new reduced load. In practice, however, production machinery is not always replaced so systematically, and existing motors often have to drive loads that are much higher, lower or more varied than those for which they were first specified. Therefore, minimising energy losses in the existing system becomes the priority.

Pumps, compressors and fans etc. operate at optimum efficiency only within a narrow range of operation (flow and pressure), and so it is important that they are kept within that range. In some systems, load control can be helpful, but the biggest improvement is often obtained simply by matching the motor size to the actual load. Energy losses in drive trains (gearboxes, drive chains etc.) can also be reduced. In some cases, effective maintenance will improve the situation.

Organisations specialising in electrical machine drive installations consider all the above factors to produce the most energy efficient system.

Panel building

Control panels are used to control electricity distribution or control systems and equipment. Most panels are built in a factory, although some companies will build their products on site. There are three types of panels: switchgear, control panels and motor control circuit systems.

Instrumentation

This involves installing process control and measurement systems for a wide variety of industries, including water companies, chemical, pharmaceutical and process plant, and manufacturing sites. System components include:

- flow meters
- pumps
- pressure and temperature gauges
- valves and level switches.

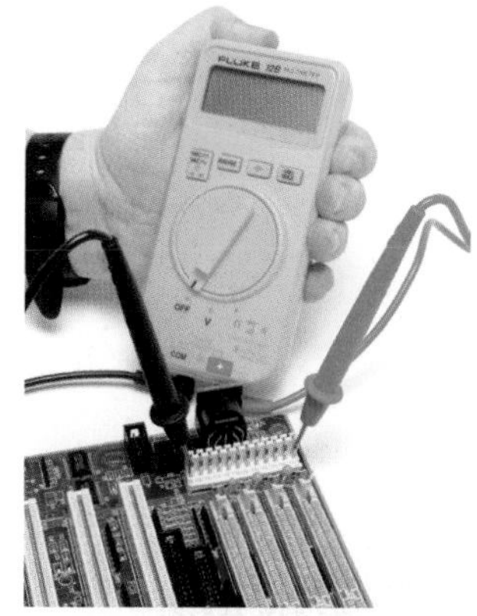

Electrical meter in use

Electrical maintenance

In the past, many installation electricians used their experience of installation to transfer to maintenance activities. Today, this job is usually seen as a specialist role, as it often has to cover a wide range of expertise related to the location, including electronics, safety issues and preventive maintenance, as well as electrical and mechanical repair and installation. Maintenance electricians are usually employed in:

- factories
- hospitals
- universities
- manufacturing plants.

They work on a wide range of systems and equipment, and do minor installation work when necessary.

The electrical contracting industry structure

Electrical contracting industry bodies

There are a variety of organisations that you should be aware of within the industry.

The Electrical Contractors' Association (ECA)

The ECA represents the interests of electrical installation companies in England, Wales and Northern Ireland and is the major association working within the electrical installation industry. It was founded in 1901 and has over 2,000 member companies, ranging in size from small traders with only a few employees to large multinational organisations. The aim of the ECA is to ensure that all electrical installation work is carried out to the highest standards by properly qualified staff. Consequently, firms that wish to become members of the ECA must demonstrate that they have procedures, staff and systems of the highest calibre.

The National Inspection Council for Electrical Installation Contracting (NICEIC)

The NICEIC is an accredited certification body set up in 1956 to protect users of electricity against the hazards of unsafe and unsound electrical installations. It is the industry's independent electrical safety regulatory body and is not a trade association.

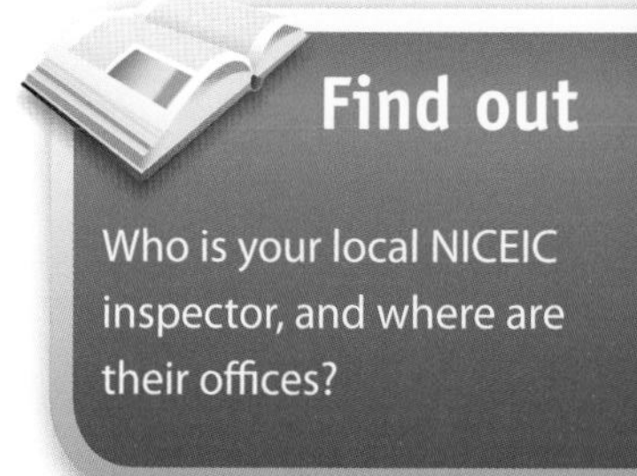

Find out

Who is your local NICEIC inspector, and where are their offices?

The NICEIC maintains a roll of approved contractors who meet the council's rules relating to enrolment and national technical safety standards, including BS 7671 (*IEE Wiring Regulations*). The roll is published annually and is regularly updated on the NICEIC website so that consumers and specifiers can select contractors who are technically competent.

The council also employs 46 inspecting engineers who make annual visits to approved contractors to assess their technical capability and to inspect samples of their work.

UNITE

Find out

Who is your local Unite representative?

For many years workers within the electrical installation industry have enjoyed good labour relationships with their employers. This was largely due to the excellent relationships and cooperation that existed between the industry's principal trade union at the time (EETPU) and the ECA. In 1992 the EETPU merged with the AEU to become the AEEU (Amalgamated Engineering and Electrical Union), subsequently merging with the MSF in 2002 to become Amicus. In 2007 Amicus then merged with the TGWU (Transport and General Workers Union) to become Unite, the UK's biggest union, with over 2 million members.

The Joint Industry Board (JIB)

Formed in 1968, the Joint Industry Board for the Electrical Contracting Industry (JIB for short) came into existence as the result of an agreement between the ECA and the EETPU. Effectively the industrial relations arm of the industry, the JIB has as its main responsibility the agreement of national working conditions and wage rates.

The Health and Safety Executive (HSE)

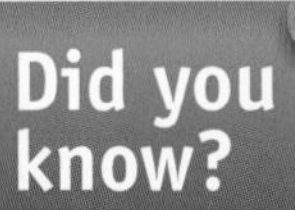

Did you know?

Local authorities are responsible to the HSC for enforcement in offices, shops and other parts of the services sector

The UK's Health and Safety Commission (HSC) and the Health and Safety Executive (HSE) are responsible for the regulation of almost all the risks to health and safety arising from work activity in the UK. Their mission is to protect people's health and safety by ensuring that risks in the workplace are properly controlled. They look after health and safety in nuclear installations and mines, factories, farms, hospitals and schools, and offshore gas and oil installations; they are further responsible for the safety of the gas grid and the movement of dangerous goods and substances, railway safety, and many other aspects of the protection both of workers and the public.

SummitSkills

SummitSkills is the Sector Skills Council for the building services engineering sector, representing the electrotechnical, heating, ventilating, air conditioning, refrigeration and plumbing industries.

SummitSkills has been created for employers to identify skills shortages and deliver action plans to address them. The organisation will also provide careers information for all industries within the sector as well as dealing with all training standards and policy matters previously handled by the former National Training Organisations (NTOs).

The Institution of Lighting Engineers

The Institution of Lighting Engineers (ILE) is the UK's most influential professional lighting association, dedicated solely to excellence in lighting. Founded in 1924 as the Association of Public Lighting Engineers, the ILE has evolved to include lighting designers, architects, consultants and engineers among its 2,500-strong membership. The key purpose of the ILE is to promote excellence in all forms of lighting. This includes interior, exterior, sports, road, flood, emergency, tunnel, security and festive lighting as well as design and consultancy services.

The Institution is a registered charity, a limited company and a licensed body of the Engineering Council.

The Institute of Electrical Engineers (IEE)

Engineering the future

The IEE was founded in 1871. It is the largest professional engineering society in Europe and has a worldwide membership of just under 130,000. As well as setting standards of qualifications for professional electrical, electronics, software, systems and manufacturing engineers, the IEE prepares regulations for the safety of electrical installations for buildings, the *IEE Wiring Regulations* (BS 7671) now having become the standard for the UK and many other countries.

Employer structure

There are about 21,000 electrical contracting companies registered in the UK. They range from one-man organisations to large multinational contractors, but the majority employ fewer than 10 people. The structure of electrical companies varies considerably between firms, depending on the number of employees and the type and size of the business. For ease we will think of them in two groups: small firms and large firms.

There are many tasks to be dealt with in electrical contracting, including:

- handling initial enquiries
- estimating cost
- issuing quotations
- dealing with suppliers and sub-contractors
- supervising the contract
- carrying out the work
- financial control
- final settlement of the account.

A large firm will have specialists who concentrate on one or two of these tasks. In a small firm, one person might do several (or even all) of them.

Small firms

In a small company, the tasks listed above (with the possible exception of actually carrying out the work) are often the responsibility of one person. This type of company structure is known as a vertical structure. The principal advantage is that lines of communication are short: everyone knows who is responsible for different aspects of a project. When someone from outside (another tradesperson, contractor or customer) needs information, they know who to talk to, and are not passed from one department to another.

However, if the person dealing with the project is ill, on holiday, or leaves the company, vital information is missing or lost and it is very difficult for someone else to continue handling the project smoothly.

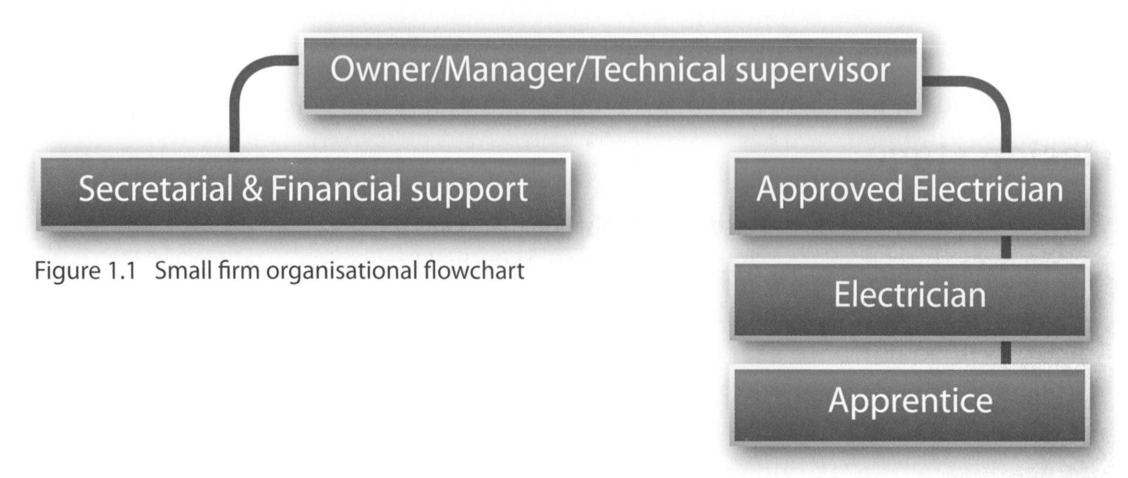

Figure 1.1 Small firm organisational flowchart

Large firms

In the larger company, the tasks are allocated to different people, for example estimators (who handle the initial enquiry and produce an estimate) and contracts engineers (who see that the work is done). This type of structure is known as a horizontal structure. The advantage is that individuals become specialists in a particular task and can work more efficiently. Companies of this size often belong to a trade organisation such as the Electrical Contractors' Association (ECA) and, because of their structure, can offer good career development prospects.

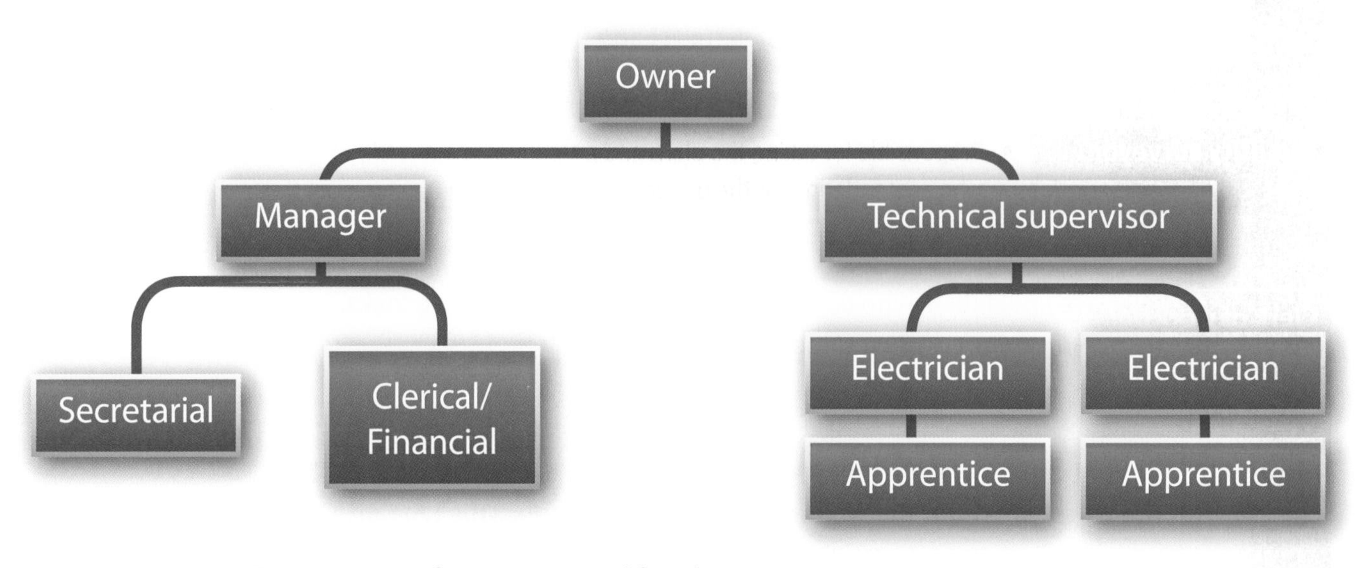

Figure 1.2 Large firm organisational flowchart

Project roles and responsibilities

Many different people are involved in a construction project, from its initial design through to construction and completion. As an electrician, you will be dealing with many of these people, and it is important that you understand their job function and how they fit into the overall project. We'll look at the people involved in the three main stages of a typical construction project:

- design
- tendering
- construction.

Design stage

Client	• Person or organisation that wants the work done and is paying for it. • Specifies the purpose of the building. • Usually gives an idea of number of rooms, size, design etc., and any specific wants. • May also give an idea of the price they are willing to pay for the work.
Architect	• Designs the appearance and construction of the building so that it fulfils its proper function. • Advises the client on the practicality of their wishes. • Ideally provides a design solution that satisfies the client and also complies with the appropriate rules and regulations for the type of building. • For small projects, an architect may draw up a complete plan. For larger, more complex buildings they will consult specialist design engineers about technical details.
Consulting engineers (Design engineers)	• Act on behalf of the architect, advising on and designing specific services such as electrical installation, heating and ventilation, etc. • Create a design that satisfies the client and architect, the supply company and regulations. • Ensure that cable sizes have been calculated properly, that the capacities of any cable trunking and conduit are adequate, and that protective devices are rated correctly. • Produce drawings, schedules and specifications for the project that will be sent out to the companies tendering for the contract. • Answer any questions that may arise from this. • Once the contract has been placed, they will produce additional drawings to show any amendments. • Act as a link between the client, the main contractor and the electrical contractor.
Quantity surveyor (QS)	• Responsible for taking the plans and preparing an initial **bill of quantities** for a project; contractors who are tendering for the project use this information to prepare their estimates. • During construction, the quantity surveyor monitors the actual quantities used, and also checks on claims for additional work and materials.
Clerk of works	• Checks that the quality of the materials, equipment and workmanship used on the project meet the standards laid down in the specification and drawings. • On big contracts there may be several clerks of work, each responsible for one aspect, such as electrical, heating or ventilation. • Effectively employed by the client. • Inspects the job at different stages. • Is also present to check any tests carried out. • May also be given the authority by the architect to sign day worksheets and to issue Architect's Instructions for alterations or additional work.

Table 1.2 Design stage

Definition

Bill of quantities – a list of all the materials required, their specification and the quantities needed

The architect, consulting engineer, quantity surveyor and clerks of works are traditionally part of the architect's design team. However, installations may not always be tackled like this.

For example, for a small job the client may approach an electrical contractor directly and ask it to carry out the work. The client provides a few basic details and requirements, and expects the electrician to ensure that the installation is properly designed and carried out.

Larger electrical contractors, and in particular those that are multi-disciplined, often offer a design service for customers which eliminates the need to use an external architect and makes things simpler (and possibly cheaper) for the client. However, if a dispute arises over design aspects of the job, the client no longer has anyone to arbitrate.

Did you know?

Some larger electrical contracting firms have their own design/consulting engineers

Tendering stage

In competition with others, tendering is the process by which a contractor works to the drawings and specifications issued by the consulting engineer and submits in writing a total cost for carrying out the work (i.e. materials, tools, equipment and labour).

Most invitations to tender have strict guidelines about the information to be supplied and a fixed deadline by which the tender must be received.

Estimator

The estimator's task is to calculate the total cost that will be given in the tender. At the start of the tendering process, a consulting engineer or building contractor usually issues an 'Enquiry to submit a tender' (or Invitation to Tender). This contains various documents, such as:

- a covering letter, giving a broad description of the work, including start and finish dates
- the form of contract that will be applicable to the project, for example, whether it is to be a fixed or fluctuating price
- drawings and specifications for the project
- a tender submission document that must be used
- a day work schedule.

Did you know?

A fixed-price contract is exactly what it says. The contractor agrees to complete the job for the price quoted in the tender, even if material or labour costs go up before the project is completed

This type of tender always has a time limit, so that the contractor is not held to a fixed price if the project is delayed. A fluctuating price contract allows the contractor to claim back the difference between costs included at the time of estimate and actual costs incurred at the time of installation.

When the Invitation to Tender is received, the estimator will check with his or her management (sometimes the contracts manager) to see whether the company wishes to tender on the basis of the submitted documents. If it does, the estimator reads the specifications carefully to understand the requirements.

Most specifications have two sections and these need to be read together.

Section 1 – General specification

This section tends to give general information about installation circumstances such as wiring systems, enclosures and equipment. The section gives generic requirements applicable to any project issued by this consulting engineer.

Section 2 – Particular specification

This section gives the specific details of the project being tendered for. For example, Section 2 may say that for this job the installation will be carried out in galvanised steel conduit. Unless there are specific conditions applicable to the project, the contractor then needs to see what the consulting engineer's general requirements are for installing galvanised steel conduit, and this information will be contained in Section 1.

Using this information and the scaled drawings, the estimator calculates the amount of materials and labour required to complete the job within the time specified by the client. This information is recorded on a 'take-off' sheet. A typical sheet is shown below.

Item	Qty	Description	Init cost	Discount	Material cost £	Hours to install	Hourly rate £	Total labour £
1	200m	20mm galvanised conduit	1.2/m	0	240.00	70	9.50	665.00
2	30	Earthing couplings	.24	0	7.20	0	0	0
3	30	20mm std. brass bushes	.10	0	3.00	0	0	0
4	150	20mm distance saddles	.48	0	72.00	0	0	0
5	150	1.5 x 8 brass screw and plugs	.03	0	4.50	0	0	0

Figure 1.3 Take-off sheet

Nowadays this work is normally done using dedicated computer software. The final tender is based on the results of these calculations.

Construction stage

The contract

Definition

Contract – a legally binding agreement between a client and a contractor

Before any work starts, **contracts** must be agreed and entered into by all the firms involved. Any failure to comply with the details of the contract by either party could result in a court action and heavy financial damages. This might happen if the contractor does not complete the work or uses sub-standard materials, or if the client does not pay.

Contracts do not have to be made in writing. A verbal agreement, even one made on the telephone, can constitute a legal contract. However, most companies use written contracts that cover all aspects of the terms and conditions of the work to be carried out.

Several conditions must be met for a contract to be binding:

1. An offer must be made which is clear, concise and understandable to the customer.
2. The customer must accept the offer and this acceptance must be received by the contractor. The acceptance must be unqualified (i.e. with no additional conditions). Up to this point there is no agreement or obligation binding either side. The contractor is free to withdraw the offer, and the customer can reject it.
3. There must be a 'consideration' on both sides. This is what each party is agreeing to do for the other.

There are several reasons why a contract may not be made. Four of them are:

1. Withdrawal – The contractor can withdraw the offer at any time until the offer is accepted. The contractor must notify the customer of the withdrawal.
2. Lapse due to time – Most offers put a time limit on acceptance. After this time, the offer expires and the contractor is under no obligation, even if the customer later accepts the offer.
3. Rejection – The customer can reject the contractor's offer; no reason has to be given. If the customer asks the contractor to submit a second offer (for additional work, or simply to lower the price), the contractor is under no legal obligation to quote again. Each quote is self-contained: the terms or conditions for a previous quote do not automatically apply.
4. Death of contractor – If the contractor dies before the offer is accepted the customer must be notified; otherwise the customer could agree (within the offer time limit) and the contract would become valid.

Definition

Breach of contract – when one party does not fulfil the contract terms

Breach of contract occurs when one of the parties does not fulfil the terms of the contract. For example if the contractor does not perform the work to the specification in the offer (it is the contractor's responsibility to ensure that the installation is in complete compliance with the specification) or if the customer refuses to pay for the work.

Contract law is very complex, and it cannot be covered fully on this course. To minimise the risks, you can use a standard form of contract. The Joint Contracts Tribunal (JCT) 'Standard Form Of Contract' (normally for projects of a complex nature or in excess of 12 months' duration) or the 'Intermediate Form Of Contract' are typical of contracts used in the industry.

If you are involved in any kind of contract it is always advisable to seek professional legal assistance before making or accepting an offer. Once contracts are agreed, construction and installation work can begin and many more people become involved.

Main contractor	• Usually the builders, because they have the bulk of the work to carry out. • Has the contract for the whole project. • Employs subcontractors to carry out different parts of the work. • In refurbishment projects, where the amount of building work is small, the electrical contractor could be the main contractor. • Paying and coordinating subcontractors.
Nominated subcontractors	• Named (nominated) specifically in the contract by the client or architect to carry out certain work. • Must be used by the main contractor. • Normally have to prepare a competitive tender. • Subcontractors will include electrical installation companies.
Non-nominated subcontractors	• Companies chosen by the main contractor (i.e. not specified by the client). • Their contract is with the main contractor.
Nominated supplier	• Supplier chosen by the architect or consulting engineer to supply specific equipment required for the project. • Main contractor must use these suppliers.
Non-nominated supplier	• Selected by main contractor or subcontractors. • For electrical supplies, this will be a wholesaler selected by the subcontractor who can provide the materials needed for the project.
Contracts manager	• Oversees the work of the contracts engineers. • May also be responsible during tendering for deciding whether a tender is to be submitted and the costs and rates to be used.
Contracts engineer	• Employed by the electrical contractor to manage all aspects of the contract and installation through to completion. • Responsible for planning labour levels, ordering and organising materials required. • Ensures the contract is completed within the contract timescales and on budget. • Liaises with suppliers to ensure planned delivery dates and builders' work programme are acceptable. • May negotiate preferential discounts with suppliers. • Attends site meetings.

Table 1.3 People involved in the construction stage (continues overleaf)

Project engineer	*Role definitions vary within the industry but generally the role is similar to that of a contracts engineer.* • Responsible for day-to-day management of on-site operations relative to a specific project. • Often based on the site.
Site supervisor	• Contractor's representative on site. • Oversees normal day-to-day operations on site. • Experienced in electrical installation work, normally an Approved Electrician. • Responsible for the supervision of the approved electricians, apprentices and labourers. • Uses the drawings and specification to direct the day-to-day aspects of the installation. • Liaises with contracts engineer to ensure that the installation is as the estimator originally planned it. Ensures materials are available on site when required. • Liaises with the contracts engineer where plans are changed or amended to ensure additional costs and labour/materials are acceptable and quoted for.
Electricians, apprentices and labourers	• The people who actually carry out the installation work. • Work to the supervisor's instructions.
Electrical fitter	• Usually someone with mechanical experience. • Involved in varied work including panel building and panel wiring and the maintenance and servicing of equipment.
Electrical technician	*Job definition varies from company to company.* • Can involve carrying out surveys of electrical systems, updating electrical drawings and maintaining records, obtaining costs, and assisting in the inspection, commissioning, testing and maintenance of electrical systems and services. • May also be involved in recommending corrective action to solve electrical problems.
Service manager	*Similar role to contracts manager (and in some cases the roles are combined) but focuses on customer satisfaction rather than contractual obligations.* • Monitors the quality of the service delivered under contract. • Checks that contract targets (e.g. performance, cost and quality) are met. • Ensures customer remains fully satisfied with the service received.
Maintenance manager	*Once the building has been completed.* • Keeps installed electrotechnical plant working efficiently. • May issue specifications and organise contracts for a programme of routine and preventive maintenance. • Responsible for fixing faults and breakdowns. • Ensures legal requirements are met. • Carries out maintenance audits.

Table 1.3 People involved in the construction stage

Sources of technical information

We've seen that lots of different people may be involved in an electrical installation project. For them to work together they must communicate with one another, using or transmitting technical information.

There are many sources of this information, including drawings, diagrams, charts and data, British Standards, Codes of Practice, specifications, manufacturers' instructions and manuals, catalogues and reports.

The information can come from hundreds of different sources, including the British Standards Institute (BSI), BSEN standards, equipment and component manufacturers, consulting engineers, trade associations, HMSO book stores, libraries and the Internet. The material can be on paper, microfilm, CD-ROM or electronically downloaded from a remote source.

In the next section we are going to look at the sort of visual information you will be dealing with on a day-to-day basis.

Drawings, diagrams and symbols

A technical diagram is simply a means of conveying information more easily or clearly than can be expressed in words. In the electrical industry drawings and diagrams are used in different forms. Those most frequently used are:

- block diagrams
- circuit diagrams
- wiring diagrams
- schematic diagrams
- assembly drawings
- record (as fitted) drawings
- layout/location diagrams
- site plans
- electrical symbols (BS EN 60617).

In this section we will be looking at these diagrams, and we will also be looking at the electrical symbols known as BS EN 60617 (formally BS 3939) which are used to help understand and draw these diagrams.

Block diagrams

A block diagram can be used to relate information about a circuit without giving details of components or the manner in which they are connected. In block diagrams the various items are represented by a square or rectangle clearly labelled to indicate its purpose. This type of diagram shows the sequence of control for installations in its simplest form, as shown in Figure 1.4.

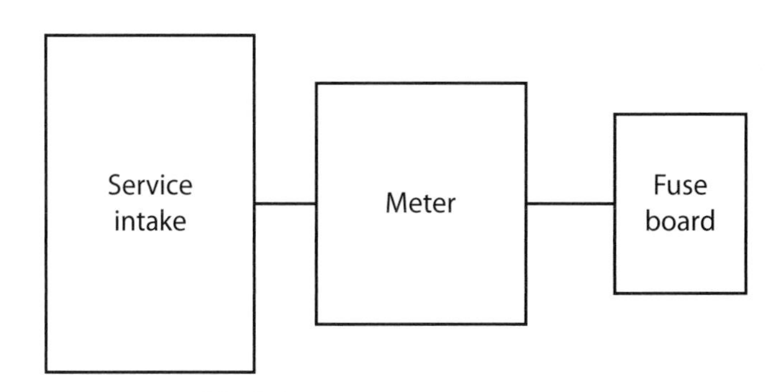

Figure 1.4 A block diagram

Circuit diagrams

A circuit diagram uses symbols to represent all circuit components and shows how these are connected. The circuit diagram should be as clear as possible and should follow a logical progression route from supply to output. In all other respects the circuit diagram cannot be regarded as a direct source of information.

For example, the shape of the diagram does not represent the physical outline of the circuit; it has no dimensions; and the symbols that are used need not bear the slightest resemblance to the components they represent. The symbols used are BS EN 60617 circuit diagram symbols. Some of the more common symbols are included at the end of this section. Figure 1.5 illustrates the use of a circuit diagram in relation to a rectification circuit.

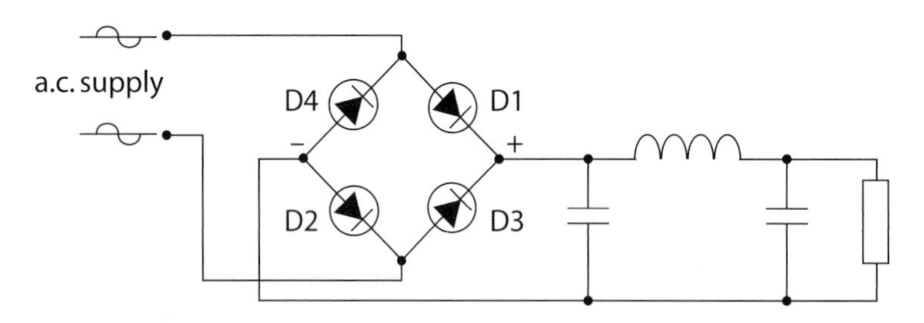

Figrure 1.5 A circuit diagram

Wiring diagrams

In a wiring diagram the physical layout is taken into consideration. The components and connections show a pictorial version of those found in the actual circuit. Wiring diagrams can be used to carry information of a specific nature regarding the wiring or connection of components. Wiring diagrams do not, as a rule, use circuit or location symbols, but there is no hard-and-fast rule about this. Since the object of the diagram is to relate information, this point should be borne in mind when drawing up the wiring or circuit diagram.

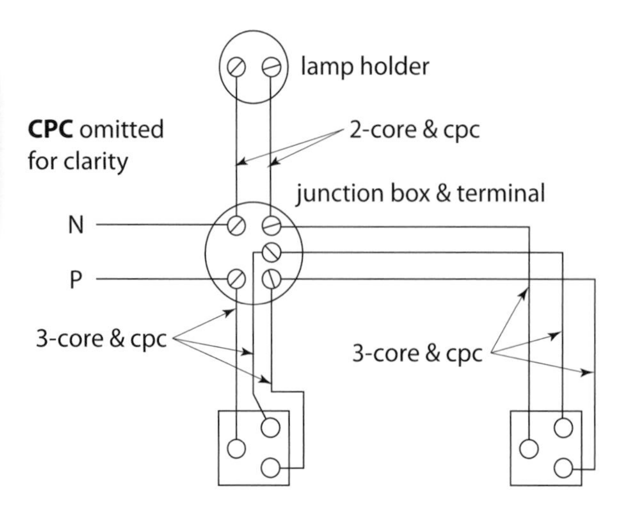

Figure 1.6 A wiring diagram

Schematic diagrams (working diagrams)

Schematic diagrams like Figure 1.7 are similar in concept to the circuit diagram on page 26. They do not show how to wire components but they do show you how the circuit is intended to work. The diagram is a control circuit of an automatic star/delta starter. These types of diagrams tend to be used for larger, more complicated electrical diagrams such as control systems for motor starters and heating systems.

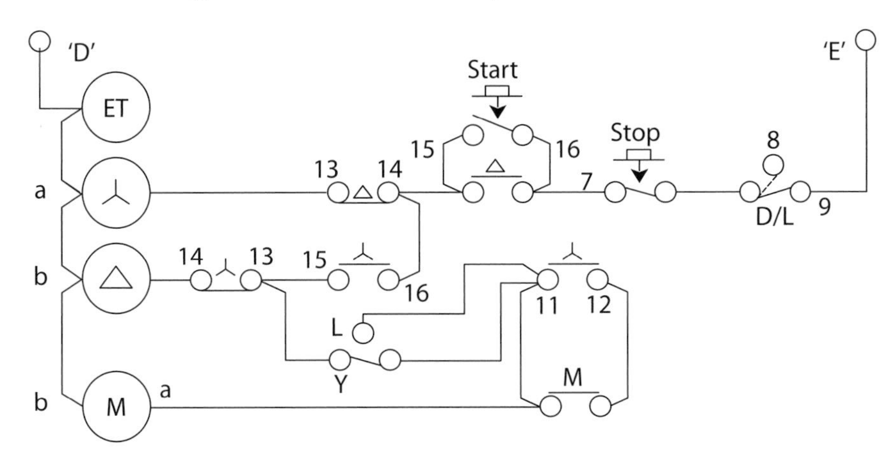

Figure 1.7 A schematic diagram

Assembly drawings

Assembly drawings should show how the individual parts (or modules) of a product fit together. They normally contain scale drawings of all the components shown in their correct position relative to each other, with some overall dimensions. Where there are internal components, these are shown by sectioning. Each component is listed and described on the drawing.

The example below is the assembly drawing for a push-button enclosure.

A Enclosure base with built in contact block clips

B Contact blocks/lamp holders

C Locking ring

D Enclosure lid

E Legend plate

F Captive screws (after screw in) loose in enclosure on delivery

G Actuators and lens cap

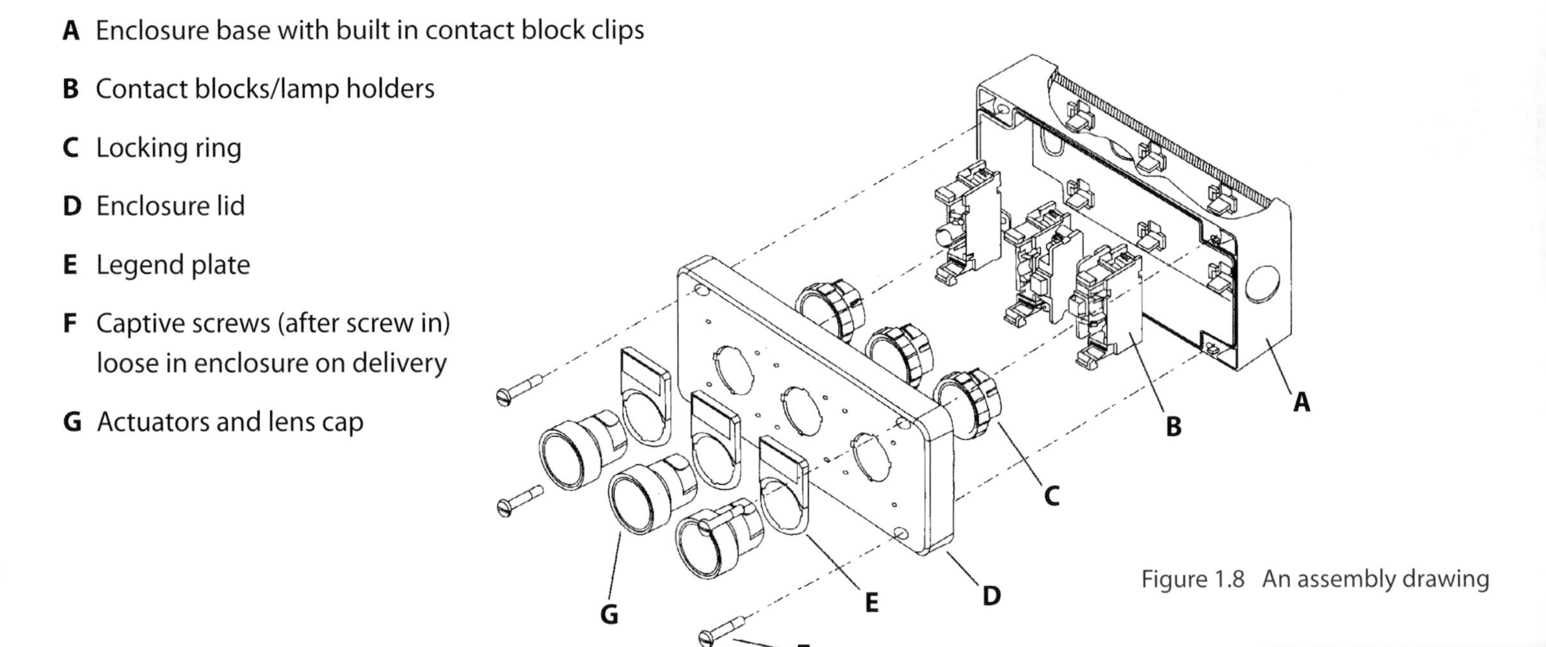

Figure 1.8 An assembly drawing

Record (as fitted) drawing

In an ideal world, every installation would follow exactly the consulting engineer's original plans. In practice, however, problems can come up which mean that conduit must be fixed in a different position or cables routed in another way. These changes must be recorded so that, in future, if maintenance or alterations are needed, the actual layout of the installation is known. This becomes very important when the installation is hidden in the walls, for example.

This is where record (as fitted) drawings come in. They are the installation drawings supplied by the consulting engineer with the final actual installed cable, conduit and trunking routes marked on them.

Layout diagrams

These are scale drawings prepared under the supervision of an electrical consultant or engineer responsible for a particular installation, and are based upon architect's drawings of the building in which the installation is to be installed. These drawings show the required position of all equipment, metering and control gear that is to be installed. They normally show the plan view of the installation, and standard BS EN 60617 location symbols are used. These diagrams are used to show the sequence of control of large installations.

Figure 1.9 shows a layout for a warehouse. Note the legend at the bottom of the page: this legend should always be included so as to help in understanding the diagram.

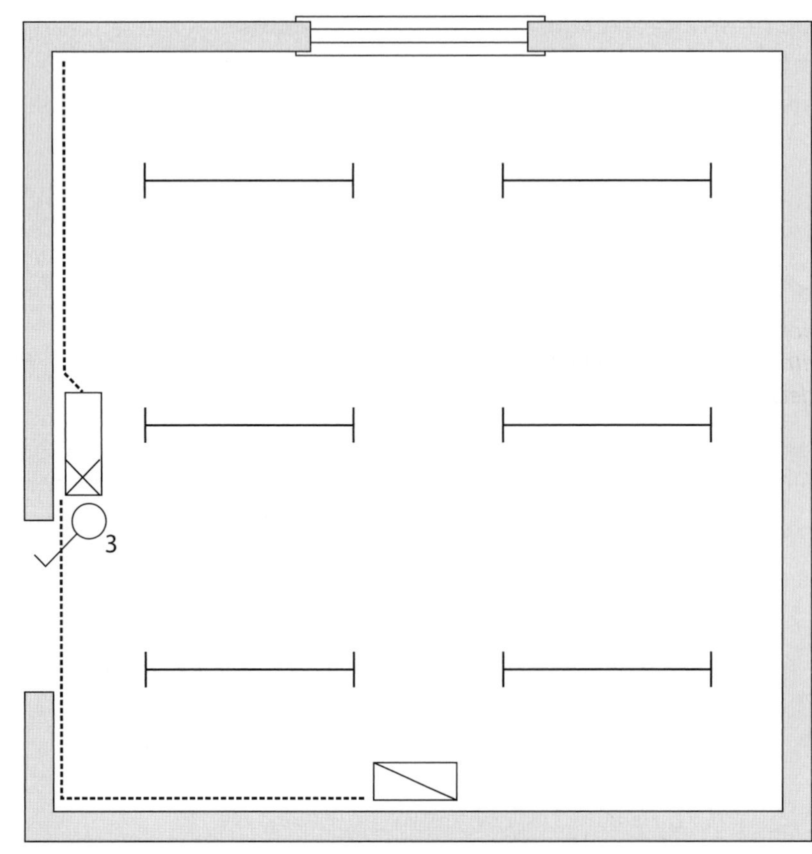

Legend

Warehouse lighting (scale 1:50)

- -------- 50 × 50mm trunking run
- 3 3-gang 1-way switch
- lighting distribution board
- main control
- fluorescent luminaire

Figure 1.9 A layout diagram

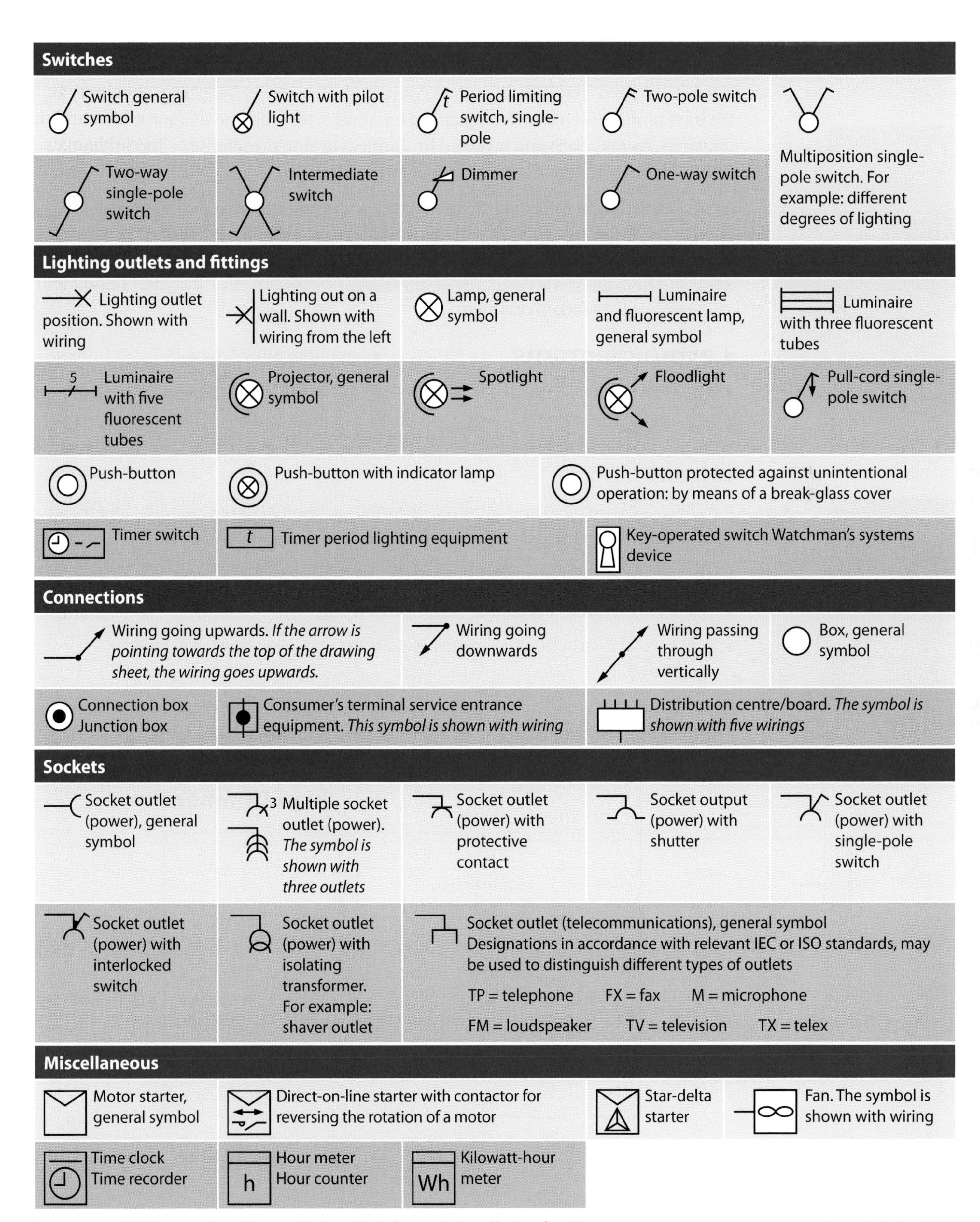

Figure 1.10 BS EN 60617 lists the standard symbols for use in installation drawings

Site documentation

Remember

The main contractor acts as coordinator of all the individual subcontractors, checks that they carry out their work properly and on time, and pays them for the work done

We have already seen how technical information is recorded on diagrams and drawings, so that everyone involved in a project during installation – and in the future – knows what to do and what has been done.

As well as technical information, there is plenty of other information that people need to communicate to others in the workplace, and various types of documents are used for this. You will have to deal with all sorts of documents and forms during your work, but here are some common examples:

- job sheets
- variation order
- day worksheets
- time sheets
- delivery notes
- manufacturers' data
- site reports and memos.

Job sheets

Remember

Prompt delivery and competitive pricing will be important factors in the choice of non-nominated suppliers

Job sheets give detailed and accurate information about a job to be done. Electrical contracting companies issue them to their electricians. They will include:

- the customer's name and address
- a clear description of the work to be carried out
- any special instructions or special conditions (e.g. pick up special tools or materials).

Sometimes extra work is done which is not included in the job sheet. In this case it is recorded on a day worksheet so that the customer can be charged for it.

Job Sheet

Evan Dimmer
Electrical contractors

Customer Dave Wilkins

Address 2 The Avenue
Townsville
Droopshire

Work to be carried out
Install 1 x additional 1200 mm
fitting to rear of garage

Special conditions/instructions
Exact location to be specified
by client

Figure 1.11 A typical job sheet

Variation orders

A **variation order** (VO) is issued ('raised') when the work done varies from the original work agreed in the contract and listed in the job sheet. If this situation arises, it is important for the site electrician to tell his supervisor immediately. A variation order can then be made out to enable the new work to be done without breaking any of the terms of the contract. The purpose of the VO is to record the agreement of the client (or the consulting engineer representing the client) for the extra work to be done, as well as any alteration that this will make to the cost and completion date of the project.

Find out

When would you need to use a variation order?

Day worksheets

Work done outside the original scope of the contract, perhaps as a result of a VO initiated by the architect, engineer or main contractor, is known as day work. When the work is completed, the electrician or supervisor fills out a day worksheet and gets a signature of approval from the appropriate client representative. Day work is normally charged at higher rates than the work covered by the main contract, and these charges are usually quoted on the initial tender. Typical day work charges are:

- labour: normal rates plus 130 per cent
- materials: normal costs plus 25 per cent
- plant: normal rate plus 10 per cent.

Disputes over day work can easily arise, so it is important that the installation team on site records any extra time, plant and materials used when doing day work.

Day Worksheet

Evan Dimmer
Electrical contractors

Customer — Job No.

Date	Labour	Start time	Finish time	Notes

Quantity	Description	Office use

Supervisor's signature — Customer's signature

Date

Figure 1.12 A typical day worksheet

Remember

Accurate time sheets are essential to make sure you get paid correctly and that your customer is charged the right amount for the job

Time sheets

Time sheets are very important to you and your company. They are a permanent record of the labour on a site, and include details of:

- each job
- travelling time
- overtime
- expenses.

This information allows the company to track its costs and make up your wages. If you work on several sites during a week, you may need to fill in a separate time sheet for each job.

Time Sheet

Evan Dimmer
Electrical contractors

Employee Project/site

Date	**Job No.**	**Start time**	**Finish time**	**Total time**	**Travel time**	**Expenses**
Mon						
Tue						
Wed						
Thu						
Fri						
Sat						
Sun						
Totals						

Employee's signature Supervisor's signature

Date

Figure 1.13 Typical time sheet

Purchase orders

Before a supplier will dispatch any materials or equipment, it will require a written purchase order. This will include details of the material, quantity required, and sometimes the manufacturer; it may also specify a delivery date and place. In many cases the initial order is made on the telephone or using email or the Internet, and a written confirmation is sent immediately afterwards. The company keeps a copy of the original order in case there are any problems.

Usually the Purchasing Department sends out these orders, but sometimes an order is raised directly from site when there is a need for immediate action.

Delivery notes

Delivery notes are usually forms with several copies that record the delivery of materials and equipment to the site. Materials delivered directly to the site will arrive with a delivery note. As the company representative on site this is the form you are most likely to deal with.

The delivery note should give the following information:

- the name of the supplier
- to whom the materials are being sent
- a list of the type, quantity and description of materials that are being delivered to the site in this particular load
- the time period allowed for claims for damage.

When materials arrive on site you should:

1. Ensure that they are unloaded and stored correctly.
2. Check each item against the delivery note.
3. Check for obvious signs of damage.
4. If everything is OK, sign the note. If not, note any missing/rejected items on the delivery note, then you and the delivery driver should both sign it.
5. Store your copy of the note safely.
6. Check the materials thoroughly for damage within the time given stated in the delivery note (usually three days), and inform the supplier immediately if there are any problems.

A delivery won't always contain all the materials listed in the purchase order. Sometimes the material is not all needed on site at the same time and it is delivered in several loads. This helps to reduce the need for on-site storage and minimises the risk of damage or loss.

Delivery note — **A. POWERS** *Electrical wholesalers*

Order No. Date

Delivery address
2 The Avenue
Townsville
Droopshire

Invoice address
Evan Dimmer
Electrical Contractors

Description	Quantity	Catalogue No.
Thorn PP 1200 mm fit fitting	1	
1.5 mm T/E cable	50 m	

Comments

Date and time of receiving goods

Name of recipient Signed

Figure 1.14 Typical delivery note

Incomplete deliveries may also occur if the supplier is out of stock, or if some of the order is coming direct from the manufacturer.

Completion orders record that all the material on an original purchase order has been delivered.

On the job: Receiving deliveries on site

Alex works for Gurnards Electrical Services. He is left to man the site while his colleagues go to lunch. A delivery arrives while he is alone. He accepts the delivery and signs the delivery note without checking the contents. When his boss arrives back, he checks the delivery and finds that some items are missing; he calls the company, but he is told that all items were delivered as specified.

1. How could this situation have been avoided?
2. Outline the steps Alex should have taken to receive the delivery.
3. What would have happened if Alex had lost the delivery note?

Manufacturers' data and service manuals

Almost all equipment will have the manufacturer's fitting instructions and other technical data or information sheets. These should be read and understood before fitting the item concerned.

Once installation of the item is complete, they should be kept safely in a central file so that they can be given to the customer in a handover manual when the project is completed. Sometimes you may need more details about an item. You can get this from the manufacturer's catalogue or datasheets, website, or by speaking to the manufacturer directly. It may be useful to talk to the contracts engineer to make sure that he or she knows what is happening.

Site reports and memos

Most companies require regular reports about the progress on site. The site foreman, supervisor or engineer in charge usually compiles them. Site reports contain details of work progress, defects, problems and delays. Sometimes other reports into specific problems or incidents will be made. The reports enable the company to:

- provide evidence when they make claims for progress payments
- take prompt action to avoid any difficulties that may be building up
- spot problems that keep being repeated and take action to eliminate them.

A memo is usually a short document about a single issue – for example, a problem installing a piece of equipment, or materials not being delivered on time.

Storing and retrieving information

Nowadays, technical information is stored and communicated using a wide variety of methods. Many of these require special equipment (e.g. computers) to get at the information. The same information is often available in several different formats.

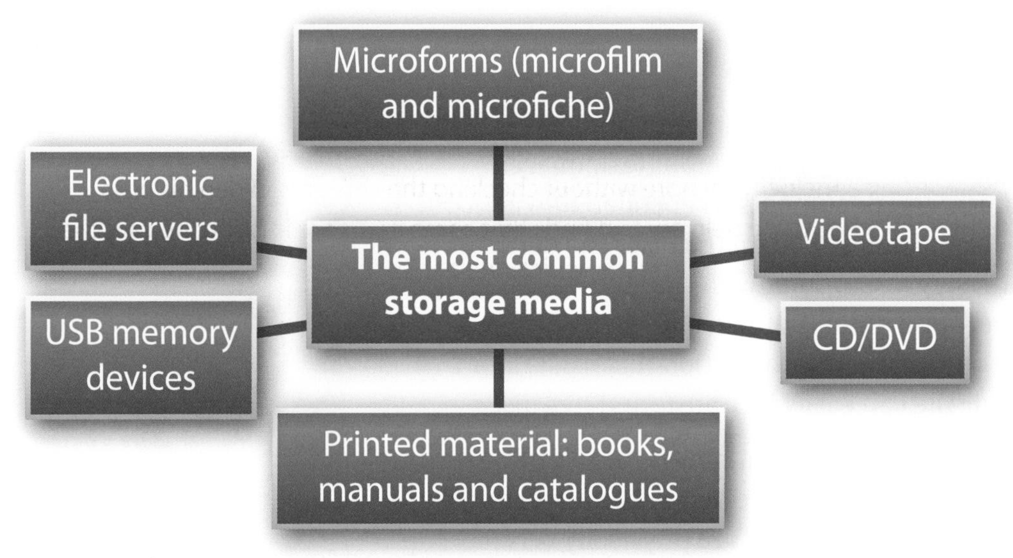

Figure 1.15 Storage media

Storage media

Printed materials

This is still the most common form of storage for general information – for example, books, newspapers, leaflets, datasheets and catalogues. Their major advantage is that they need no special equipment for reading. Increasingly, many of these items are also available in the other formats described below. This is particularly true of technical information, where electronic methods have the advantage of lower costs and increased delivery speed.

Microforms

Microforms (also known as microfiches) store large amounts of information such as text, documents and photographs using photographic techniques. Their main advantage is long life, and therefore they are popular for storing information that needs to be preserved for many years or even centuries, for example in libraries or newspaper offices. Microforms are based on polyester film, and have a life expectancy of about 500 years when stored correctly. Microfilm is a length of 16 mm or 25 mm film; microfiche is a larger piece of photographic film (around 150 mm × 100 mm). Both formats store information as tiny pages, and each type can hold hundreds of pages on one microform. Special optical reading devices are used to read the images. They have a temperature-controlled light source and a magnifying device to show the images on a screen similar to a computer display.

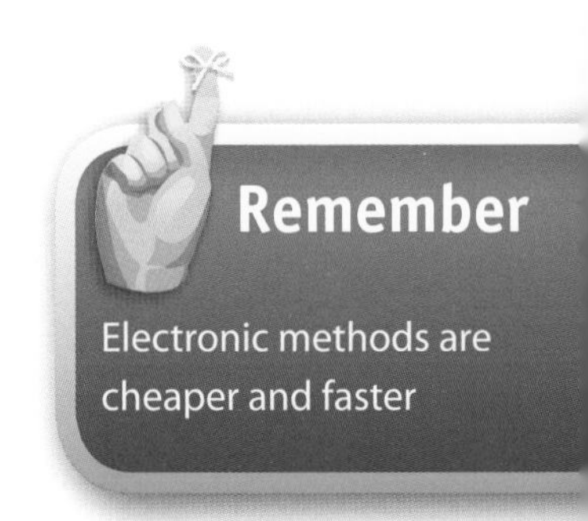

Videotape

Videotape is a polyester film on which television signals are stored magnetically. The most common format is VHS. Its main advantages are low cost and the widespread availability of video recorders for storing and reading material. Videotape is not normally used for storing technical documents, but it has a useful role where visual information is important, such as for product demonstrations and tours.

CD/DVD/USB device

CDs and DVDs are among the most popular storage media today. Information (text, images, film, music and documents) is encoded into a digital format and stored on the disc. The data can be read using a computer with suitable drive and software, or on a stand-alone device. The discs are low cost, and any information that exists as an electronic file can easily be transferred to a disc.

CDs can store a vast amount of information

Perhaps even more popular now is the use of USB memory sticks or similar devices. These can store data in a digital format and are physically very small, about 50 mm long. They can typically be used on computers without a CD/DVD drive and can therefore easily transfer information between machines.

CD	700 MB
DVD	4.7 GB
USB device	4 GB

Table 1.4 Storage capacity of media devices

Each type of information can be encoded using a number of different formats, each having certain advantages for different applications. Some of these are more common than others. To read the information, the reading device must be equipped with software to handle it.

Some typical file types (denoted by their normal file-name suffix) are:

Images	jpg, bmp, tif, pcx, gif, png
Movies	wmv, mpg
Text	txt
Documents	doc, pdf

Table 1.5 Table of file types

USB storage device

Electronic file servers

The storage methods described so far need the reader to have a copy of the information, whether it is a book, CD or video; in effect, there is one copy per reader (although one copy may be shared among several readers, like a library book). The increasing use of electronic networking to connect computers together means that a single copy of some information can be stored and then accessed by many users at the same time. The information is held on a file server located anywhere in the world, provided that it is connected to the network. The Internet is the ultimate example of this. The information is stored in digital format, using the same formats as a CD or DVD. Access to the information can be limited to certain individuals or groups, or may be open to anyone.

Information sharing

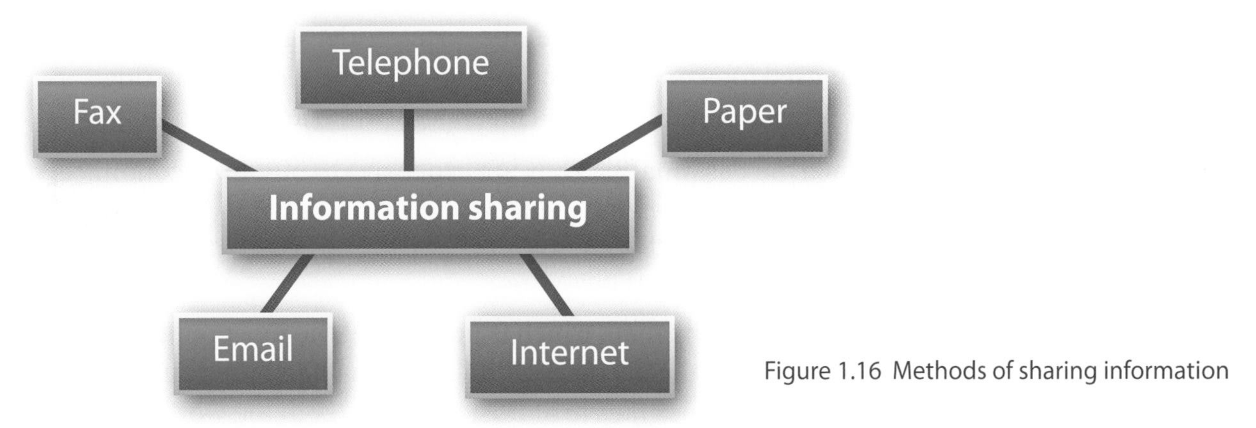

Figure 1.16 Methods of sharing information

Written material

This is one of the simplest ways of communicating, using printed material or by hand written notes and memos.

Telephone

Speaking directly to someone is sometimes the quickest way to obtain answers and information. Manufacturers nearly all have telephone helplines.

Facsimile (fax)

The fax machine enables written or printed material to be sent to another fax machine over a normal telephone line. Although email has greatly reduced the use of fax, it is still useful for sending handwritten documents and printed text without needing a computer, and it is quite fast. However, the end result is a sheet of paper, so it cannot be edited electronically. There are several different types of stand-alone fax machine, which vary as to how they print out the received document.

A fax machine is basically a scanner combined with a printer. Therefore a computer equipped with these two items, along with a fax modem and suitable software, can send and receive faxes without the need for a stand-alone fax machine.

Faxes have been around in one form or another for about 100 years

Printers

There are several types of printers in use with computers today. These include dot matrix, thermal paper, thermal ribbon, and laser.

Dot matrix are rather old fashioned. They used patterns of pins pressing against an ink ribbon to create type.

Thermal paper printers use special paper that is coated with chemical to turn black when heated. They are cheap, have few moving parts, and do not need ink or ribbon supplies. However, they cannot work with ordinary paper, and over time, the image fades or disappears, particularly if it is left in a hot place.

Thermal ribbon printers have a heat-sensitive ribbon or film. As the paper passes over the ribbon (which also moves), the ink is transferred to the paper to make a mark. They do not need special paper, but the ribbon or film has to be replaced regularly.

Inkjet printers work with ordinary paper, but the ink cartridges need to be replaced. If the machine is not used for some time, the ink may dry and clog the cartridge, giving a poor image. They can produce good black-and-white and colour images.

Laser printers produce the best images. They are expensive to buy, but economical to run, and use ordinary paper. They come in black-and-white, or colour models.

Did you know?

The Internet started as a defence project by the Advanced Research Projects Agency (ARPA) for the US government in 1969 and was first known as the ARPANET

Email

Most of you will be familiar with email, which is now one of the most popular methods of communicating in the business world and elsewhere. You can transmit simple text messages and complex files to almost any country in the world in seconds, and you can also attach other files, such as documents, images and movies. To send and receive email messages you need:

- a computer running a software application called a 'mail client' (e.g. Outlook Express or Pegasus)
- connection to the Internet (via modem or broadband)
- an account with an Internet service provider (ISP).

Did you know?

Web pages are coded using HyperText Markup Language (HTML). This enables the same source document to be read using any compatible browser, although the appearance of a particular Web page may vary slightly

The Internet

The Internet, sometimes called simply 'the Net', is a worldwide system of computer networks.

In practice, the Internet is really a 'network of networks'. Today, the Internet is accessible to hundreds of millions of people across the world. It enables users to get information from other computers on the Net, and is used to transmit voice, radio and video as well as data, particularly email. More recently, Internet telephony hardware and software has made it possible to use the Internet for normal voice conversations.

Using the World Wide Web (often abbreviated 'www', or simply 'the Web'), you have access to billions of pages of information. Web browsing is done using web-browser software, of which Microsoft Internet Explorer and Netscape Navigator are the most popular. Most companies now have websites that allow you to browse products or technical information.

Working with documentation, drawings and specifications

It should be clear by now that documents, especially drawings and specifications, are very important for the success of a project. Without them, everyone would have a different idea about how things should be done, and the result would be chaotic. Documents provide the technical information that everyone needs to do their job properly. However, they are only useful if the information they provide is clear and the users can understand them. Unfortunately, this doesn't always happen.

In our own industry the estimator uses the drawings and specification provided by the consulting engineer during the tendering stage. The electrician will also need the drawings during the installation stage. In the project specifications (which usually come in two parts: General and Particular) there might be hundreds of different drawings, particularly if it is a large project like a school or hospital.

There are several types of drawing that will be used in a project; each one has a specific purpose, and no single document will contain all the information needed, except possibly on the simplest level. Different people will need different documents, depending on their task. You will often have to use several documents, reading them alongside each other to get all the information you need.

Scaled drawings

Layout and assembly drawings give information about physical objects, such as the floor layout in a building, or a mechanical object. If we were to make the drawing the same size as the object, the drawings would often be far too big to handle.

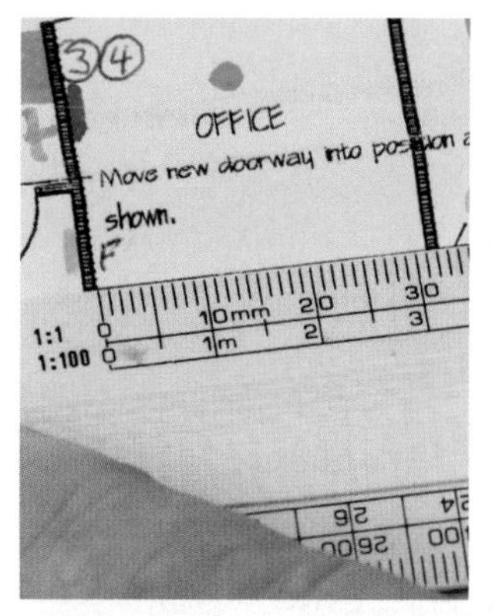

Using a scaled drawing

To make the drawing a sensible size, we use **scaled drawings**. You may, for example, have built model aeroplanes from a kit. Quite often these are described as 1/32nd scale: in other words, every part in the model is 32 times smaller than the real thing. This type of scale is known as a ratio scale, and it makes the drawings easy to use. To find a measurement on the actual object, you measure the distance on the drawing and multiply it by the scale. It doesn't matter what unit of measurement you choose, because you are simply going to multiply it by a number (the scale).

For example, on most construction projects, the scale used to show the floor layout of a building is 1:100. So 10 mm on the drawing represents something that is 100 times bigger in reality (i.e. $10 \times 100 = 1{,}000$ mm (1 metre).

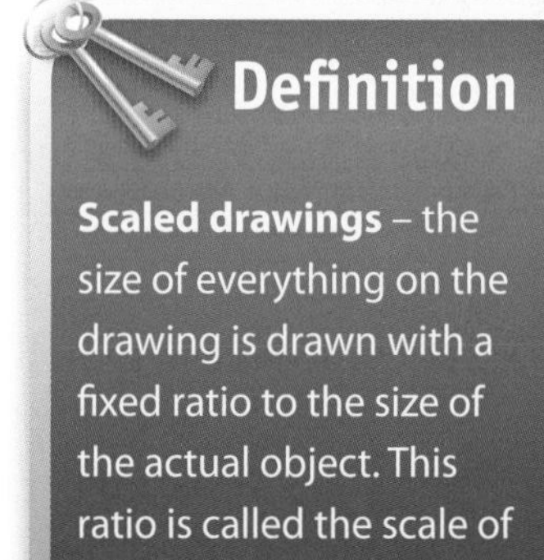

The drawing scale is chosen to make the drawing a reasonable size, according to its purpose. Although a scale of 1:100 may be fine for the layout of a building, it would be impractical for a road map, because you would only be able to get a few miles on each sheet. A scale of 1:500,000 (1 cm = 5 km) would be better.

In the same way, an assembly drawing for a wristwatch would be too small to read if we used 1:100; a better scale might be 20:1 (20 mm on the drawing represents 1 mm on the actual watch).

Preparing a materials list

Take a good look at the drawing: it shows the layout of the building, the location of specific services and how some items are to be installed and connected. As it is to scale (1:50), you can measure it to find the actual dimensions of the building and prepare a materials list for the job. You can also scale up positions shown on the drawing and mark them for real inside the building itself. However, what the drawing doesn't tell us is anything about the fitments or how they should be installed. This information will be in the specification, so you'll need a copy of that too. Opposite is an extract from the specification, which tells us what we need to know.

> the warehouse lighting and sub-main installation will be completed in PVC single core cable contained within galvanised steel conduit and galvanised steel trunking where required. All fluorescent lighting fittings will be 1700 mm x 58 W of type Tamlite TM58 and suspended on chain from back boxes. All light switches will be of MK type Metalclad Plus with an aluminium front plate. The installation will be fed from the new lighting distribution board (MEM) which shall be fed from the existing MCB DB located in the warehouse.

Extract from specification

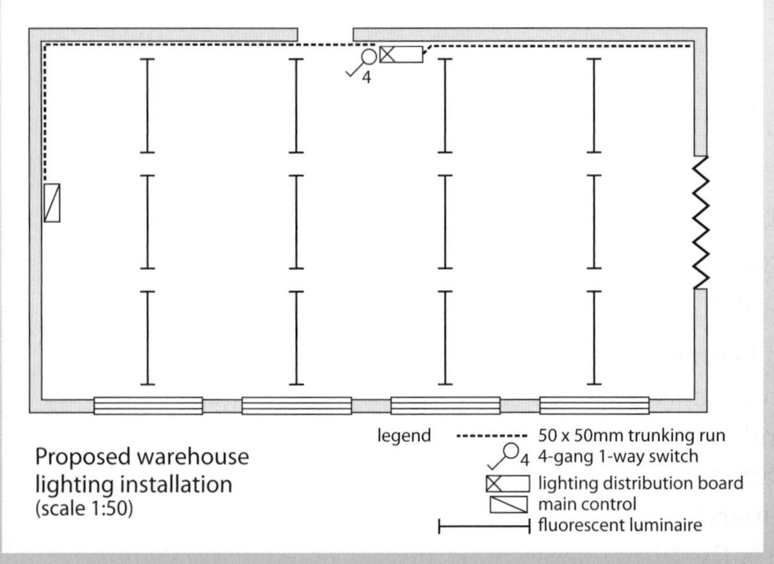

Scale layout drawing

Step 1: Count up all the major pieces of equipment needed

Looking at the plan and referring to the specification, we can see that we need:

- 12 Tamlite TM58 fittings, plus tubes
- 1 MK Metalclad Plus 4-Gang 1-way surface switch with aluminium front plate
- 1 MEM 3-way SPN surface mounted DB (distribution board).

Step 2: Decide the best runs of conduit and trunking

A logical way would be to install the 50 x 50 mm trunking as shown on the drawing and then install separate conduits from the trunking up to each row of lights and then along each row of lights, fixing the conduit to the roof structure.

Step 3: Calculate lengths of trunking, conduit and cable required

We measure these from the drawing and calculate the actual distance using the scale provided (1:50). You will also need to know the height of the building, which is not shown on the layout diagram.

Step 4: Include accessories, fixings etc.

We will need to allow back boxes and hook and chain arrangements at each luminaire and order sufficient fixings (screws, bolts, saddles for conduit etc.) of various types for the installation.

Step 5: Consider special access equipment

Do you need special ladders, scaffolding etc? It's no use having the materials if you can't get to the right places!

As well as compiling a materials list, you will need to work out what wiring is required to control the installation and feed the lighting distribution board.

Preparing a materials list

At some stage, you are going to have to put together a materials list for a project or part of a project. The estimator and contracts engineer will usually have ordered most of the systems and equipment, but there will always be smaller projects, or parts of a big project, where you will be asked to do it for yourself.

Using the example opposite, this is how you go about preparing a materials list. The drawing is of a warehouse that is part of a large hospital project. It's going to be used for storing information and leaflets for medical staff. You'll probably be able to order the materials you need from a central store or, if not, directly from a local wholesaler. Either way, the process of working out what you need is just the same. The first thing to do is to get a copy of the layout drawing, like the one on page 40. It *must* be a scaled drawing.

Remember

As well as compiling a materials list, you will need to work out what wiring is required to control the installation and feed the lighting distribution board

Using charts and reports

In addition to drawings and specifications, charts and reports are two other methods of showing and communicating technical information and data. For this Technical Certificate, you must be able to interpret the data contained in a chart.

Charts

Charts can often make information easier to understand and allow the user to see clearly what they need to know. The most popular chart used within construction work is the bar chart. When it shows activities against time, it is sometimes referred to as a **Gantt chart**, after its inventor, Robert Gantt.

In Figure 1.17, the bar chart shows several activities and when they are due to happen. This helps the supervisor to keep an eye on how the contract is actually progressing when compared to the original plan. Main contractors often use this sort of bar chart to show when individual trades should be on site at any time during the contract.

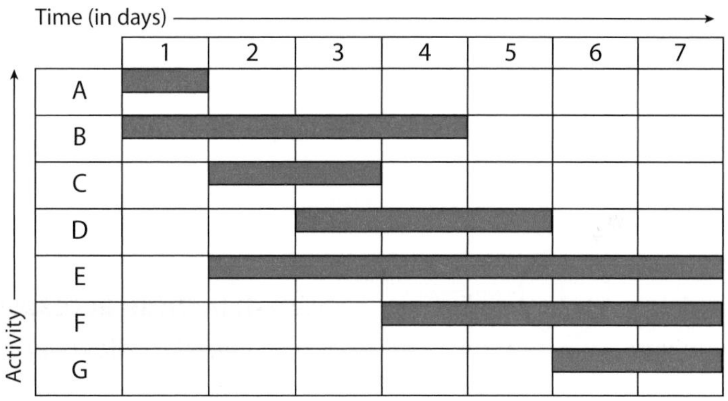

Time (in days)
Activity
1 2 3 4 5 6 7
A B C D E F G

Figure 1.17 Bar chart 1

Looking at the chart we can see that:

- Activity A should take one day
- Activity B starts on the same day, and lasts four days
- Activity C lasts two days, but doesn't start until Day Two... and so on.

Bar charts can be used to show additional information by adding colours, codes and symbols.

A further example of a bar chart is shown in Figure 1.18. The activities are the same as before but with the actual progress against each one shown by the shaded blue area beneath the original bars. The chart shows progress to the end of Day 3. It is easy to see which activities have been completed, and which ones are lagging behind.

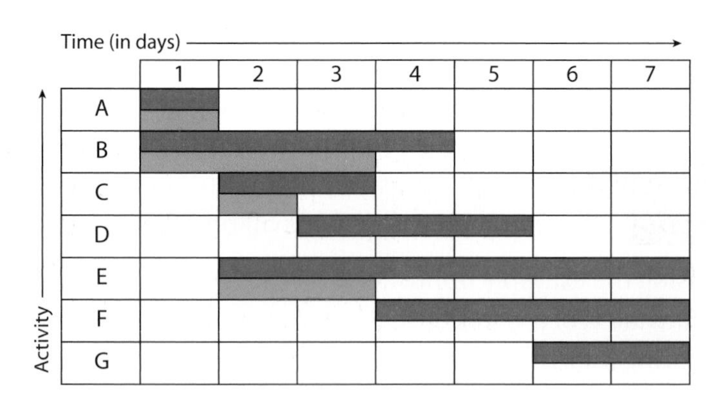

Figure 1.18 Bar chart 2

Critical path analysis

In larger projects, many tasks must be completed before the project is finished. Not all of these activities can be done at the same time, and some can't begin until others are completed. We need a way of working out the best way to organise the project efficiently, and critical path analysis (CPA) is one solution.

CPNs were developed in the 1950s by the chemical industry

Critical path networks (CPNs) are diagrams that represent each task and how they relate to one another. If we know the time needed for each activity, the overall project completion time can be calculated. The chart also helps us to see what happens if a task is delayed unexpectedly.

The critical path is the sequence of activities that fix the duration of a project. In many projects there will be some activities that aren't on the critical path. The first step in constructing a critical path network is to list all the activities, what must be done before they can start (the sequence) and expected time needed.

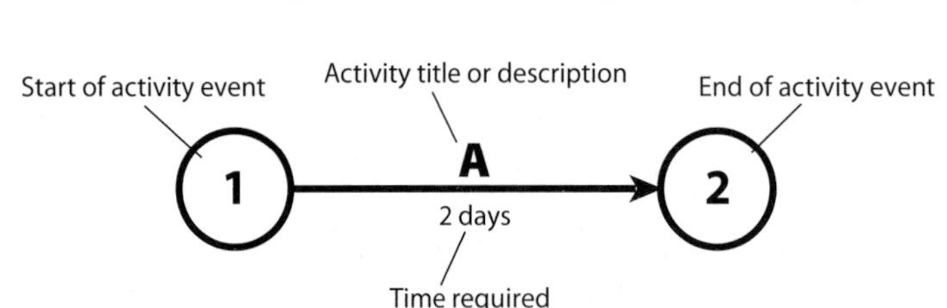

Figure 1.19 Critical path diagram

The circles are 'events'; they have no duration, but represent the time at which the activity starts or finishes. The arrow represents the activity, and always goes from left to right. It starts and ends with an event. The diagram above represents an activity (A), which lasts for two days.

We build the network by linking the activities from left to right at their start and end events. One rule of CPN is that no two activities can begin and end on the same two events. To explain this, we'll use an example.

We have four activities (A, B, C and D) to complete. Activities A and B can run at the same time, but C cannot begin until Activities A and B have been completed, and Activity D cannot begin until Activity B has been completed.

Activity	Activities that need to be done before this activity
A	None
B	None
C	A and B
D	B

Table 1.6 Sequence of activities for a Critical path analysis

As the first two activities (A and B) can begin at the same time, the temptation could be to show them as:

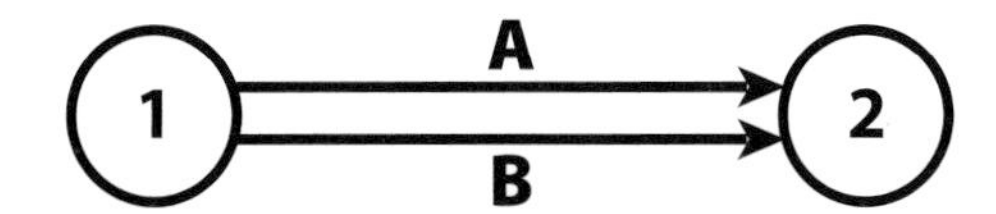

Figure 1.20 Timing of Activities A and B

However, to comply with the rule that no two activities can start and end on the same events, we introduce a 'dummy activity'. Shown as a dotted line, a dummy activity doesn't take any time. Our drawing becomes:

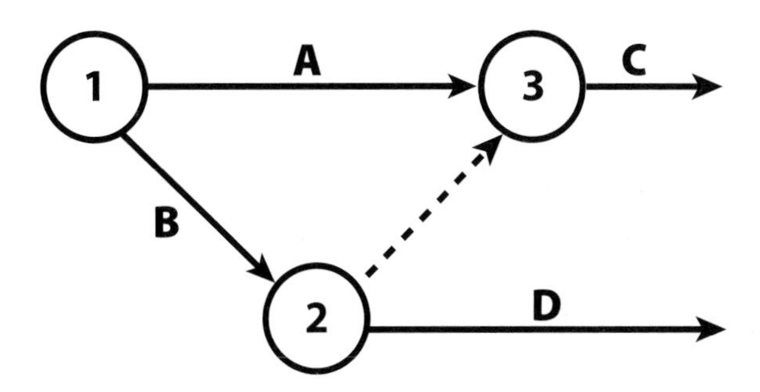

Figure 1.21 Introducing a dummy activity

This shows that Activity C can only start at Event 3, when both A and B are complete. Now let's look at a more detailed example.

The information we have about the activities is:

- **Activity A** starts on Day one and lasts two days.
- **Activity B** starts on Day one and lasts three days.
- **Activity C** can only begin once Activity B is complete and lasts 3 days.
- **Activity D** can only begin once Activity C is complete and lasts 3 days.
- **Activity E** cannot start until Activity A is complete; it will take 5 days.
- **Activity F** can start at any time and lasts 2 days.

Time in days →

Acctivity	1	2	3	4	5	6	7	8	9
A	■	■							
B	■	■	■						
C				■	■	■			
D							■	■	■
E			■	■	■	■	■	■	
F	■	■							

Figure 1.22 Activities shown as a bar chart

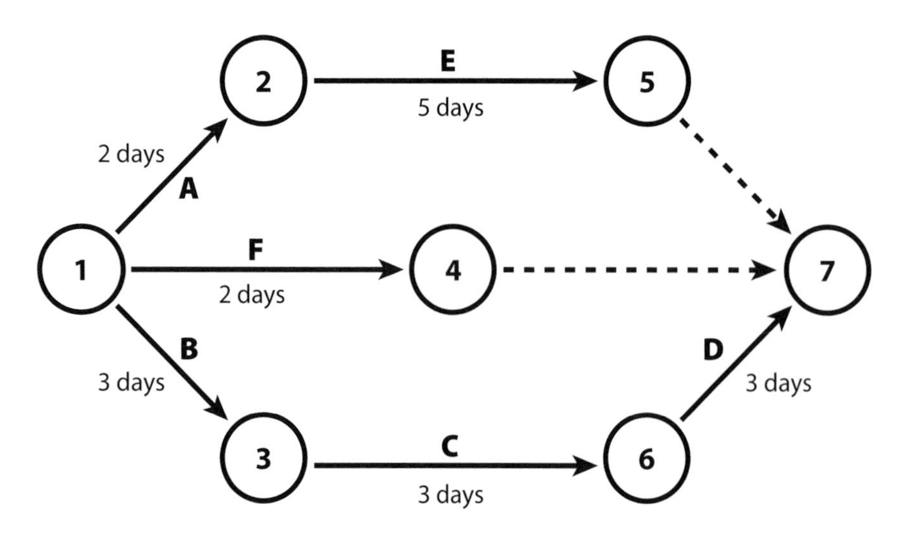

This may need a little careful study to see how it all fits together.

The minimum time for project completion is taken as the longest time path through the network. In this example it is through points 1, 3, 6 and 7, giving us a project time of nine days.

Figure 1.23 Critical path network

The bar chart may seem easier to understand, but it can't readily show when several activities have to be completed before another can begin. With practice, using the critical path method will give you greater control over establishing when activities may start.

Remember

Both the bar chart and the critical path network are management planning tools; their purpose is to help a contract to run smoothly

Bar charts and critical path networks will help you to see which activities affect your work and how it fits in with everyone else's, so you can:

- plan which areas to work in
- see when you can start each activity
- make sure you have the correct materials and equipment ready at the right time
- avoid causing delays to others and the overall contract.

INTERNAL MEMO

To: D Boss, Contracts Engineer

From: A Foreman, Site Supervisor

Project: The New Hospital

Following the arrival of the new essential services generator on site, we have found that it is too big for the entrance to the existing generator house. I have spoken to the Main Contractor, and we believe that a section of roof could be removed easily and the generator craned into position.

Please advise.

A Foreman

Figure 1.24 A short memo from a site supervisor to a contracts engineer

Reports

A report is simply a written document that tells someone about something. In a construction project, a report may be needed for many reasons. For example:

- progress made since the last report
- description of an equipment installation problem
- an investigation into the different ways of solving a problem (with a recommendation for action)
- explanation of why there is an attendance problem on site.

Reports need to be factual, to the point and not long-winded. Often a report is requested by someone needing information about an issue. At other times, a report will be written by someone who has information to share.

The memo shown in Figure 1.24 is really a short report. It may be that the contracts engineer will ask for more detail from the site supervisor to understand how the situation has arisen. When preparing a report, an electrician may have to refer to other documents, such as the site diary. Other documents that might be helpful include letters from other contractors, inspection and test results sheets, and drawings and specifications.

Project management

You should now be able to see that many different skills and activities are needed to complete a successful project. Every project is unique, but a lot of tasks are common to them all. We need to have a sensible plan for pulling all these together. This is called a **work plan**.

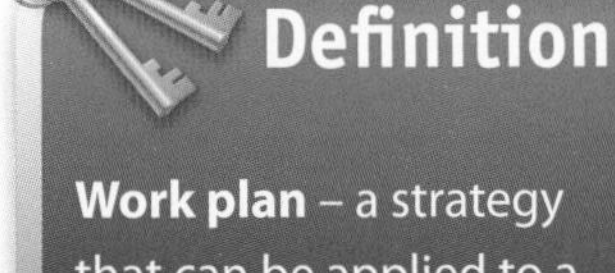

Work plan – a strategy that can be applied to a whole project (or part of one) to ensure that the installation is carried out safely and efficiently

Typically, a work plan will include the following activities:

- checking the drawings, instructions and specifications
- identifying the tasks to be done
- checking that the work area and environment are suitable and safe at all times
- listing the tools, materials and equipment needed for the project, and making sure they are available at the right time
- finding out what skills are required
- creating a logical sequence for all the activities
- co-ordinating with other contractors
- managing the installation process
- making sure that the site and installation comply with all appropriate legislation and codes of practice
- inspecting and testing of the installation
- ensuring the project has been completed satisfactorily to specified requirements.

Depending on the size of the project, one person or many may have the responsibility for devising and monitoring the work plan. On a large project, Contract Managers and Engineers, Project Engineers, Safety Officers and Site Supervisors may all be involved. For a small project, many of these tasks (if not all) may fall to the Electrician on site. This might include producing drawings and specifications.

All the tools we have covered in this chapter, including:

- communication skills
- time sheets
- delivery records
- letters and reports
- bar charts
- critical path networks
- day work sheets
- variation orders

will help you complete a project successfully.

Relationships and teamworking

Customer relations

Remember

Be polite to everyone you come into contact with during your work projects, not just the main 'customer'

Who is your **customer**? At first, you'll probably think of the person who wants the work done and is paying for it. But does this mean you can ignore everyone else you will meet in your job? A wider definition of a customer is: 'Anyone who has a need or expectation of you.'

Using this definition, almost everyone you work with becomes your customer, and you will, in turn, be theirs. They may be architects, consultants, clients, other tradespeople or a member of the public. They may ask you to do some complex task, or simply ask you a question. In either case, they will expect a good, polite response. If you treat everyone as a valuable customer, and always try to give them your best service, it will bring you many benefits. Dealing with people is an important part of your job. It is never wise to upset people, if only because it may cause problems later for you or your firm.

DO	DON'T
• be honest • be neat and tidy in your personal appearance, and look after your personal hygiene • learn how to put people at ease, and be pleasant and cheerful • show enthusiasm for the job • try to maintain friendly relationships with customers, but don't get over-familiar • know your job and do it well – good knowledge of the installation and keeping to relevant standards gives the customer confidence in you and your company • explain what you are going to do, and how long it will take • if you are not sure about something – ask!	• 'badmouth' your employer • use company property and materials to do favours for others • speak for your employer when you have no authority to do so • use bad language • smoke on customer premises • gossip about the customer or anyone else • tell lies – the customer will find out eventually if he or she is being misled or ripped off • assume that you know what your employer wants without bothering to ask.

Table 1.7 Some dos and don'ts to improve customer relations

When working in someone's home or office:

- take care to protect their property. Use dust covers, and ask them to remove objects that might get broken or damaged
- if there are pets or small children around, ask for them to be kept well away from the working area

- make sure you inform the customer before using hazardous substances. Take the correct precautions and respect any instructions you get from the customer
- before starting work, make sure that you understand exactly what the client expects
- if you have recommendations for improvements or alterations, take time to discuss these with the client and give clear technical information if it is needed.

Always try to provide answers to any questions about the work. You might be asked:

- Is this the right product for the job?
- Will it cost a lot to buy and install?
- Will it do what I need?
- How reliable is it?
- Will I be able to use it?
- How easy is it to repair?
- How long does the guarantee last?
- Will you be finished on time?

Before answering the question, try to understand why the customer is asking it: what do they really want to know? Are they worried that they cannot afford the installation? Have they booked a holiday that starts just after you are scheduled to finish?

If you don't know the answer, don't guess. Promise to find out – and then do so! When the job is completed, make sure the client fully understands how to use the installation, and leave behind any manufacturers' user guides or installation manuals. Invite the customer to contact you if there are any problems in the future.

If you follow this advice you will have excellent relationships with your customers. In the unlikely event that a dispute arises, you will both need to seek the help of an independent mediator to settle your differences.

Remember

Remember that you are representing your company. People will judge the company by the way you behave. If you do well, your company could get more work from the client. On the other hand, if you don't, contracts may be lost and you may be out of a job

On the job: Cleaning up mess

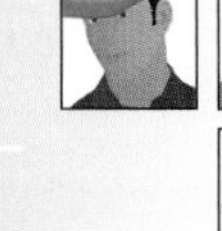

Ellie is an apprentice electrician who always cleans up any mess she has made. She is working with Joe, the senior electrician. While installing an extra socket outlet in an old people's home Joe has left a mess in the kitchen; he has also burnt the kitchen work surface with a cigarette.

1. How could this situation have been avoided?
2. What steps could Ellie take to explain the situation to the home's manager?
3. What should happen to Joe when he gets back to the company premises?

Promoting good relationships with fellow workers

A site where everyone works together to finish the job is much happier and more productive than one where people are at loggerheads. Here is a checklist of the things you can do to help make things run smoothly:

- ✔ cooperate with other trades – it's always better than conflict
- ✔ be patient and tolerant with others
- ✔ attend site meetings regularly – this helps liaison with other trades
- ✔ keep to the agreed work programme
- ✔ do your work in a professional manner
- ✔ finish your work on time; don't hold others up if you can help it
- ✔ respond cheerfully to reasonable requests from colleagues
- ✔ don't leave the site for long periods of time
- ✔ don't borrow tools and materials unless it is necessary, and return them promptly and in good condition if you do
- ✔ tell your employer if you have personal or work difficulties – don't be too proud
- ✔ take good care of your and others' property
- ✔ keep noise down, especially from your radio
- ✔ show respect for everyone on site – make an effort to learn their names
- ✔ make sure everyone, including visitors to the site, has the right PPE
- ✔ report any breakdown in discipline or disputes between co-contractors promptly to the site supervisor
- ✔ keep a current edition of the Wiring Regulations or the Amicus guide book with you on site
- ✔ always do your best to answer questions from visitors or other tradespeople
- ✔ never play practical jokes on colleagues (for example hiding tools, lunch boxes, car keys). This can cause bad feeling and may result in injury or accident.

Teamworking

Successful projects require good teamwork. But who exactly is 'the team'? In one sense, everyone working on the project is part of the team. More commonly, though, contractors will see their own group of people as a team, who will be working with other teams of tradespeople to complete the whole project.

Working well with others is all part of the job

Here is one way to look at the development of a team.

1. Forming
2. Storming
3. Norming
4. Performing.

The model gives useful insights whether we take a team as being just the squad of electricians, or everyone working on the site. As a team develops, relationships between team members shift, and the team leader changes leadership style. As you read through it, try to apply the model to teams you have been involved with: a gang of mates at school, perhaps, or sports team.

Team development

1. Forming

This is the first stage, where the team has just come together. The members probably don't know one another very well, and individual roles and responsibilities are unclear. There is little agreement between members about what the team is trying to do. Some will feel confused and won't know what they should be doing. At this stage the team relies heavily on the team leader for guidance and direction. The leader must provide lots of answers about the team's purpose and objectives and relationships with groups outside the team.

2. Storming

During this stage, team members jockey for position as they try to find themselves a role in the team. The leader might receive challenges to his or her authority from other team members. The team's purpose becomes clearer, but there remains plenty of uncertainty, and decisions are hard to achieve because members will argue a lot. Small groups or factions may form, and there may be power struggles. The team needs to be focused on its goals to avoid being broken up by relationship and emotional issues. Some compromises will be needed to make any progress. The leader has to become less bossy and more of a coach.

3. Norming

This is a more peaceful stage, when team members generally reach agreements easily. Roles and responsibilities are clear and accepted by all, and the team works together. Members develop ways and styles of working by discussion and agreement together. Big decisions are made by the whole team or the team leader, but smaller decisions are left to individuals or small groups inside the team. The team may also enjoy fun and social activities together. The leader now acts to guide the team gently and enable it to do its job, and has no need to enforce decisions. The team may share some leadership roles.

4. Performing

In this final stage, the team knows clearly what it is doing and why. It has a shared vision and needs little or no input from the leader. If disagreements occur they are tackled positively by the team itself. The team works together towards achieving the goal, and also copes with relationship, style and process issues along the way. Team members look after each other. The leader's role is to delegate and oversee tasks and projects, and there is no need for instruction or assistance, except for individual personal development.

Did you know?

During a recent survey, employers were asked: 'During a half-hour interview with a prospective employee, when do you make your mind up?' Most replied: 'Within 10 seconds. If I like them, they have 30 minutes to lose the job. If I don't, they have 30 minutes to persuade me to change my mind'

Communication

All the ideas we have looked at so far in this section involve communication. Communication is about much more than just speaking or writing. You communicate an enormous amount by how you look, the gestures and facial expressions you use, and the way you behave. Even something as simple as a smile can make a big difference to your communication.

The benefits of good communication are huge. Good attitude, appearance and behaviour will immediately put your customers at ease and earn you respect from others. Clearly conveyed thoughts and ideas, when backed up with good working practices and procedures, will invariably improve productivity, increase profitability and produce satisfied customers.

Speaking, writing, appearance, attitude and behaviour all combine to make up your personal 'communication package'. It affects your relationships with everyone you deal with – client, client representative, architect, main contractor, other site trades, your supervisor and your work colleagues.

Letter and report writing

Good written communication skills are vital for avoiding confusion, especially in an industry that relies heavily on documentation. Writing good letters and reports is a part of the communication process. Letters can be used to:

- request information, e.g. delivery dates from a wholesaler, or progress updates from a main contractor
- inform people of situations.

Reports can be used to:

- summarise investigations into causes and effects of problems or trends, and recommend solutions, e.g. Why has there been an increase in absenteeism? Why is the work below standard?
- provide statistical or financial summaries (e.g. end-of-week materials costs, annual accounts)
- record decisions made at meetings
- supply information for legal purposes, e.g. accident reports
- monitor progress, e.g. negotiations, construction work, implementation of a new system
- look into the feasibility of introducing new procedures, processes or products, or changing company policy.

Remember

Try to use simple language, and don't try to impress the reader with your huge vocabulary - it will only frustrate or annoy them. Don't use jargon or abbreviations unless you are sure your reader is familiar with them

The writing process

Writing isn't too difficult if you break it down into steps and then focus on one step at a time. You can tackle almost any writing project if you follow this method. The main steps are:

Step 1: Have a clear idea about the overall purpose of your letter or report. Think about exactly what you want to say, and why. Most people get stuck because they have not thought carefully about what they are trying to achieve.

Step 2: Gather all the information you will need.

Step 3: Plan the logical order for presenting your ideas. Write a list of headings, and check that each one follows on from the previous one. If it is a report, the first paragraph should be a summary of the rest of the document, and if you want your reader to take some action, say so in your last paragraph. Don't start writing in detail until you have completed this step.

Step 4: Now that you know what each section or paragraph will contain, go ahead and write each one. Then read it through again and, if necessary, edit what you have written. Finally, present it as a properly formatted and typed document.

Hints and tips for good writing

- Avoid 'wordy' and complicated sentences.
- Read this sentence: 'I found out that I should take an investigative look at the new plant room in order to establish a plan to help us re-evaluate the installation methods and techniques that we intend to use.' It could be written as: 'I will be checking our proposed installation in the new plant room to see if there is a better way of doing things.' Much easier to understand, isn't it?
- Communication always involves two people: the writer/speaker and the reader/listener. You should think about your reader and write for them, clearly conveying your ideas so that the reader can easily understand them. For example, you wouldn't expect to write the same for a five-year-old as you would for a college lecturer.

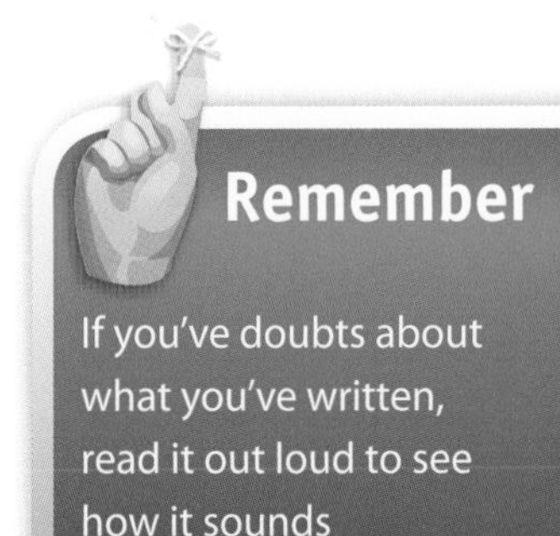

Make your writing interesting

Don't let your document become boring and repetitive. Use a **thesaurus** to find alternative words to use. A thesaurus contains lists of words with similar meanings grouped together. For example, instead of the word 'wherewithal' in the sentence 'I have the wherewithal to pay the bill', the thesaurus gives 'means', 'resources', or 'ability' (among others) as possible alternatives that you could use.

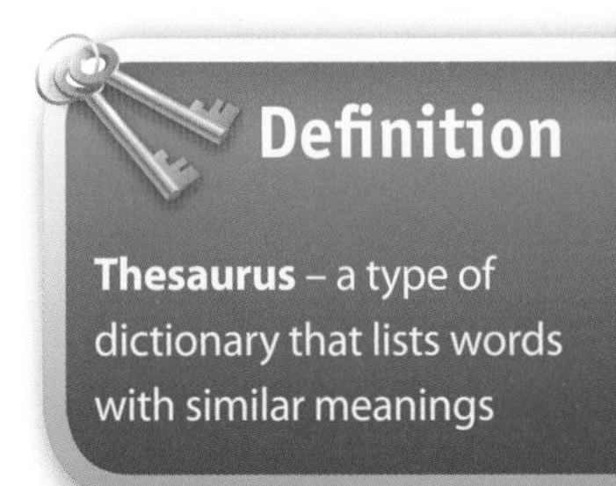

Check your work before sending it

We all make mistakes! Checking your work (known as proofreading) is a very important aspect of writing. Pay attention to grammar, spelling and punctuation. If you are using word-processing software to write your letter or report, it will often indicate when something might be wrong. Get into the habit of finding out what the problem is and correcting it. However, always remember that the software cannot detect every mistake. The following sentence has several errors, but they won't show up on the computer screen:

> *'Two many electricians were note iced to be erecting the too extract fans in 2 the to staff toilets.'*

A dictionary can help: if you aren't sure how to spell a word, look it up in a dictionary. If possible, don't proofread something immediately after you have written it. You are more likely to spot any mistakes and find better ways of saying things if you go back to it later.

Report writing

Reports usually describe a problem or an investigation. Their purpose is to help someone, or a team, to make a decision, by 'reporting' all the relevant facts and perhaps making some recommendations for action. Some reports will present a solution to the problem, others may simply list historical and factual data.

Remember

Always proofread written material before sending it out

Many companies will have a standard structure or format for reports, and you should use this whenever you can. However, a lot depends on the subject and type of the report, and there is no merit in slavishly following a format if it makes it difficult to understand what the report is about. As with any writing, it is important to know clearly what you are trying to say, and then to put that into a style and format that will be acceptable to the reader.

Style, although hard to define, can make a big difference to the success of a report. A good style can help to convince the reader of the merits of the report and its recommendations. A bad style may put the reader off, even if the content of the report is perfectly OK.

Computers and calculators can help when you're doing the paperwork

Report structure

Companies may have a required format, but they will probably include most or all of the elements described below.

For a business report or memo:

1. Basic identification data (who it is to, who is writing it, the date, subject, reference number).
2. Summary – the project or problem and the purpose of the report.
3. Background – the history of the issue being reported.
4. Relevant data – the evidence you have gathered.
5. Conclusions and recommendations.

For a more formal technical report:

1. Title page
2. Acknowledgement
3. Change history
4. Contents list
5. Report summary
6. Introduction
7. Technical chapters
8. Conclusions and recommendations
9. References
10. Appendices (these contain additional specialised information).

Whatever style and format is used, the principles of good writing still apply.

Be clear	• The reader must be able to understand the report easily. • Explain symbols, tables or diagrams. • Never assume that your reader has prior knowledge of the issues. • Don't use jargon or abbreviations unless you know that the reader will be familiar with them.
Be concise	• Don't waffle. • Be succinct without making the report hard to understand. • Think about what the reader wants to know, i.e. evidence and conclusions, not a vivid description of the valiant effort you made to compile them.
Be logical	• Ensure you have a beginning, middle and end. • There should be a sensible flow between sections, chapters, paragraphs and sentences. • Avoid jumping randomly from idea to idea. • Take out unnecessary distractions such as pretty graphics that don't add any extra information.
Be accurate and objective	• Base your report on honest facts – an inaccurate report is at best pointless and at worst harmful. • Remember that someone may make an important decision based on your recommendations. • If you have to make assumptions, make it clear that you have done so. • Be tactful. What you say may contradict others or what the reader thinks. • If you feel very strongly about something, don't send the report until you have had some time to think about it, and make sure it contains only facts. • Don't use reports as a sounding-off platform. It is very easy to destroy a relationship by sending an angry, ill-conceived letter or report.

Table 1.8 Good writing practice

FAQ

Q Do I work in the construction or engineering industry?

A You are really employed within the electrotechnical industry, providing whatever electrical services are required. There are many people who work in one area only, but equally there are many more that work across the complete sector.

Q Why do I need to know about delivery notes, day work sheets etc, my foreman deals with all these?

A Today he does, but what about the future? If you are not aware of how important and necessary all the paperwork is you will not be employed for very long. For example, by checking the delivery against the delivery note you are able to ensure that both the correct quantity and correct type of materials have been delivered to site. When the delivery note is forwarded to your employer it triggers payment to the supplier. On larger jobs the regularity of the delivery notes arriving at the office, recording the materials delivered to site, gives an accurate picture of progress against the programme for the job. I'm sure you know the value of completing a time sheet!

FAQ

Q Do I really have to know about critical path analysis?

A If in the future you become a design/project engineer then it will be essential if you are to successfully estimate and run projects. However complicated it sounds with its dummy activities, event times and nodes, it is only about breaking a task down into segments and deciding the order in which it is to be done. You will be doing this every day on every task you are given, but without writing it out on paper.

Knowledge check

1. List 10 areas that qualified electricians may find themselves working in.
2. Describe in your own words 'electrical services'.
3. What does the NICEIC stand for? Write a brief description of this organisation.
4. Which body controls and produces the Wiring Regulations (BS 7671)?
5. Describe the role of the clerk of works in relation to a works project.
6. Describe how a contract can be made.
7. What's the difference between a nominated subcontractor and a non-nominated subcontractor?
8. Why are time sheets, and the information on them, very important?
9. Who would normally compile a site report or site memo?
10. With the aid of diagrams describe a Gantt chart.
11. Describe the four main steps in the writing process.

chapter 2

Health and safety

Overview

All employers, including the self-employed, have duties under the Health and Safety at Work Act 1974 to ensure the health and safety of themselves and others affected by what they do. This includes people working for employers and the self-employed (e.g. part-time workers, trainees and subcontractors), those who use the workplace and equipment they provide, those who visit their premises, and people affected by their work (e.g. neighbours or the general public).

This chapter looks at UK health and safety legislation and its effects, and best practices to ensure a safe working environment. It will cover:

- **Safety signs**
- **Personal Protective Equipment (PPE)**
- **Legislation**
- **Implementing and controlling health and safety**
- **Risk assessment**
- **Fire safety**
- **Site safety**
- **Electrical safety**
- **Safe isolation of electrical supplies**
- **First aid**
- **Other safety topics**
- **General hazards**

Did you know?

242 people were killed in workplace accidents in Britain in 2006/07. There were also 28,267 reported major injuries and 113,083 reported injuries that caused three or more days absence from work (source: The Health and Safety Executive)

Safety signs

Working on construction sites, in factories or elsewhere, you will see a variety of signs and notices. It's up to you to learn and understand what they mean and to take notice of them. Most warn of a possible danger and must not be ignored.

Figure 2.1 Types of safety sign

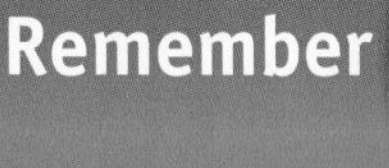

Remember

Accidents don't just happen: they are caused, usually by human failing. However, with knowledge and the application of common sense and tested procedures, there is no reason why you should ever become one of the statistics

Each type can be recognised by its shape and colour. Some signs are just symbols while others have words or other information such as heights and distances.

	Prohibition signs	Mandatory signs
Shape:	Circular	Circular
Colour:	Red borders and cross bar. Black symbol on white background	White symbol on blue background
Meaning:	Shows what must NOT be done	Shows what must be done
Example:	No smoking	Wear eye protection

	Warning signs	Information or safe condition signs
Shape:	Triangular	Square or rectangular
Colour:	Yellow background with black border and symbol	White symbols on green background
Meaning:	Warns of hazard or danger	Indicates or gives information on safety provision
Example:	Danger: electric shock risk	First aid post

Figure 2.2 Safety signs

Personal Protective Equipment (PPE)

Updated in 2002, the Personal Protective Equipment at Work Regulations 1992 set out principles for selecting, providing, maintaining and using personal protective equipment (**PPE**). This is equipment designed to be worn or held to protect against a risk to health or safety. It includes most types of protective clothing and equipment such as eye, hand, foot and head protection.

Employers have a duty to reduce the workplace risk to as low as is reasonably practicable; where some risk remains then PPE is issued as a last resort. PPE should then be issued free of charge by the employer and must be suitable to protect against the risk for which it is issued.

Employers	Employees
Must train employees and give information on maintaining, cleaning and replacing damaged PPE	Must use PPE provided by their employer, in accordance with any training in the use of the PPE concerned
Must provide storage for PPE	Must inform employer of any defects in PPE
Must ensure that PPE is maintained in an efficient state and in good repair	Must comply with safety rules
Must ensure that PPE is properly used	Must use safety equipment as directed

Table 2.1 PPE responsibilities of employers and employees

Areas requiring protection, together with related examples of PPE, are shown on the following pages. However, as formal risk assessment defines the type of PPE required in any given situation, no attempt has been made to quantify the circumstances in which they should be used.

Remember

Your eyes are two of your most precious possessions. They are among the most vulnerable parts of your body to injury at work

Eye protection

Every year thousands of workers suffer eye injuries, which result in pain, discomfort, lost income and even blindness. Following safety procedures correctly and wearing eye protection can prevent these injuries. There are many types of eye protection equipment available, for example safety spectacles, box goggles, cup goggles, face shields, welding goggles etc. To be safe you have to have the right type of equipment for the specific hazard you face.

Safety goggles

Helmet with visor face screen

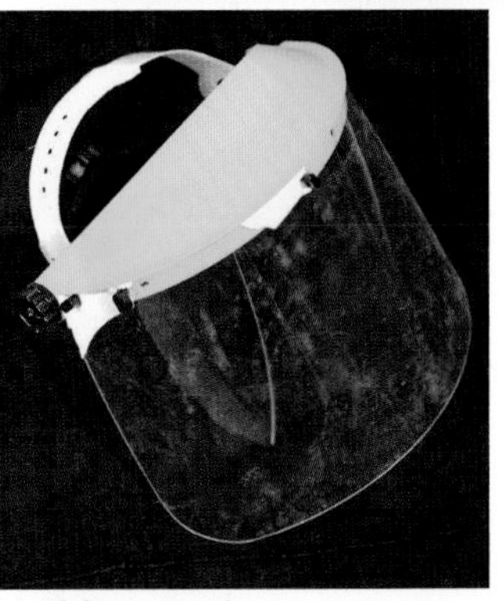

Half-face mask

Foot and leg protection

Did you know?

Toe and foot accidents account for a large proportion of reported accidents in the workplace each year, and many more accidents go unrecorded

Guarding your toes, ankles, feet and legs from injury also involves protecting your whole body from injury caused by improper footwear, for example an injury caused by electric shock.

Protective footwear can help prevent injury and reduce the severity of injuries that do occur.

PPE		Usage
Safety shoe, boot or trainer 		Basic universal form of foot protection Safety clogs can also be worn. These are used particulary as protection against hot asphalt.
Spats	Often made of leather and worn over the shoe	Protect the feet from stray sparks during welding
Gaiters	Worn over the lower leg and top of the shoe	Used to give protection against foul weather/ splashing water
Leggings 		Protect general clothing

Figure 2.3 Types of leg and foot PPE

Hand protection

This involves the protection of two irreplaceable tools – your hands, which you use for almost everything: working, playing, driving, eating etc. Unfortunately hands are often injured. One of the most common problems other than cutting, crushing or puncture wounds is **dermatitis**.

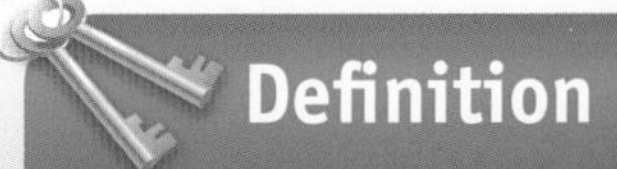

Dermatitis
– inflammation of the skin normally caused by contact with irritating substances

Skin irritation may be indicated by sores, blisters, redness or dry, cracked skin that is easily infected.

To protect your hands from irritating substances you need to:

- keep them clean by regular washing using approved cleaners
- wear appropriate personal protection when required
- make good use of barrier creams where provided.

Rigger gloves

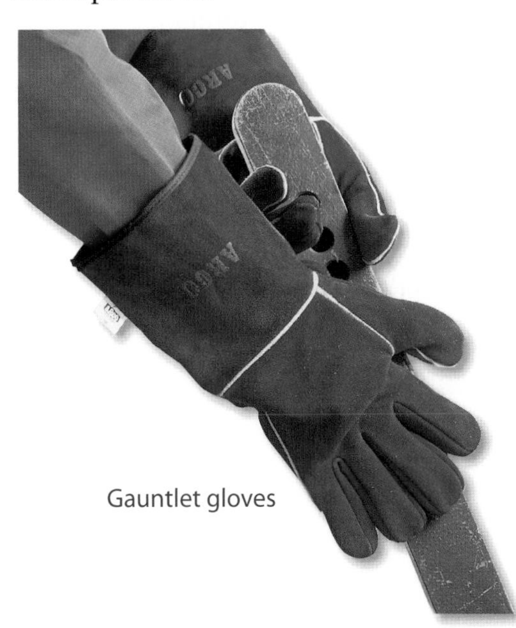

Gauntlet gloves

Head protection

Head protection is important because it guards your most vital organ – your brain! Head injuries pose a serious threat to your brain and your life. Head protection can help to prevent such injuries.

A single injury can handicap a person for life or even be fatal

Here is a list of good safety practice:

- know the potential hazards of your job and what protective gear to use
- follow safe working procedures
- take care of your protective headgear
- notify your supervisor of unsafe conditions and equipment
- get medical help promptly in the case of head injury.

There are several types of protective headwear for use in different situations; use them correctly and wear them whenever they are required.

Safety helmets

Did you know?

An estimated 80% of industrial head injuries are sustained by people who are not wearing any protective equipment

Here are a few important rules:

- adjust the fit of your safety helmet so it is comfortable
- all straps should be snug but not too tight
- don't wear your helmet tilted or back to front
- never carry anything inside the clearance space of a hard hat, e.g. cigarettes, cards and letters
- never wear an ordinary hat under a safety helmet
- do not paint your safety helmet as this could interfere with electrical protection or soften the shell
- only use approved types of identification stickers on your safety helmet, e.g. First Aider

Safety helmet

- do not use sticky tape or Dymo tape as the adhesive could damage the helmet
- handle the helmet with care: do not throw it or drop it etc.
- regularly inspect and check the helmet for cracks, dents or signs of wear, and if you find any, get your helmet replaced
- check the strap for looseness or worn stitching and also check your safety helmet is within its 'use-by' date.

Safety tip

If you have long hair and don't want it ripped off by something, tie it up or use a hair net

Bump caps

For less dangerous situations, where there is a risk of bumping your head rather than things falling, or where space is restricted, bump caps, which are lighter than safety helmets, may be acceptable.

If you have to work outside in poor conditions, and a safety helmet is not a requirement, consider using a Souwester and cape.

On the job: Head protection

Fazal is a first-year apprentice who is starting his first day on a construction site. The electrician in charge gives Fazal a hard hat to wear. Fazal notices that the hat has a crack down one side and reports this to his supervisor. The supervisor tells Fazal that he must wear this hat because there are no more in the store and he needs to get on with his work.

1. What should Fazal do in this situation?
2. If an object hit Fazal on the head and he was injured because of the crack in the hat, who would be responsible for this injury?

Hearing protection

You may sometimes have to work in noisy environments, and if you do, your employer should provide you with suitable hearing protection. Like any other sort of PPE it is important that you wear hearing protection properly and check it regularly to make sure it is not damaged. Typical types are ear plugs that fit inside the ear, or ear defenders that sit externally such as headphones.

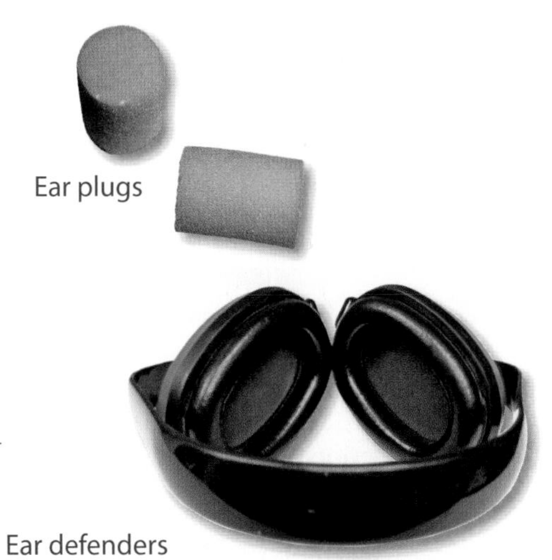

Ear plugs

Ear defenders

Did you know?

Failure to use ear protection when required can result in permanent damage to your hearing

Lung protection

We all need clean air to live, and we need correctly functioning lungs to allow us to inhale that air. Fumes, dusts, airborne particles such as asbestos or just foul smells, such as in sewage treatment plants, can all be features of construction environments.

A range of respiratory protection is available, from simple dust protection masks to half-face respirators, full-face respirators and powered breathing apparatus. To be effective these must be carefully matched to the hazard involved and correctly fitted. You may also require training in how to use them properly.

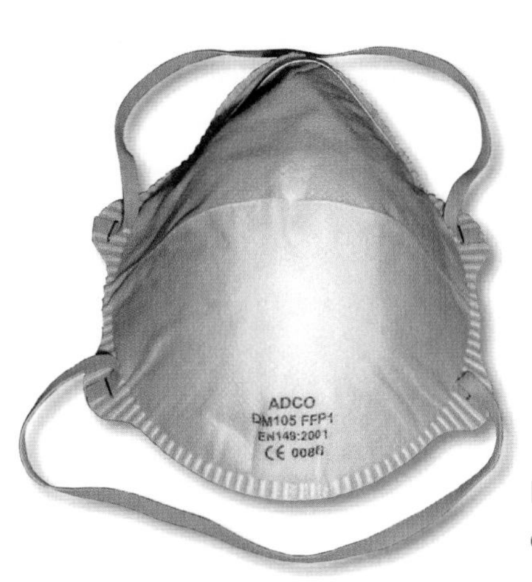

Disposable dust respirator

Whole body protection

To complete our 'suit of armour' against all things harmful, we need to protect the rest of our body. Usually this will involve overalls, donkey jackets or similar, to protect against dirt and minor abrasions. However, specialist waterproof or thermal clothing will be needed in adverse weather conditions; high visibility clothing is required on sites or near traffic; and chemical resistant clothing, such as neoprene aprons, are necessary if working near or with chemical substances. When working outdoors, sunscreen is now also a consideration.

High-visibility jacket

High-visibility waistcoat

Overalls

Remember

Remember: all items of PPE that are provided for your protection must be worn and kept in good condition

When should I wear PPE?

Risk assessment is the key here, and we'll talk more about that later. However, all construction sites that you work on will require you to wear a similar basic level of PPE.

PPE	When worn
Hard hats	• where there is a risk of you either striking your head or being hit by falling objects
Eye protection	• when drilling or chiselling masonry surfaces • when grinding or using grinding equipment • when driving nails into masonry • when using cartridge-operated fixing tools • when drilling or chiselling metal • when drilling any material that is above your head
Ear protection	• when working close to noisy machinery or work operations
Gloves	• whenever there is a risk to the hands from sharp objects or surfaces • when handling bulky objects to prevent splinters, cuts or abrasion • when working with corrosive or other chemical substances
Breathing protection	• when working in dusty environments • when working with asbestos • when working where noxious odours are present • when working where certain gases are present

Table 2.2 When to wear PPE

Other items

When working involves long periods of kneeling or having to take your weight on your elbows, you may be issued with specialist protectors for these areas. Other items you may use could include face masks, safety harnesses or breathing apparatus.

FAQ

Q Do I really need to wear a hard hat? They look so silly.

A Simply put, if the risk assessment says so, then yes, you do need to wear it. Better to look silly than to look dead!

Q Do I have to remember all the safety signs?

A No, there are far too many for you to be able to remember them all. What is important is to remember the four different types and colours used, so you know if it's just advising you or telling you something you must or must not do. For example, warning notices are triangular and have black symbols on a yellow background. Mandatory notices show what must be done and are round with a blue background and a white symbol.

Here are some guidelines for safe working. Take a minute to look at these and think about how you work. Are you taking any unnecessary risks?

Work tidily and cleanly. Do not leave objects lying on the floor where they may cause accidents. Clean all materials and debris away from the site at the end of the working day and ensure that when working overhead on scaffolds, trestles, ladders and steps, you do not lay anything down in such a position that it may fall on anyone or anything below.

Observe all rules and work instructions provided. When you start work your employer will make you aware of the company's rules and expectations with regard to safety. Most employers will have a health and safety policy or statement and it is part of your job to acquaint yourself with the contents and to ensure that all of your activities comply with the stated requirements.

Running or hurrying can cause accidents. Never run or take short cuts, even if you are in a hurry. You may collide with someone, trip over an obstruction or run into a protruding object causing an injury. It is always better to walk and arrive safely. Construction sites are particularly dangerous places to work if you do not take common-sense precautions.

Keep all machinery and equipment well maintained and in good condition. Never use damaged machinery, tools or equipment, and make sure that any damage that you may cause is reported or repaired promptly so that it does not endanger the next user.

Secure all loose clothing and repair any torn articles immediately. Overalls should always be fastened with no flaps or torn pieces hanging off that may become tangled with rotating machinery etc. If your hair is long, it should be covered by a 'snood' cap (hair net) or tied up so that it is not a hazard, and all jewellery should be removed where a safety hazard exists.

Advise supervisors immediately if you observe any unsafe practices or notice any defects in any of the equipment that is provided for use by yourself and others.

Follow all manufacturers' instructions and recommendations when using items of equipment.

Examine all electrical tools and equipment very carefully before use to ensure that they are in good working order and show signs of having been recently inspected.

Let others know when you are working overhead or nearby when your activities may pose a particular danger to them.

You are responsible for the safety of yourself and others with whom you work or who may be affected by your work. Don't leave things lying around. Everything that you do must be of the very highest commercial and safety standards so that it DOES NOT present any significant danger to you or other people who may be affected by your actions.

Remember

Accidents don't just happen; they are caused and are invariably the result of human failing. However, with knowledge and the application of sensible thoughts and procedures, there is no reason why you should ever have an accident

Legislation

This section looks at the regulatory requirements and responsibilities laid down by Acts and Regulations. All of the following legislation is law and enforceable. As a worker, these Regulations provide a legal framework that protects you. This includes the building in which you are working, the electrical supply that you are working with and they also ensure that you do not cause harm to yourself or others by using bad practices.

There are many sites and projects and consequently many Acts and Regulations. We cannot predict every site that you will ever work on, but we will look at the most important legislation.

Health and Safety at Work Act 1974 (HASAWA)

The Health and Safety at Work Act 1974 (HASAWA) is what is called an 'Enabling Act'. It sets out the basic principles by which health and safety at work is regulated. Attached to the Act is a series of Regulations covering the practical detail of how employers and employees create, maintain and operate a safe working environment. They are also used by government authorities to control the standard of working conditions throughout industry.

The Act is **statutory**, that means it is binding in law, and criminal penalties can be imposed on people found guilty of malpractice and misconduct.

All employers are covered by the HASAWA. The Act places certain specific duties on both employers and employees which must be complied with by law.

Each employee, and this includes *you*, is also required by law to assist and co-operate with their employer and others in making sure that safe working environments are maintained, that all safety equipment is fully and correctly used and that all safety procedures are followed.

There are many sections to the Act. The main ones that will affect you are:

Section 1 Gives the general purposes of Part 1 of the Act, which are to maintain or improve standards of health and safety at work, to protect other people against risks arising from work activities, to control the storage and use of dangerous substances and to control certain emissions into the air.

Section 2 Contains the duties placed upon employers with regard to their employees. We'll talk more about this later.

Section 3 Places duties on employers and the self-employed to ensure their activities do not endanger anybody (with the self-employed that includes themselves) and to provide information, in certain circumstances, to the public about any potential hazards.

Section 4 Places a duty on those in control of premises, which are non-domestic and used as a place of work, to ensure they do not endanger those who work within them. This extends to plant and substances, means of access and egress as well as to the premises themselves.

Section 6 Places duties on manufacturers, suppliers, designers, importers etc. in relation to articles and substances used at work. Basically they have to research and test them and supply information to users.

Section 7 Places duties upon employees.

Section 8 Places a duty on everyone not to intentionally or recklessly interfere with or misuse anything provided in the interests of health, safety and welfare.

Section 9 Provides that an employer may not charge his employees for anything done for, or equipment provided for, health and safety purposes under a relevant statutory provision.

Duties of employers to their employees (section 2)

The Act requires employers to ensure, so far as is reasonably practicable, the health, safety, and welfare of their employees at work. The things it covers include:

- the provision and maintenance of plant and systems of work that are safe and without risk to health
- safety in the use, handling, storage and transport of articles and substances
- the provision of information, instruction, training and supervision as is necessary to ensure the health and safety at work of employees
- the provision of access to and egress from the place of work that are safe and without risk
- the provision of adequate facilities and arrangements for welfare at work.

The employer must also:

- draw up a health and safety policy statement if there are five or more employees, including the organisation and arrangements for bringing this policy to your attention
- carry out an assessment of risks associated with all the company's work activities
- identify and implement control measures
- inform employees of the risks and control measures
- review periodically the assessment
- record the assessment if over five persons are employed
- consult a safety representative if one is appointed by a recognised trade union about matters affecting your health and safety
- if requested in writing to do so by any two safety representatives, establish a safety committee within three months of the request being made.

Duties of employees (section 7)

As an employee you have legal duties to:

- take reasonable care at work for the health and safety of yourself and others who may be affected by what you do or do not do
- not intentionally or recklessly interfere with or misuse anything provided for your health and safety
- co-operate with your employer on health and safety matters and assist the employer in meeting their statutory obligations
- bring to your employer's attention any situation you consider a serious and imminent danger
- bring to your employer's attention any weakness in their health and safety arrangements.

Electricity at Work Regulations 1989 (EAWR)

Made under the HASAWA 1974 (covered in the last section), the Electricity at Work Regulations (EAWR) came into force on 1 April 1990. They impose general health and safety requirements to do with electricity at work, on employers, self-employed people and employees.

Every employer and self-employed person has a duty to comply with the provisions of the Regulations, in so far as they relate to matters that are within their control.

Every employee has the duty to co-operate with their employer, so far as is necessary to enable the Regulations to be complied with. Because these are statutory Regulations, penalties can be imposed on people found guilty of malpractice or misconduct.

The Regulations refer to a person as a 'duty holder' in respect of systems, equipment, and conductors. The Regulations clearly define the various duty holders. It is the duty of every duty holder to comply with the provisions of these Regulations in so far as they relate to matters that are within their control.

Employer: any person or body who:

- **a.** employs one or more individuals under a contract of employment or apprenticeship or
- **b.** provides training under the schemes to which the HASAWA applies.

Self-employed: an individual who works for gain or reward other than under contract of employment, whether or not he or she employs others.

Employee: Regulation 3(2)(b) repeats the duty placed on employees by the HASAWA which are equivalent to those placed on employers and self-employed persons where these matters are within their control. This includes trainees like you who are considered employees under the Regulations.

The Regulations are designed to take account of the responsibilities that many employees in the electrical trades and professions have to take on as part of their job.

The level of responsibility you hold to make sure the regulations are met depends on the amount of control you have over electrical safety in any particular situation. A person may find himself or herself responsible for causing danger to arise elsewhere in an electrical system, at a point beyond his or her own installation. This situation may arise, for example, if you energise a circuit while somebody is working in a different room on that circuit. This is obviously a dangerous situation. Because such circumstances are 'within their control', the effect of regulation 3 is to bring responsibilities for compliance with the rest of the Regulations to that person, thus making them a duty holder.

Absolute/reasonably practicable

Duties in some of the regulations are either regarded as 'absolute' meaning they absolutely *have* to be met, or if they have a qualifying term applied to them, 'reasonably practical'. The meaning of reasonably practical has been well established in law. The interpretations below are given as only a guide.

Absolute

If the requirement in a regulation is 'absolute', for example if the requirement is not qualified by the words 'so far as is reasonably practicable', the requirement must be met regardless of cost or any other consideration.

Reasonably practicable

a. Someone who is required to do something 'so far as is reasonably practicable' must think about the amount of risk of a particular work activity or site and, on the other hand, the costs in terms of the physical difficulty, time, trouble and expense which would be involved in taking steps to reduce the risks to health and safety of a particular work process. For example in your own home you would expect to find a fireguard in front of a fire to prevent young children from touching the fire and being injured. This is a cheap and effective way of preventing accidents; this would be a reasonably practicable situation. If the cost or technical difficulties of taking certain steps to prevent those risks are very high, it might not be reasonably practicable to take those steps. The greater the degree of risk, the less weight that can be given to the cost of measures needed to prevent that risk.

b. In the context of the Regulations where the risk is very often that of death from electrocution and where the nature of the precautions which can be taken are so often very simple and cheap e.g. insulation surrounding cables, the level of duty to prevent that danger approaches that of an absolute duty.

Here is a summary of the regulations you are most likely to have to comply with and whether they are regarded as 'absolute' or 'reasonably practicable'.

Regulation 4
Standard of duty: *reasonably practicable*
All electrical systems shall be constructed and maintained to prevent danger. All work activities are to be carried out so as not to give rise to danger.

Regulation 5
Standard of duty: *absolute*
No electrical equipment is to be used where its strength and capability may be exceeded so as to give rise to danger.

Regulation 6
Standard of duty: *reasonably practicable*
Electrical equipment sited in adverse or hazardous environments must be suitable for those conditions.

Regulation 7
Standard of duty: *reasonably practicable*
Permanent safeguarding or suitable positioning of live conductors is required.

Regulation 8
Standard of duty: *absolute*
Equipment must be earthed or other suitable precautions must be taken e.g. the use of residual current devices, double insulated equipment, reduced voltage equipment etc.

Regulation 9
Standard of duty: *absolute*
Nothing is to be placed in an earthed circuit conductor that might, without suitable precautions, give rise to danger by breaking the electrical continuity or introducing a high impedance.

Regulation 10
Standard of duty: *absolute*
All joints and connections in systems must be mechanically and electrically suitable for use.

Regulation 11
Standard of duty: *absolute*
Suitable protective devices should be installed in each system to ensure all parts of the system and users of the system are safeguarded from the effects of fault conditions.

Regulation 12
Standard of duty: *absolute*
Where necessary to prevent danger, suitable means shall be available for cutting off the electrical supply to any electrical equipment. (Note: drawings of the distribution equipment and methods of identifying circuits should be readily available. Ideally, mains signed isolation switches should be provided in practical work areas.)

Regulation 13
Standard of duty: *absolute*
Adequate precautions must be taken to prevent electrical equipment, which has been made dead in order to prevent danger, from becoming live whilst any work is carried out.

Regulation 14
Standard of duty : *absolute*
No work can be carried out on live electrical equipment unless this can be properly justified. This means that risk assessments are required. If such work is to be carried out, suitable precautions must be taken to prevent injury.

Regulation 15
Standard of duty: *absolute*
Adequate working space, adequate means of access and adequate lighting shall be provided at all electrical equipment on which or near which work is being done in circumstances that may give rise to danger.

Regulation 16
Standard of duty: *absolute*
No person shall engage in work that requires technical knowledge or experience to prevent danger or injury, unless he or she has that knowledge or experience, or is under appropriate supervision.

Construction (Design and Management) Regulations 1994 (CDM)

The CDM Regulations are aimed at improving the overall management and co-operation of health, safety and welfare throughout all stages of a construction project and so reducing the large number of serious and fatal accidents and cases of ill health that happen every year in the construction industry.

The CDM Regulations place duties on all those who can contribute to the health and safety of a construction project. Duties are placed upon clients, designers and contractors and the Regulations create a new duty holder – the planning supervisor. They also introduce new documents – health and safety plans and the health and safety file.

As the person requiring construction work, the client has to appoint:

- a planning supervisor to co-ordinate and manage health and safety during the design and early stages of preparation
- a principal contractor to co-ordinate and manage health and safety issues during the construction work.

One of the duties placed on the planning supervisor is to ensure that a pre-tender stage health and safety plan is prepared before arrangements are made for the principal contractor to carry out or manage construction work. The principal contractor is then required to develop the health and safety plan before work starts on site and keep it up to date throughout the construction phase.

The degree of detail required in the health and safety plan for the construction phase and the time and effort in preparing it should be in proportion to the nature, size and level of health and safety risks involved in the project. Projects involving minimal risks will call for simple, straightforward plans. Large projects or those involving significant risks will need more detail.

The health and safety plan should set out the arrangements for securing the health and safety of everyone carrying out the construction work and all others who may be affected by it. It should deal with:

- the arrangements for the management of health and safety of the construction work
- the monitoring systems for checking that the health and safety plan is being followed
- the health and safety risks to those at work, and others, arising from the construction work, and from other work in premises where construction work may be carried out.

Exceptions to CDM

The CDM Regulations apply to most construction projects. However, there are a number of situations where the Regulations do not apply. These include:

- construction work other than demolition that does not last longer than 30 days and does not involve more than four people
- construction work for a domestic client
- construction work carried out inside offices and shops or similar premises without interrupting the normal activities in the premises and without separating the construction activities from the other activities
- the maintenance or removal of insulation on pipes, boilers or other parts of heating or water systems.

Workplace (Health, Safety and Welfare) Regulations 1992

These Regulations apply to all workplaces, where a workplace is defined as being any premises or part of premises which are not domestic premises, and are made available to any person as a place of work, and including:

- any place within the premises to which such person has access while at work; and
- any room, lobby, corridor, staircase, road or other place used as a means of access to or egress from that place of work or where facilities are provided for use in connection with the place of work other than a public road

but shall not include a modification, an extension or a conversion of any of the above until such modification, extension or conversion is completed.

The Regulations do not apply to places of work that relate to ships, aircraft, locomotive or rolling stock, construction sites and sites where extraction of mineral resources is carried out, temporary work sites and workplaces in agricultural or forestry land.

The requirements of these Regulations are imposed upon every employer or any person who has to any extent control of a workplace, in respect of the following matters:

- working environment
- safety
- facilities.

Working environment

The following regulations apply.

Ventilation

- Regulations 5 to 11 require that effective and suitable provision is made to ensure that every enclosed workplace is ventilated by a sufficient quantity of fresh air.

Temperature

- The temperature in all workplaces inside buildings is reasonable within the first hour of the building being occupied, i.e. 16°C for offices and 13°C where manual work is carried out.
- There should be a sufficient number of thermometers provided for monitoring.

Lighting

- Every workplace should have suitable and sufficient lighting
- If possible every workplace should have natural lighting.

Cleanliness

- All furniture, furnishings, fittings, surfaces of floors, walls and ceilings of all workplaces must be kept clean.

Space

- Every room where persons work should have sufficient floor area, height and unoccupied space for purposes of health, safety and welfare.
- No room must be so overcrowded as to cause risk to the health and safety of persons at work in it.
- The number of persons at work at any one time must be so that the amount of space allowed for each is not less than 11 cubic metres.

Ergonomics

- Every workstation should be arranged so that it is suitable for both any person at that workstation and any person who is likely to use that workstation.

Outdoor protection

- In the case of outdoor workstations protection must be provided from adverse weather (this should include protection from exposure to UVA and UVB from the sun's rays).

Safety

The following regulations apply.

Floors and surfaces

- Regulations 12 to 19 require that every floor and surface area be suitable for the purpose for which it is used.
- No floor should be slippery, uneven or have holes or a slope that may expose any person to a risk to their health and safety.
- Every floor where necessary must have a means of drainage.

Falling hazards

Measures are to be taken to ensure:

- persons are protected from falling objects
- persons are protected from falling from height.

Traffic

- Workplaces must be organised in such a way that pedestrians and vehicles are separated and can circulate in a safe manner.

Escalators

- Escalators and moving walkways are to be equipped with any necessary safety devices
- one or more emergency stop controls must be fitted, which are to be easily identifiable and readily accessible.

Facilities

The following regulations apply.

Toilets

- Regulations 20 to 25 require that suitable toilets be provided at readily accessible places.

Water supply

- Hot and cold water, soap and towels must be supplied.
- Clean drinking water should be provided for all persons at the workplace.

Electricity Safety, Quality and Continuity Regulations 2002

The Electricity Safety, Quality and Continuity Regulations 2002 replace the Electricity Supply Regulations 1988, and specify safety standards that are aimed at protecting the general public and consumers from danger. In addition, the Regulations specify power quality and supply continuity requirements to ensure an efficient and economic electricity supply service for consumers.

The duty holders identified in the Regulations are generators, distributors, suppliers and meter operators, including licensed and non-licensed duty holders. It should be noted that contractors and agents of duty holders, parties constructing networks, persons installing connections, persons operating small-scale embedded generation, and consumers may also have duties under the Regulations.

The requirements of the Regulations apply to public and private operators and to electricity networks used to supply consumers in England, Wales and Scotland.

The Regulations are mainly concerned with the electricity transmission and distribution systems, the overhead service lines to consumers' premises and the apparatus therein up to the consumers' terminals, as all of these are accessible to the public. However, street furniture falling within the scope of the Regulations would include streetlights, traffic signals, bollards, advertising hoardings, bus shelters, public telephones, etc. situated on or adjacent to roads, streets and footpaths.

Management of Health and Safety at Work Regulations 1999

The Management of Health and Safety at Work Regulations 1999 generally make more explicit what employers are required to do to manage health and safety under the Health and Safety at Work Act. Like the Act, they apply to every work activity.

The main requirement on employers is to carry out a risk assessment. Employers with five or more employees need to record the significant findings of the risk assessment. Risk assessment can be straightforward in a simple workplace such as a typical office. However, it can become complicated if it deals with serious hazards such as those on a construction site, nuclear power station, chemical plant or an oil rig. We will look at Risk Assessment later in this chapter.

Besides carrying out a risk assessment, employers also need to:

- make arrangements for implementing the health and safety measures identified as necessary by the risk assessment
- appoint competent people (often themselves or company colleagues) to help them to implement the arrangements
- set up emergency procedures
- provide clear information and training to employees
- work together with other employers sharing the same workplace.

Provision and Use of Work Equipment Regulations 1998 (PUWER)

PUWER 1998 replaced the Provision and Use of Work Equipment Regulations 1992 and carried forward those existing requirements with a few changes and additions, for example the inspection of work equipment and specific new requirements for mobile work equipment.

The Regulations require risks to people's health and safety from equipment that they use at work, be prevented or controlled.

In addition to the requirements of PUWER, lifting equipment is also subject to the requirements of the Lifting Operations and Lifting Equipment Regulations 1998.

In general terms, the Regulations require that equipment provided for use at work is:

- suitable for the intended use
- safe for use, maintained in a safe condition and, in certain circumstances, inspected to ensure this remains the case
- used only by people who have received adequate information, instruction and training; and accompanied by suitable safety measures, e.g. protective devices, markings, warnings.

Generally, any equipment which is used by an employee at work is covered, for example hammers, knives, ladders, drilling machines, power presses, circular saws, photocopiers, lifting equipment (including lifts), dumper trucks and motor vehicles.

If an employer allows employees to provide their own equipment, it will be covered by PUWER and the employer will need to make sure it complies

The Regulations cover places where the HASAWA applies – these include factories, offshore installations, offices, shops, hospitals, hotels, places of entertainment etc. PUWER also applies in common parts of shared buildings and temporary places of work such as construction sites.

While the Regulations cover equipment used by people working from home, they do not apply to domestic work in a private household.

Control of Substances Hazardous to Health Regulations 2002 (COSHH)

Using hazardous substances can put people's health at risk. COSHH therefore requires employers to control exposures to hazardous substances to protect both employees and others who may be exposed from work activities.

Hazardous substances are anything that can harm your health when you work with them if they are not properly controlled e.g. by using adequate ventilation.

Hazardous substances are found in nearly all work places e.g. factories, shops, mines, construction sites and offices and can include:

- substances used directly in work activities, e.g. glues, paints, cleaning agents
- substances generated during work activities, e.g. fumes from soldering and welding
- naturally occurring substances, e.g. grain dust, blood, bacteria.

Employers must ensure employees' and others' safety wherever practicable. COSHH

Employees must therefore:

- work out what hazardous substances are used in your workplace and find out the risks to people's health from using these substances
- decide what precautions are needed before starting work with hazardous substances
- prevent people being exposed to hazardous substances, but where this is not reasonably practicable, control the exposure. If it is reasonably practicable, exposure must be prevented by changing the process or activity so that the hazardous substance is not required or generated, or by replacing it with a safer alternative or using it in a safer form, e.g. pellets instead of powder. If prevention is not reasonably practicable, exposure should be adequately controlled by one or more of the measures outlined in the Regulations, e.g. total enclosure of the process. For a carcinogen special requirements apply. Only as a last resort should personal protective equipment be provided
- make sure control measures are used and maintained properly and that safety procedures are followed. Engineering controls and respiratory protective equipment have to be examined and, where appropriate, tested at suitable intervals
- if required, monitor exposure of employees to hazardous substances. Occupational Exposure Limits or OELs (threshold limits for concentrations of hazardous substances in the air) are approved by HSC and have legal status under COSHH
- carry out health surveillance where your assessment has shown that this is necessary or COSHH makes specific requirements. This might involve examinations by a doctor or trained nurse. In some cases trained supervisors could, for example, check employees' skin for dermatitis, or ask questions about breathing difficulties where work involves substances known to cause asthma
- if required, prepare plans and procedures to deal with accidents, incidents and emergencies
- provide their employees with suitable information, instruction and training about the nature of the substances they work with or are exposed to and the risks created by exposure to those substances and the precautions they should take
- provide employees with sufficient information and instruction on control measures (their purpose and how to use them), how to use personal protective equipment and clothing provided, results of any exposure monitoring and health surveillance (without giving people's names) and emergency procedures.

Control of Major Accidents and Hazards Regulations 1999 (COMAH)

COMAH applies mainly to the chemical industry, but also to some storage activities, explosives and nuclear sites and other industries where threshold quantities of dangerous substances identified in the Regulations are kept or used.

Their main aim is to prevent and mitigate the effects of those major accidents involving dangerous substances such as chlorine, liquefied petroleum gas, explosives and arsenic pentoxide, which can cause serious damage/harm to people and/or the environment.

The general duty on all operators that underpins all the regulations is to take all measures necessary to prevent major accidents and limit their consequences to people and the environment and applies to all establishments within scope.

By requiring measures both for prevention and mitigation, there is recognition that all risks cannot be completely eliminated. Thus, the phrase 'all measures necessary' is taken to include this principle and a judgement must be made about the measures in place. Where hazards are high, then high standards will be required to ensure risks are acceptably low.

Control of Noise at Work Regulations 2006

The Control of Noise at Work Regulations 2006 state:

- Every employer shall reduce the risk of damage to the hearing of his employees from exposure to noise to the lowest level reasonably practicable. The first action level is a daily personal noise exposure of 80 dB(A), the second action level is a daily personal noise exposure of 85 dB(A), and the peak action level is a peak sound pressure of 137 dB(A).
- When any employee is likely to be exposed to the first action level or above, or to the peak action level or above, every employer must ensure that a competent person makes a noise assessment. This noise assessment must identify which employees are exposed and give sufficient details of the situation to allow the employer to conform with these Regulations.
- Every employer must ensure, so far as is practicable, that when any employee is likely to be exposed to the first action level or above in circumstances where the daily personal noise exposure of that employee is likely to be less than 85 dB(A), the employee is provided, at his or her request, with suitable and efficient personal ear protectors.
- If an employee is likely to be exposed to the second action level or above or to the peak action level or above, then the employer must reduce, so far as is reasonably practicable (other than by the provision of personal ear protectors), the exposure to noise of that employee.
- Every employer must ensure, so far as is practicable that, when any employee is likely to be exposed to the second action level or above or to the peak action level or above, that employee is provided with suitable personal ear protectors which, when properly worn, can reasonably be expected to keep the risk of damage to that employee's hearing to below that arising from exposure to the second action level or, as the case may be, to the peak action level.
- Every employer must also ensure that for any premises under his control, so far as is reasonably practicable, that each ear protection zone is separate and identified by means of a specified sign including text indicating that it is an ear protection zone, that employees wear personal ear protectors whilst in such a zone and that no employees enter any such zone unless they are wearing personal ear protectors.

Health and Safety (First Aid) Regulations 1981

The Health and Safety (First-Aid) Regulations 1981 require employers to provide adequate and appropriate equipment, facilities and personnel to enable first aid to be given to employees if they are injured or become ill at work. These Regulations apply to all workplaces including those with five or fewer employees and to the self-employed.

What is adequate will depend on the circumstances in the workplace. This includes whether trained first aiders are needed, what should be included in a first aid box and if a first aid room is needed. Employers should carry out an assessment of first aid needs to determine this.

That said, the minimum first aid provision on any work site is a suitably stocked first aid box and an appointed person to take charge of first aid arrangements. As a guide, and where there is no special risk in the workplace, a minimum stock of first aid items would be:

- a leaflet giving general guidance on first aid e.g. HSE leaflet Basic advice on first aid at work
- 20 individually wrapped sterile adhesive dressings (assorted sizes)
- two sterile eye pads
- four individually wrapped triangular bandages (preferably sterile)
- six safety pins
- six medium sized (approximately 12 cm x 12 cm) individually wrapped sterile unmedicated wound dressings
- two large (approximately 18 cm x 18 cm) sterile individually wrapped unmedicated wound dressings
- one pair of disposable gloves.

You should not keep tablets or medicines in the first aid box.

The above is a suggested contents list only; equivalent but different items will be considered acceptable.

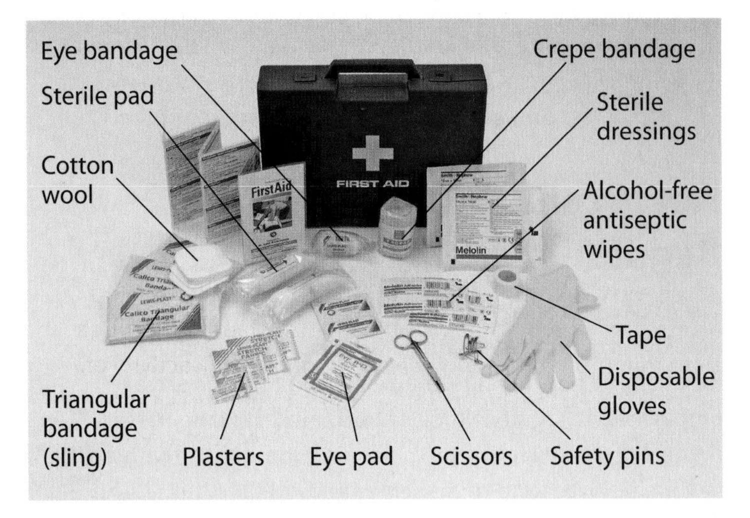

Figure 2.4 A typical first aid kit

Each employer must also inform employees of the first aid arrangements. Putting up notices telling staff who and where the first aiders or appointed persons are and where the first aid box is located is usually regarded as sufficient.

Manual Handling Operations Regulations 1992

As amended in 2002, the Manual Handling Operations Regulations 1992 seek to reduce the very large incidence of injury and ill-health arising from the manual handling of loads at work. More than one in four of all reportable injuries are caused by manual handling. These accidents do not include cumulative injuries, particularly to the back, which can lead to physical impediment or even permanent disablement.

The Regulations extend to the manual handling of people and animals and include '… any transporters or supporting of a load (including the lifting, putting down, pushing, pulling, carrying or moving thereof) by hand or by bodily force'.

Manual handling implies that an attempt is being made to move a load. Therefore, if a girder being moved manually is dropped and fractures an employee's foot, it is a manual handling accident. If the girder is inadvertently knocked over and causes a similar injury this would not be due to manual handling.

An important exception is that a tool or machine being used for its normal purpose is not a load. Therefore chainsaws being unloaded from a vehicle would be regarded as a 'load' and subject to these Regulations, but they would not be a 'load' in normal use.

In essence, the Regulations set out three steps:

- avoid manual handling operations which involve a risk of injury, so far as is reasonably practicable
- suitably and sufficiently assess all manual handling operations that cannot be avoided
- take steps to reduce the risk of injury during those operations to the lowest level reasonably practicable.

Employees have a duty to follow a safe system of work where it has been properly provided and to use any machinery or equipment for which they have been adequately trained, when provided as a lifting aid.

Work at Height Regulations 2005 (amended 2007)

The Work at Height Regulations 2005 came into force on 6 April 2005 and apply to all work at height where there is a risk of a fall liable to cause personal injury. The regulations place a duty on employers, the self-employed, and any person that controls the work of others (for example facilities managers or building owners who may contract others to work at height). The Work at Height (Amendment) Regulations 2007 came into force on 6 April 2007. These also added a duty for those who work at height providing instruction or leadership to one or more people engaged in caving or climbing by way of sport, recreation, team building or any other similar activity in Great Britain.

If you are an employee or working under someone else's control, regulation 14 says you must:

- report any safety hazard to them
- use the equipment supplied (including safety devices) properly, following any training and instructions (unless you think that would be unsafe, in which case you should seek further instructions before continuing).

There is a simple hierarchy for managing and selecting equipment for work at height shown in Figure 2.4.

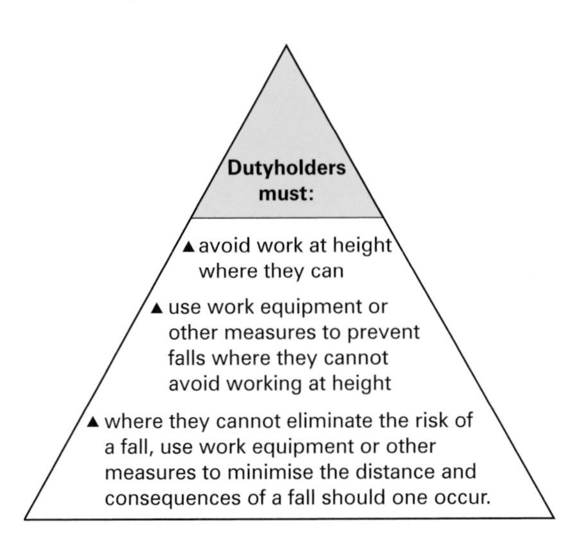

Figure 2.5 The role of duty holders

As part of the Regulations, duty holders must ensure:

- all work at height is properly planned and organised
- all work at height takes account of weather conditions that could endanger health and safety
- those involved in work at height are trained and competent
- the place where work at height is done is safe
- equipment for work at height is appropriately inspected
- the risks from fragile surfaces are properly controlled
- the risks from falling objects are properly controlled
- that no work is done at height if it is safe and reasonably practicable to do it other than at height
- that the work is properly planned, appropriately supervised, and carried out in as safe a way as is reasonably practicable
- a plan exists for emergencies and rescue
- they have taken account of the risk assessment carried out under regulation 3 of the Management of Health and Safety at Work Regulations
- that when selecting equipment for work at height the most suitable equipment is used
- collective protection measures (e.g. guard rails) are provided as a priority over personal protection measures (e.g. safety harnesses)
- they have taken account of working conditions and risks to the safety of all those at the place where the work equipment is to be used.

The Regulations also include schedules giving requirements for existing places of work and means of access for work at height, collective fall prevention (e.g. guardrails and working platforms), collective fall arrest (e.g. nets, airbags), personal fall protection (e.g. work restraints, fall arrest, rope access) and ladders.

Control of Vibration at Work Regulations 2005

The Control of Vibration at Work Regulations 2005 (the Vibration Regulations), came into force on 6 July 2005 and aim to protect workers from risks to health from vibration. The regulations introduce action and limit values for hand–arm and whole-body vibration.

Whole body vibration

Whole-body vibration means mechanical vibration which is transmitted into the body, when seated or standing, through the supporting surface. This can be during a work activity or, as described in regulation 5.3.f, beyond normal working hours, including exposure in rest facilities supervised by the employer.

Hand–arm vibration

Hand–arm vibration is vibration transmitted from work processes into workers' hands and arms. It can be caused by operating hand-held power tools, such as road breakers, and hand-guided equipment, such as powered lawnmowers, or by holding materials being processed by machines, such as pedestal grinders.

Regular and frequent exposure to hand–arm vibration can lead to permanent health effects. This is most likely when contact with a vibrating tool or work process is a regular part of an employee's job. Occasional exposure is unlikely to cause ill health.

There are many types of tools and equipment that can cause ill health from vibration, with some of the more relevant being: hammer drills, hand-held grinders and jigsaws.

Consequently, the regulations introduce:

- an exposure action value of 2.5 m/s^2 A(8) at which level employers should introduce technical and organisational measures to reduce exposure
- an exposure limit value of 5.0 m/s^2 A(8) which should not be exceeded
- a transitional period from the exposure limit value for hand–arm vibration until 2010 to allow work activities where the use of older tools and machinery cannot keep exposures below the exposure limit value, to continue in certain circumstances.

Health and Safety Information for Employees Regulations 1989

The Health and Safety Information for Employees Regulations 1989 (HSIER) require employers to provide their employees with certain basic information concerning their health, safety and welfare at work. This information is contained in both a poster and a leaflet approved by HSE. Employers can comply with their duty by either displaying the poster or providing employees with a copy of the leaflet.

The revised versions are substantially different from the previous poster and leaflet, reflecting changes in health and safety legislation. Some of the main changes are:

- removal of references to repealed duties under the Factories Act 1961 and the Offices, Shops and Railway Premises Act 1963
- reference to the main duties under some of the more recent legislation
- drawing attention to the duty to consult employees or their representatives on health and safety
- amended references to Employment Medical Advisory Service (EMAS) to reflect its current role
- the inclusion on the poster of two further information boxes which the employer is encouraged to complete as appropriate in order to personalise the information. These give details of the names and locations of employee health and safety representatives and the names of competent persons appointed by the employer together with their health and safety responsibilities.

Employers have two principal duties under the Regulations:

a. either to display the poster OR to distribute the leaflet (HSIER, Reg. 4)

b. provide further information giving details of the enforcing authority for the premises and the local address for EMAS (HSIER, Reg. 5).

Dangerous Substances and Explosive Atmospheres Regulations 2002 (DSEAR)

The Dangerous Substances and Explosive Atmospheres Regulations 2002 (DSEAR) came into force on 9 December 2002. DSEAR applies to all dangerous substances at nearly every workplace in the United Kingdom and sets minimum requirements for the protection of workers from fire and explosion risks arising from dangerous substances and potentially explosive atmospheres. DSEAR complements the requirement to manage risks under the Management of Health and Safety at Work Regulations 1999.

Employers and the self-employed must:

- carry out a risk assessment of any work activities involving dangerous substances
- provide technical and organisational measures to eliminate or reduce as far as is reasonably practicable the identified risks
- provide equipment and procedures to deal with accidents and emergencies
- provide information and training to employees
- classify places where explosive atmospheres may occur into zones, and mark the zones where necessary.

For DSEAR, the definition of workplace is very wide and means any premises or part of premises used for work. Premises include all industrial and commercial premises and also land-based and offshore installations as well as vehicles and vessels. Common parts of shared buildings, private roads and paths on industrial estates, and business parks are also 'premises' – as are houses and other domestic dwellings.

DSEAR applies to any substance or preparation (mixture of substances) with the potential to create a risk to persons from energetic (energy-releasing) events such as fires, explosions, thermal runaway from exothermic reactions etc. Known in DSEAR as dangerous substances, these include petrol, liquefied petroleum gas (LPG), paints, varnishes and certain types of combustible and explosive dusts produced in, for example, machining and sanding operations, flour mills and distilleries.

It should be noted that many of these substances will also create a health risk as well. For example many solvents are toxic as well as being flammable.

DSEAR does not address these health risks. They are dealt with by the Control of Substances Hazardous to Health Regulations (COSHH), which have now been amended to implement the health side of the Chemical Agencies Directive.

DSEAR is concerned with harmful physical effects from thermal radiation (burns), over-pressure effects (blast injuries) and oxygen depletion effects (asphyxiation) arising from fires and explosions.

Some examples and their industries might be:

- storage of petrol as a fuel for cars
- motor boats
- horticultural machinery
- use of flammable gases, such as acetylene for welding

- handling and storage of waste dusts in a range of manufacturing industries
- handling and storage of flammable wastes including fuel oils, and work activities that could release naturally occurring methane
- dusts produced in the mining of coal
- use of flammable solvents in pathology and school laboratories
- storage/display of flammable goods, such as paints in the retail sector
- filling, storage and handling of aerosols with flammable propellants such as LPG
- transport of flammable liquids in containers around the workplace
- deliveries from road tankers, such as petrol or bulk powders and chemical manufacture, processing and warehousing.

Reporting of Injuries, Diseases and Dangerous Occurrences Regulations 1995 (RIDDOR)

The Reporting of Injuries, Diseases and Dangerous Occurrences Regulations 1995 (RIDDOR) place a legal duty on employers, the self-employed and those in control of premises to report some work-related accidents, diseases and dangerous occurrences to the relevant enforcing authority for their work activity. This can be the Health and Safety Executive (HSE) or one of the local authorities (LAs).

The law requires the following work-related incidents to be reported:

- deaths
- injuries where an employee or self-employed person has an accident which is not major, but results in the injured person being away from work or unable to do the full range of their normal duties for more than three days (including any days they wouldn't normally be expected to work such as weekends, rest days or holidays) not counting the day of the injury itself
- injuries to members of the public where they are taken to hospital
- work-related diseases
- dangerous occurrences where something happens that does not result in a reportable injury but which could have done.

Incident Contact Centre

Reporting accidents and ill health at work is a legal requirement. The information enables the enforcing authorities to identify where and how risks arise and to investigate serious accidents. The enforcing authorities can then advise you on preventive action to reduce injury, ill-health and accidental loss – much of which is uninsurable.

Previous systems required employers to complete an accident report form and notify the relevant authorities. However, sometimes people were unsure to which authority they should be reporting.

To make things easier and to remove all confusion, in April 2001 the HSE set up the Incident Control Centre, which provides a central point for employers to report incidents irrespective of whether their business is HSE or Local Authority enforced.

As employers can now report incidents over the telephone to the Incident Control Centre, the need for the old F.2508 Accident Report Form has been largely removed, as the Incident Control Centre now completes the forms on an employer's behalf.

The centre is based in Caerphilly, and allows individuals to make their report to a single point irrespective of their location. The centre will then send the report to the correct enforcing authority. The centre will also send a copy of their final report on the incident to the employer. This meets the statutory obligation for employers to keep records of all reportable incidents for inspection and also allows the employer to correct any errors or omissions.

The quickest way to notify the centre is by telephone. The number is: 0845 300 9923.

Reporting requirements

In each case, provided that the incident is reported through the Incident Centre, there is no need to complete an F.2508 Accident Report Form or F.2508A Disease Report Form, although the centre will ask for brief details about the business, the affected person or persons and the incident or disease.

A report needs to be made in the following circumstances:

Death or major injury

If there is an accident connected with work and:

- an employee or a self-employed person working on your premises is killed or suffers a major injury (including as a result of physical violence) or
- a member of the public is killed or taken to hospital

the enforcing authority must be informed by telephone via the Incident Centre (see feature) without delay (normally within 24 hours).

A list of the major injuries covered by the legislation can be found on page 77.

Over-three-day injury

If there is an accident connected with work (including an act of physical violence) and an employee or self-employed person working on the premises suffers an over-three-day injury, then the enforcing authority must be informed by telephone via the Incident Centre without delay (but no longer than 10 days).

Disease

If a doctor notifies an employer that an employee suffers from a reportable work-related disease, then the enforcing authority must be informed by telephone via the Incident Centre as soon as possible after notification (but no longer than 10 days).

A list of the diseases covered by the legislation can be found on page 77.

Dangerous occurrence

If something happens which does not result in a reportable injury, but which clearly could have done, it may be a dangerous occurrence that must be reported.

The enforcing authority should be informed by telephone via the Incident Centre immediately (but no longer than 10 days).

A list of the dangerous occurrences covered by the legislation can be found on page 77.

Accident reporting

Employers legally have to keep a record of any reportable injury, disease or dangerous occurrence. This must include the date and method of reporting, the date, time and place of the event, personal details of those involved and a brief description of the nature of the event or disease along with any supporting evidence (photographs etc.). Employers can keep the record in any form they wish.

Incident Control Centre forwards a copy of every report form it completes to the respective employer for it to check and hold as a record. The importance of checking this and any other document cannot be over emphasised.

Accident report book

To deal with small scale, non-reportable incidents and injuries, most employers hold an accident book on site. Should an employer wish to continue with the paper-based system, the Health and Safety Executive (HSE) has launched a new Accident Book, which will help organisations comply with the Data Protection Act 1998. Approved by the Information Commissioner, it has been revised because most existing accident books store personal details and information, which can then be seen by anyone reading or making an entry.

The new book allows for accidents to be recorded, but individuals' details can be held separately in a secure location. It contains information on first aid and how to manage health and safety information to help prevent accidents from happening in the first place.

The previous version, produced by the Department for Work and Pensions (DWP), and other similar books, are not compliant with the Data Protection Act, and DWP has given responsibility for the publication to HSE. The Information Commissioner ruled that businesses had to change their accident book to comply with the Data Protection Act by 31 December 2003.

Definitions of reportable major injuries, diseases and dangerous occurrences under the requirements of these Regulations.

Reportable major injuries

- Fracture other than to fingers, thumbs or toes
- Amputation
- Dislocation of the shoulder, hip, knee or spine
- Loss of sight (temporary or permanent)
- Chemical or hot metal burn to the eye or any penetrating injury to the eye
- Injury resulting from an electric shock or electrical burn leading to unconsciousness or requiring resuscitation or admittance to hospital for more than 24 hours
- Any other injury leading to hypothermia, heat-induced illness or unconsciousness; or requiring resuscitation; or requiring admittance to hospital for more than 24 hours
- Unconsciousness caused by asphyxia or exposure to a harmful substance or biological agent
- Acute illness requiring medical treatment, or loss of consciousness arising from absorption of any substance by inhalation, ingestion or through the skin
- Acute illness requiring medical treatment where there is reason to believe that this resulted from exposure to a biological agent or its toxins or infected material.

Reportable diseases

- Certain poisonings.
- Some skin diseases such as occupational dermatitis, skin cancer, chrome ulcer, oil folliculitis and acne.
- Lung diseases including occupational asthma, farmer's lung, pneumoconiosis, asbestosis, mesothelioma.
- Infections such as leptospirosis, hepatitis, tuberculosis, anthrax, legionellosis and tetanus.
- Other conditions such as: occupational cancer, certain musculoskeletal disorders, decompression illness and hand–arm vibration syndrome.

Reportable dangerous occurrences

- Collapse, overturning or failure of load-bearing parts of lifts and lifting equipment.
- Explosion, collapse or bursting of any closed vessel or associated pipework.
- Failure of any freight container in any of its load-bearing parts.
- Plant or equipment coming into contact with overhead power lines.
- Electrical short circuit or overload causing fire or explosion.
- Any unintentional explosion, misfire, failure of demolition to cause the intended collapse, projection of material beyond a site boundary, injury caused by an explosion.
- Accidental release of a biological agent likely to cause severe human illness.
- Failure of industrial radiography or irradiation equipment to de-energise or return to its safe position after the intended exposure period.
- Malfunction of breathing apparatus while in use or during testing immediately before use.
- Failure or endangering of diving equipment, the trapping of a diver, an explosion near a diver, or an uncontrolled ascent.
- Collapse or partial collapse of a scaffold over five metres high, or erected near water where there could be a risk of drowning after a fall.
- Unintended collision of a train with any vehicle.
- Dangerous occurrence at a well (other than a water well).
- Dangerous occurrence at a pipeline.
- Failure of any load-bearing fairground equipment, or derailment or unintended collision of cars or trains.
- A road tanker carrying a dangerous substance overturns, suffers serious damage, catches fire or the substance is released.
- A dangerous substance being conveyed by road is involved in a fire or released.

The following dangerous occurrences are reportable except in relation to offshore workplaces:

- The unintended collapse of any building or structure under construction, alteration/demolition where over five tonnes of material falls, a wall or floor in a place of work and any false-work.
- Any explosion or fire causing suspension of normal work for over 24 hours.
- The sudden, uncontrolled release in a building of 100 kg or more of a flammable liquid; 10 kg or more of a flammable liquid above its boiling point; or 10 kg or more of a flammable gas; or 500 kg of these substances if the release is in the open air.
- The accidental release of any substance that may damage health.

Implementing and controlling health and safety

We can see from the previous section that there is an awful lot of legislation to be read and interpreted correctly if the workplace is to be made as safe as possible. Keeping up to date with and understanding the implications of this legislation can be daunting, so how do we manage it?

Strangely, there is no set method for distributing new or amended health and safety requirements to employers. It is up to every employer to ensure that they are meeting legal requirements; lack of knowledge is not regarded as an excuse in a court of law.

The means of keeping up to date vary from company to company and often depend on their size and structure. In smaller companies the owner/manager is likely to have responsibility, but pressure on their time might be very high. Many smaller companies choose to hire the services of an external health and safety consultant.

The consultant will normally visit the employer and carry out an inspection of the main premises and types of site and activities undertaken. The consultant will then produce the necessary paperwork – e.g. risk assessments, safety policy, Control of Substances Hazardous to Health regulations (COSHH) assessments etc. – and provide a manual for the employer to refer to and implement.

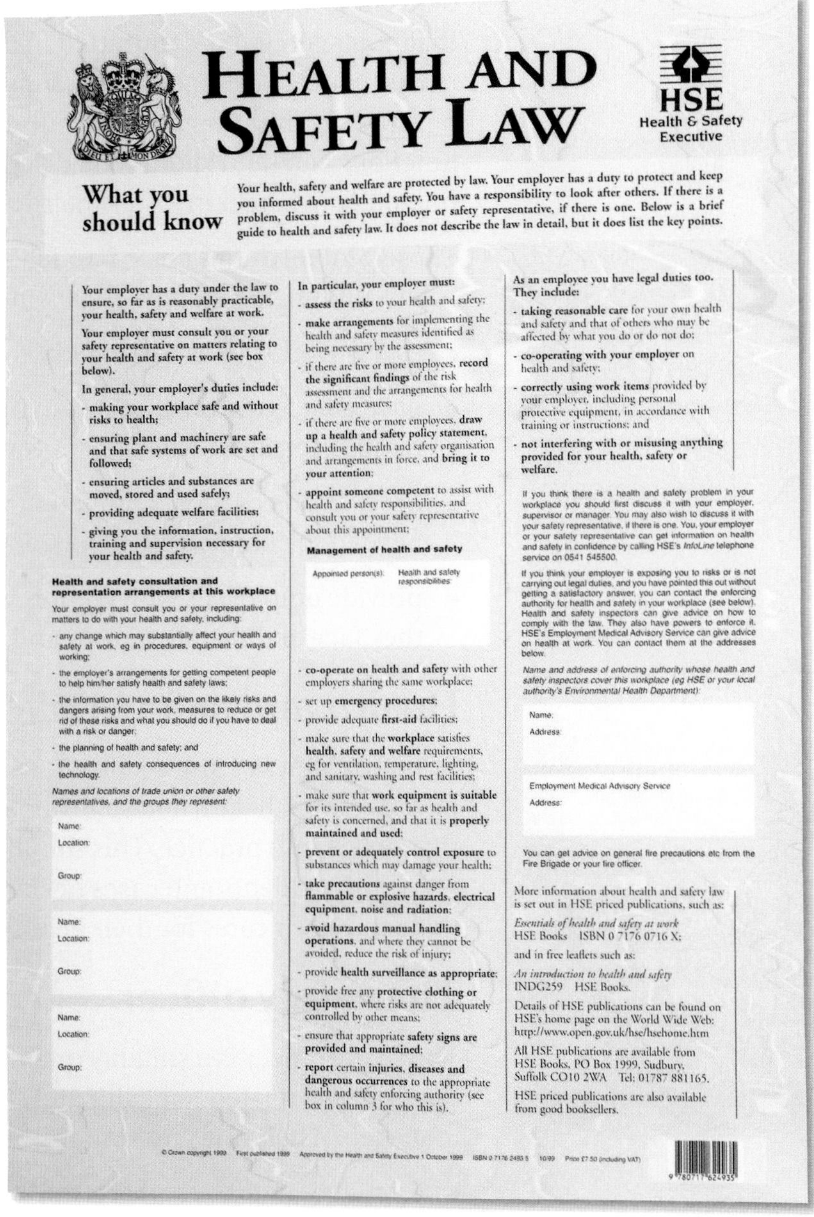

HEALTH AND SAFETY LAW

HSE Health & Safety Executive

What you should know

Your health, safety and welfare are protected by law. Your employer has a duty to protect and keep you informed about health and safety. You have a responsibility to look after others. If there is a problem, discuss it with your employer or safety representative, if there is one. Below is a brief guide to health and safety law. It does not describe the law in detail, but it does list the key points.

Your employer has a duty under the law to ensure, so far as is reasonably practicable, your health, safety and welfare at work.

Your employer must consult you or your safety representative on matters relating to your health and safety at work (see box below).

In general, your employer's duties include:

- **making your workplace safe and without risks to health;**
- **ensuring plant and machinery are safe and that safe systems of work are set and followed;**
- **ensuring articles and substances are moved, stored and used safely;**
- **providing adequate welfare facilities;**
- **giving you the information, instruction, training and supervision necessary for your health and safety.**

Health and safety consultation and representation arrangements at this workplace

Your employer must consult you or your representative on matters to do with your health and safety, including:

- any change which may substantially affect your health and safety at work, eg in procedures, equipment or ways of working;
- the employer's arrangements for getting competent people to help him/her satisfy health and safety laws;
- the information you have to be given on the likely risks and dangers arising from your work, measures to reduce or get rid of these risks and what you should do if you have to deal with a risk or danger;
- the planning of health and safety; and
- the health and safety consequences of introducing new technology.

Names and locations of trade union or other safety representatives, and the groups they represent:

Name:
Location:
Group:

Name:
Location:
Group:

Name:
Location:
Group:

In particular, your employer must:

- **assess the risks** to your health and safety;
- **make arrangements** for implementing the health and safety measures identified as being necessary by the assessment;
- if there are five or more employees, **record the significant findings** of the risk assessment and the arrangements for health and safety measures;
- if there are five or more employees, **draw up a health and safety policy statement**, including the health and safety organisation and arrangements in force, and **bring it to your attention**;
- **appoint someone competent** to assist with health and safety responsibilities, and consult you or your safety representative about this appointment;

Management of health and safety

Appointed person(s):	Health and safety responsibilities:

- **co-operate on health and safety** with other employers sharing the same workplace;
- set up **emergency procedures**;
- provide adequate **first-aid** facilities;
- make sure that the **workplace** satisfies **health, safety and welfare** requirements, eg for ventilation, temperature, lighting, and sanitary, washing and rest facilities;
- make sure that **work equipment is suitable** for its intended use, so far as health and safety is concerned, and that it is **properly maintained and used**;
- **prevent or adequately control exposure** to substances which may damage your health;
- **take precautions** against danger from **flammable or explosive hazards, electrical equipment, noise and radiation**;
- **avoid hazardous manual handling operations**, and where they cannot be avoided, reduce the risk of injury;
- provide **health surveillance as appropriate**;
- provide free any **protective clothing or equipment**, where risks are not adequately controlled by other means;
- ensure that appropriate **safety signs are provided and maintained**;
- **report** certain **injuries, diseases and dangerous occurrences** to the appropriate health and safety enforcing authority (see box in column 3 for who this is).

As an employee you have legal duties too. They include:

- **taking reasonable care** for your own health and safety and that of others who may be affected by what you do or do not do;
- **co-operating with your employer** on health and safety;
- **correctly using work items** provided by your employer, including personal protective equipment, in accordance with training or instructions; and
- **not interfering with or misusing anything provided for your health, safety or welfare.**

If you think there is a health and safety problem in your workplace you should first discuss it with your employer, supervisor or manager. You may also wish to discuss it with your safety representative, if there is one. You, your employer or your safety representative can get information on health and safety in confidence by calling HSE's *InfoLine* telephone service on 0541 545500.

If you think your employer is exposing you to risks or is not carrying out legal duties, and you have pointed this out without getting a satisfactory answer, you can contact the enforcing authority for health and safety in your workplace (see below). Health and safety inspectors can give advice on how to comply with the law. They also have powers to enforce it. HSE's Employment Medical Advisory Service can give advice on health at work. You can contact them at the addresses below.

Name and address of enforcing authority whose health and safety inspectors cover this workplace (eg HSE or your local authority's Environmental Health Department):

Name:
Address:

Employment Medical Advisory Service
Address:

You can get advice on general fire precautions etc from the Fire Brigade or your fire officer.

More information about health and safety law is set out in HSE priced publications, such as:

Essentials of health and safety at work
HSE Books ISBN 0 7176 0716 X;

and in free leaflets such as:

An introduction to health and safety
INDG259 HSE Books.

Details of HSE publications can be found on HSE's home page on the World Wide Web: http://www.open.gov.uk/hse/hsehome.htm

All HSE publications are available from HSE Books, PO Box 1999, Sudbury, Suffolk CO10 2WA Tel: 01787 881165.

HSE priced publications are also available from good booksellers.

© Crown copyright 1999 First published 1999 Approved by the Health and Safety Executive 1 October 1999 ISBN 0 7176 2493 5 10/99 Price £7.50 (including VAT)

9 780717 624935

Displaying a health and safety notice in the workplace is compulsory

Some employers use a consultant on a one-off basis and then choose to maintain the information and systems provided themselves. However, there are risks associated with this, in that they must ensure that the manual remains current with all developments. Other employers engage a consultant on an ongoing basis where the consultant monitors all legislation, Codes of Practice and relevant information and continually provides the employer with updated information, documentation and services.

Large companies often have their own specialist member(s) of staff with responsibility for health and safety.

Find out

What is your employer's approach to health and safety? Can you see any ways in which it could be improved?

Did you know?

Employers are covered by sections 2 and 3 of the Health and Safety at Work Act 1974. Employers need to prepare – and make sure that workers know about – a written statement of the company health and safety policy and the arrangements in place to put it into effect

Remember

Writing a health and safety policy statement should be more than just a legal requirement – it should show the commitment of an employer to planning and managing health and safety

Remember

What is written in the policy has to be put into practice. The true test of a health and safety policy is the actual conditions in the workplace, not how well the statement has been written

Health and safety responsibility

Responsibility begins with senior management. Strong leadership is vital in delivering effective health and safety risk control. Everyone should know and believe that management is committed to continuous improvement in health and safety performance. Consequently, management should explain its expectations, and how the organisation and procedures will deliver them. Although health and safety functions can (and should) be delegated, legal responsibility for health and safety (including where it involves members of the public) rests with the employer.

These general duties are expanded and explained in the Management of Health and Safety at Work Regulations 1999. Employers must:

- assess work-related risks for both employees and those not in their employ
- have effective arrangements in place for planning, organising, controlling, monitoring and reviewing preventive and protective measures
- appoint one or more competent persons to help in undertaking the measures needed to comply with health and safety law
- provide employees with comprehensible and relevant information on the risks they face and the preventive and protective measures that control those risks.

Health and safety policy statement

By law (HASAWA 1974, section 2(3)), anyone employing five or more people must have a written health and safety policy which includes arrangements for putting that policy into practice. This should be the key to achieving acceptable standards, reducing accidents and cases of work-related ill-health, and it shows employees that their employer cares for their health and safety.

The safety officer

The safety officer is a suitably qualified member of staff with delegated responsibility for all things related to health and safety, and is answerable to the company managers. Their role and responsibilities are likely to include:

- arranging internal and external training for employees on safety issues
- monitoring and implementing codes of practice and regulations
- the update and display of information
- liaison with external agencies such as the HSE
- carrying out and recording regular health and safety inspections and risk assessments
- advising on selection, training, use and maintenance of PPE
- maintenance of accident reports and records.

Should an accident occur, the safety officer would lead the investigations, identify the causes and advise on any improvements in safety standards that need to be made.

The safety representative

The Safety Representatives and Safety Committees Regulations 1977 were originally made under section 15 of the Health and Safety at Work Act 1974, although they have been amended since. The Regulations and associated Codes of Practice provide a legal framework for employers and trade unions to reach agreement on arrangements for safety representatives and safety committees to operate in their workplace. The safety representative is often a trade union member, and their role is similar to that of the safety officer:

- making representations to the employer on behalf of members on any health, safety and welfare matter
- representing members in consultation with the Health and Safety Executive (HSE) inspectors or other enforcing authorities
- inspecting designated workplace areas at least every three months
- investigating any potential hazards, complaints by members and causes of accidents, dangerous occurrences and diseases
- requesting facilities and support from the employer to carry out inspections and receive legal and technical information
- paid time off to carry out the role and to undergo either TUC or union-approved training.

The Health and Safety Commission (HSC) has issued guidance for employers who do not recognise independent trade unions. For all but the smallest companies, they recommend setting up a safety committee of members drawn from both management and employees. This should help employers comply with their legal duties, particularly under section 2 (Part 4) of the Health and Safety at Work Act 1974 and the Health and Safety (Consultation with Employees) Regulations 1996.

Under the Consultation With Employees Regulations, any employees not in groups covered by trade union safety representatives must be consulted by their employers; the employer can choose to consult with them directly or through the elected representatives.

Health and Safety Commission (HSC)and Executive (HSE)

Britain's Health and Safety Commission (HSC) and the Health and Safety Executive (HSE) are responsible for the regulation of almost all the risks to health and safety arising from work activity in Britain. Their mission is to protect people's health and safety by ensuring risks in the changing workplace are properly controlled.

Among other things, they look after health and safety in nuclear installations and mines, factories, farms, hospitals and schools, offshore gas and oil installations, the gas grid, railways, and the movement of dangerous goods and substances.

The HSC members are appointed by the appropriate Secretary of State, and the HSC in turn appoints members of the HSE.

The HSE has about 4,000 employees throughout the UK including advisers, scientific and medical experts and inspectors

HSE inspectors

To help enforce the law, HSE inspectors have a range of statutory powers. Consequently, they can, and sometimes do, visit and enter premises without warning. If there is a problem the inspector can:

- issue an **improvement notice**. This notice will say what needs to be done, why and by when. The time period within which to take the remedial action will be at least 21 days to allow the duty holder time to appeal to an Employment Tribunal if they so wish. The notice also contains a statement that in the opinion of an inspector an offence has been committed
- issue a **prohibition notice**. Where an activity involves, or will involve, a risk of serious personal injury, the inspector can serve a prohibition notice prohibiting the activity immediately, or after a specified time period, and not allowing it to be resumed until remedial action has been taken. The notice will explain why the action is necessary
- take legal action. In certain circumstances, the inspector may also wish to start legal proceedings. Improvement and prohibition notices, and written advice, may be used in court proceedings.

Before deciding on a course of action, HSE inspectors apply a concept known as 'proportionality'. Effectively, this means that the degree of enforcement action to be taken will be in proportion to the degree of risks discovered.

Environmental health officers

Environmental health officers are employed by local authorities and they inspect commercial businesses such as warehouses, offices, shops, pubs and restaurants within a borough area.

They have the right to enter any workplace without giving notice, though they may in practice give notice. Normally, the officer looks at the workplace, work activities and management of health and safety, and checks that the business is complying with health and safety law.

They may offer guidance or advice to help businesses. They may also talk to employees and their representatives, take photographs and samples, serve improvement notices and take action if there is a risk to health and safety that needs to be dealt with immediately.

FAQ

Q **Can I be prosecuted by the HSE?**

A Yes, we all can if we are employed. Prosecution would only happen as a result of blatantly or continuously failing to implement safe working practices where such failure has resulted in injury or death.

Penalties for health and safety offences

(As at January 2002. These penalties can change from time to time.)

The Health and Safety at Work Act 1974 (HASAWA), section 33 (as amended), sets out the offences and maximum penalties under health and safety legislation.

Failing to comply with an improvement or prohibition notice, or a court remedy order (issued under HASAWA, sections 21, 22 and 42 respectively) will incur the following penalty:

- Lower court maximum: £20,000 and/or six months' imprisonment
- Higher court maximum: unlimited fine and/or two years' imprisonment.

Breach of sections 2–6 of the HASAWA, which set out the general duties of employers, self-employed persons, manufacturers and suppliers to safeguard the health and safety of workers and members of the public who may be affected by work activities, will incur the following penalty:

- Lower court maximum: £20,000
- Higher court maximum: unlimited fine.

There can be other breaches of the HASAWA and breaches of 'relevant statutory provisions' under the Act, which include all health and safety regulations. The legislation imposes both general and more specific requirements, such as requirements to carry out a suitable and sufficient risk assessment or to provide suitable personal protective equipment. Breaches incur the following penalties:

- Lower court maximum: £5,000
- Higher court maximum: unlimited fine.

Remember

The more severe the offence, the more severe the punishment

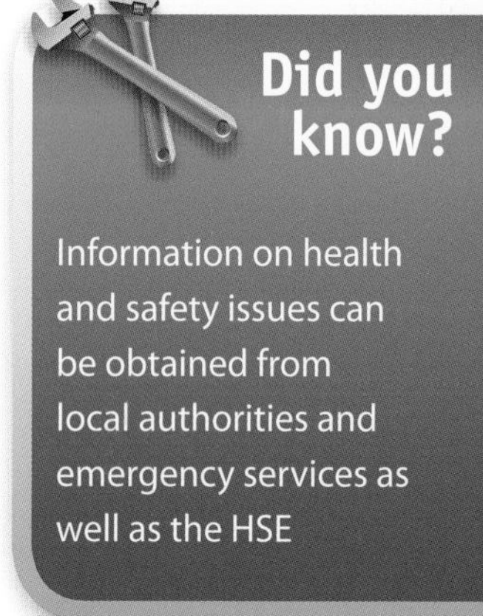

Did you know?

Information on health and safety issues can be obtained from local authorities and emergency services as well as the HSE

Sources of information

Apart from via the HSE itself, information can also be sought from local authorities and emergency services.

Approved Codes of Practice (ACOPs)

ACOPs give practical advice on how to comply with the law and are approved by the HSC, with the consent of the Secretary of State.

A failure by an employer to follow an ACOP doesn't make them liable to any civil or criminal proceedings. However, in any criminal proceedings where an employer is said to have committed an offence by contravening any requirement or prohibition imposed for which there was an Approved Code of Practice at the time of the alleged contravention, the court will use the ACOP to demonstrate the employer's non-compliance. In such circumstances the employer's only defence would be to prove that it was using another appropriate method/system.

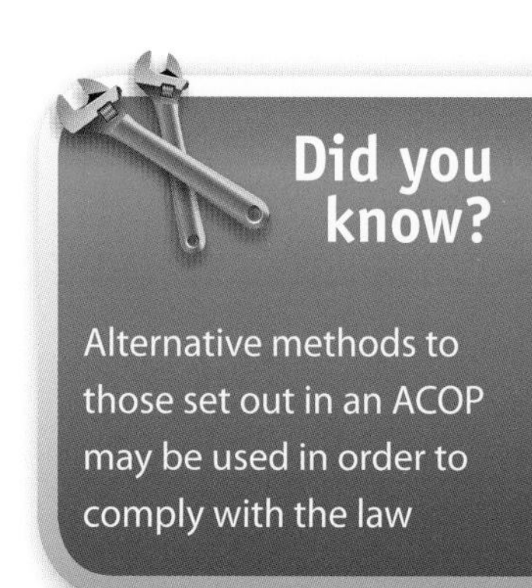

Did you know?

Alternative methods to those set out in an ACOP may be used in order to comply with the law

Risk assessment

Accidents and ill health can ruin lives, and affect business too if output is lost, machinery is damaged, insurance costs increase or employers have to go to court. Employers are therefore legally required to assess the risks in their workplace.

A risk assessment is nothing more than a careful examination of what, during working activities, could cause harm to people, so that an employer can weigh up whether it has taken enough precautions or should do more to prevent harm. The aim is to make sure that no one gets hurt or becomes ill.

The important things to decide are:

- What are the hazards?
- Are they significant?
- Are hazards covered by satisfactory precautions so that the risk is small?

Employers need to check this when assessing risks. For example, we know that electricity is a hazard that can kill, but the risk of it doing so in a tidy, well-run office environment is remote, provided that the installation is sound, 'live' components are insulated and metal casings are properly earthed.

What is a hazard and what is a risk?

A hazard means anything that can cause harm (e.g. chemicals, asbestos, electricity, working from ladders or scaffolding, etc.).

A risk is the chance, high or low, that somebody will be harmed by the hazard.

We'll look more at hazards at the end of this section.

How to assess risks in the workplace

Some years ago the HSE produced guidance for employers to help with the process. Known as the *5 Steps to Risk Assessment*, it has become an invaluable tool for grasping the essentials of risk assessment and comprises:

- **Step 1** Look for the hazard.
- **Step 2** Decide who might be harmed and how.
- **Step 3** Evaluate the risks and decide whether the existing precautions are adequate or whether more should be done.
- **Step 4** Record your findings.
- **Step 5** Review your assessment and revise it if necessary.

To help understand the concepts, when you read on, try to put yourself in the position of an employer carrying out a risk assessment. We've worded it in that way.

(See two sample forms from *5 Steps to Risk Assessment* on pages 86–87)

Step 1: Look for the hazards

When you are doing the assessment, walk around your workplace and look afresh at what could reasonably be expected to cause harm. Ignore the trivial and concentrate on significant hazards that could result in serious harm or affect several people.

Some typical examples are use of tools, machinery and equipment, falling and tripping, lifting and handling, fire, storage of materials, electricity and working at height.

Ask your employees or their representatives what they think. They may have noticed things that are not immediately obvious to you. Manufacturers' instructions or data sheets can also help you spot hazards and put risks in their true perspective. So can accident and ill-health records.

Remember to think about long term hazards, such as noise.

Step 2: Decide who might be harmed and how

Don't forget categories such as:

- young workers, apprentices and trainees, new and expectant mothers etc. who may all be at particular risk
- cleaners, visitors, contractors, maintenance workers etc. who may not be in the workplace all the time
- members of the public, or people you share your workplace with, if there is a chance they could be hurt by your activities.

Ask other staff if they can think of anyone that you may have missed.

Step 3: Evaluate the risks

Evaluate the risks and decide whether existing precautions are adequate or more should be done. Consider how likely it is that each hazard could cause harm. This will determine whether or not more needs to be done to reduce the risk. Even after all precautions have been taken, some risk usually remains. Decide for each hazard whether this remaining risk is high, medium or low.

First, ask yourself whether you have done all the things that the law says you have to do. For example, there are legal requirements on prevention of access to dangerous parts of machinery. Then ask yourself whether generally accepted industry standards are in place.

But don't stop there – think for yourself, because the law also says that you must do what is reasonably practicable to keep your workplace safe.

Your real aim is to make all risks small by adding precautions as necessary. If something needs to be done, draw up an 'action plan' and give priority to any remaining risks which are high and/or those which could affect most people. In taking action ask yourself:

- Can I get rid of the hazard altogether?
- If not, how can I control the risks so that harm is unlikely?

In controlling risks apply the principles below, where possible in the following order:

- try a less risky option
- prevent access to the hazard (e.g. by guarding and using barriers and notices)
- organise work to reduce exposure to the hazard
- issue personal protective equipment (e.g. clothing, goggles etc.)
- provide welfare facilities (e.g. washing facilities for removal of contamination; first aid).

Improving health and safety need not cost a lot. For instance, placing a mirror on a dangerous blind corner to help prevent vehicle accidents, or putting some non-slip material on slippery steps, are inexpensive precautions considering the risks. Failure to take simple precautions can cost you a lot more if an accident does happen.

If the work you do tends to vary a lot, or you or your employees move from one site to another, identify the hazards you can reasonably expect and assess the risks from them.

Step 4: Record your findings

If you have fewer than five employees, you do not need to write anything down, though it is sensible and useful to keep a written record of what you have done. But if you employ five or more people you must record the significant findings of your assessment.

This means writing down the significant hazards and conclusions. You must also tell your employees about your findings.

Risk assessments must be suitable and sufficient. You need to be able to show that:

- a proper check was made
- you asked who might be affected
- you dealt with all the obvious significant hazards, taking into account the number of people who could be involved
- the precautions are reasonable, and the remaining risk is low
- you involved your staff or their representative in the process.

Keep the written record for future reference or use; it can help you if an inspector asks what precautions you have taken, or if you become involved in any action for civil liability.

It can also remind you to keep an eye on particular hazards and precautions and it helps to show that you have done what the law requires.

Step 5: Review your assessment

Review your assessment and revise it if necessary. Sooner or later you will bring in new machines, substances or procedures that could lead to new hazards. If there is any significant change, add to the risk assessment to take account of the new hazard. Don't amend your assessment for every trivial change, or still more, for each new job.

But if a new job introduces significant new hazards of its own, you will want to consider them in their own right and do whatever you need to keep the risks down. In any case, it is good practice to review your assessment from time to time (no later than annually) to make sure that the precautions are still working effectively.

Company name:

Step 1
What are the hazards?

Spot hazards by:

- walking around your workplace;
- asking your employees what they think;
- visiting the *Your industry* areas of the HSE website or calling HSE Infoline;
- calling the Workplace Health Connect Adviceline or visiting their website;
- checking manufacturers' instructions;
- contacting your trade association.

Don't forget long-term health hazards (such as noise).

Step 2
Who might be harmed and how?

Identify groups of people. Remember:

- some workers have particular needs;
- people who may not be in the workplace all the time;
- members of the public;
- if you share your workplace think about how your work affects others present.

Say how the hazard could cause harm.

Step 5 **Review date:**

5 Steps to Risk Assessment

Date of risk assessment:

Step 3 What are you already doing?	What further action is necessary?	Step 4 How will you put the assessment into action?		
List what is already in place to reduce the likelihood of harm or make any harm less serious.	You need to make sure that you have reduced risks 'so far as is reasonably practicable'. An easy way of doing this is to compare what you are already doing with good practice. If there is a difference, list what needs to be done.	Remember to prioritise. Deal with those hazards that are high-risk and have serious consequences first.		
		Action by whom	Action by when	Done

- Review your assessment to make sure you are still improving, or at least not sliding back.
- If there is a significant change in your workplace, remember to check your risk assessment and, where necessary, amend it.

Remember

Prevention is better than cure!

Accident prevention

How do we prevent accidents? So far everything that we have looked at has been based around the legal requirements and fairly obvious material problems. These tend to be referred to as 'environmental causes' as they relate to the environment that you are working in. Such causes can be unguarded machinery, defective tools and equipment, poor ventilation, excessive noise, poorly lit workplaces, overcrowded workplaces and untidy or dirty workplaces.

However, there are other causes of accidents and they invariably involve people. It would be nice to think that we all possess common sense, but the 'human' causes of accidents include:

- carelessness
- bad and foolish behaviour
- improper dress
- lack of training
- lack of experience
- poor supervision
- fatigue
- use of alcohol or drugs.

Remember

The first person to alert about a health and safety issue should be your Site Supervisor or Safety Officer

Health and safety in the workplace is something we each need to take personal responsibility for, and by thinking about what we are doing, or are about to do, we can avoid most potentially dangerous situations.

Equally, we can demonstrate a responsible approach by reporting potentially dangerous situations, hazards or activity to the correct people. Even if the hazard is something you can easily fix yourself, for instance you might move a brick that was causing a trip hazard, still report it to your supervisor. The fact that the brick was there at all might be indicative of someone not doing their job properly and it could happen again.

Safety tip

If the hazard is something that can be easily fixed, and as long as it is not dangerous to yourself, then you should fix it

You may also have a role to play in other circumstances. For example, at present premises are inspected by local Fire Authorities. If you are aware of something unsafe, such as a fire exit that has been blocked, then tell them about it.

As with the risk assessment process, the secret is to be aware of all possible danger in the workplace and have a positive attitude towards health and safety. Follow safe and approved procedures where they exist, and always act in a responsible way to protect yourself and others, and the construction site can be a happy and safe working environment for many years.

Fire safety

As an electrician, you may well end up working in what are known as **explosive atmospheres**, and fire is always a risk when dealing with electrical installation. So what exactly is a fire?

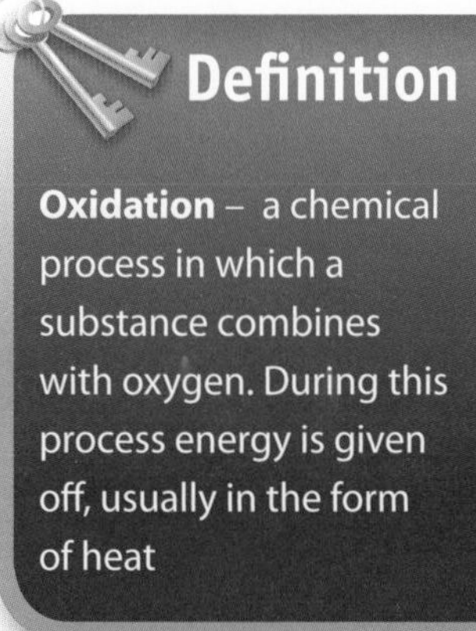

Oxidation – a chemical process in which a substance combines with oxygen. During this process energy is given off, usually in the form of heat

What is a fire?

Essentially, fire is very rapid **oxidation**.

Rusting iron and rotting wood are common examples of slow oxidation. Fire, or **combustion**, is rapid oxidation as the burning substance combines with oxygen at a very high rate. Energy is given off in the form of heat and light. Because this energy production is so rapid, we can feel the heat and see the light as flames.

How fire happens

All matter exists in one of three states: solid, liquid or gas (vapour). The atoms or molecules of a solid are packed closely together, and those of a liquid are packed loosely. The molecules of a vapour are not really packed together at all and are free to move about.

In order for a substance to oxidise, its molecules must be well surrounded by oxygen molecules. The molecules of solids and liquids are packed too tightly for this to happen, and therefore only vapours can burn.

When a solid or liquid is heated, its molecules move about rapidly. If enough heat is applied, some molecules break away from the surface to form a vapour just above the surface. This vapour can now mix with oxygen. If there is enough heat to raise the vapour to its ignition temperature, and if there is enough oxygen present, the vapour will oxidise rapidly and it will start to burn.

What we call burning is the rapid oxidation of millions of vapour molecules. The molecules oxidise by breaking apart into individual atoms and recombining with oxygen into new molecules. It is during the breaking-recombining process that energy is released as heat and light. The heat that is released is radiant heat, which is pure energy. It is the same sort of energy that the sun radiates and that we feel as heat. It radiates (travels) in all directions. Therefore, part of it moves back to the seat of the fire, to the 'burning' solid or liquid (the fuel). The heat that radiates back to the fuel is called radiation feedback.

Part of this heat releases more vapour, and part of it raises the vapour to the ignition temperature. At the same time, air is drawn into the area where the flames and vapour meet. The result is that there is an increase in flames as the newly formed vapour begins to burn.

Figure 2.6 The fire triangle

The fire triangle

The three things that are needed for combustion to take place are:

- fuel (to vaporise and burn)
- oxygen (to combine with fuel vapour)
- heat (to raise the temperature of the fuel vapour to its ignition temperature).

The fire triangle shows us that fire cannot exist without all three together:

- if any side of the fire triangle is missing, a fire cannot start
- if any side of the fire triangle is removed, the fire will go out.

Remember

Quite simply, a fire can be extinguished by destroying the fire triangle. If fuel, oxygen or heat is removed, the fire will die out. If the chain reaction is broken, the resulting reduction in vapour and heat production will put out the fire

Classes of fire

Combustible and flammable fuels have been broken down into five categories:

- Class A fires are those involving organic solids such as paper or wood
- Class B fires are those involving flammable liquids
- Class C fires are those involving flammable gases
- Class D fires are those involving metals
- Class F fires are those involving cooking oils

  (Electrical fires have no classification, as electricity is a source of ignition that will feed a fire until switched off or isolated.)

Fire legislation

With effect from 2006, most of the existing UK fire safety legislation was swept away and replaced under the 2001 Reform Act by one new, simpler piece of legislation, the Regulatory Reform (Fire Safety) Order 2005. As a result, fire certificates have been abolished and employers become solely responsible for fire safety within their respective workplaces.

Responsibility for meeting the order lies with anyone who has control of premises or who has a degree of control over certain areas or systems, such as the employer (for those parts of premises staff may go to), the occupier, such as self-employed people, or voluntary organisations if they have any control.

The order applies to virtually all premises and covers nearly every type of building, structure and open space, such as factories, offices, shops, care homes, hospitals, community halls, places of worship, pubs, clubs, restaurants, schools and sports centres. However, it does not apply to private homes, including individual flats within a block or house.

The main rules for the responsible person under the order are to:

- carry out a fire-risk assessment identifying any possible dangers and risks
- consider who may be especially at risk
- get rid of or reduce the risk from fire as far as is reasonably possible and provide general fire precautions to deal with any possible risk remaining
- make sure that everyone on the premises, or nearby, can escape safely if there is a fire

- take other measures to make sure there is protection if flammable or explosive materials are used or stored
- create a plan to deal with any emergency and, in most cases, keep a record of your findings
- review findings when necessary.

Fire authorities will be the main agency responsible for enforcing all fire-safety legislation in non-domestic premises and will target their resources and inspections at those premises that present the highest risk.

Did you know?

The new fire regulations will state that fire-fighting equipment in the workplace is there only to enable a small fire to be attacked to prevent it from spreading

Fire prevention

Fires can spread rapidly. Once established, even a small fire can generate sufficient heat energy to spread and accelerate the fire to surrounding combustible materials. Fire prevention is largely a matter of common-sense and good housekeeping. For example, keep the workplace clean and tidy. If you smoke, don't throw lit cigarettes on to the ground, and don't leave flammable materials lying around or near sources of heat or sparks.

From an electrical perspective, make sure that all leads are in good condition, that fuses are of the correct rating and that circuits are not overloaded. Any alterations or repairs to electrical installations must be carried out only by qualified personnel and must be to the standards laid down in the IEE Regulations (BS 7671).

Fire fighting

A fire safety officer once said that the only use people should think of for a portable fire extinguisher 'was to break the window so that you could escape from the building'. The point he was making is that it is dangerous to try to fight a fire, and the use of fire extinguishers should only be considered as a first-response measure, e.g. where the fire is very small or where it is blocking your only means of exit. Fire fighting is a job for the professional emergency service.

Remember

Recent European legislation has dictated that new extinguishers must be coloured red, whatever substance they contain, but will carry 5% of the colour the extinguisher would have been in the original system. So now a carbon dioxide extinguisher (for use on electrical equipment fires only) will be red with a black stripe, triangle or lettering

Fire-fighting equipment

Normally available fire-fighting equipment includes portable appliances such as extinguishers, buckets of sand or water and fire-resistant blankets. In larger premises you will find automatic sprinklers, hose reels and hydrant systems.

Fire extinguishers

There are many types of fire extinguisher, each with a specific set of situations in which they may or may not be used. Three common types of fire extinguisher are shown overleaf.

In the past, the whole extinguisher was coloured according to its type. Black ones, for instance, were carbon dioxide (CO_2). Today this colour-coding is not immediately obvious as all extinguishers are red. In the heat of the moment, be careful to pick up the right extinguisher for the type of fire you're trying to put out. Powder and foam extinguishers both come in two types, and none is totally effective on every kind of fire. Before buying one, it is vital to look carefully at what kinds of fires it can be used on.

Dry powder fire extinguisher

Water fire extinguisher

Carbon dioxide fire extinguisher

Foam fire extinguisher

Fire extinguisher types	
Standard/Multi-purpose dry powder	
Colour	Blue
Application	The powder 'knocks down' the flames. Safe to use on most kinds of fire. Multi-purpose powders are more effective, especially on burning solids; standard powders work well only on burning liquids.
Dangers	The powder does not cool the fire well. Fires that seem to be out can re-ignite. Doesn't penetrate small spaces, like those inside burning equipment. The jet could spread burning fat or oil around.
How to use	Aim the jet at the base of the flames and briskly sweep it from side to side.
Water	
Colour	Red
Application	The water cools the burning material. You can only use water on solids, like wood or paper. Never use water on electrical fires or burning fat or oil.
Dangers	The water can conduct electricity back to you. Water actually makes fat or oil fires worse – they can explode as the water hits them.
How to use	Aim the jet at the base of the flames and move it over the area of the fire.
CO_2	
Colour	Black
Application	Displace oxygen with CO_2 (a non-flammable gas). Good for electrical fires as they don't leave a residue.
Dangers	Pressurised CO_2 is extremely cold. DO NOT TOUCH. Do not use in confined spaces.
How to use	Aim the jet at the base of the flames and sweep it from side to side.
Foam/AFFF (Aqueous Film Forming Foam)	
Colour	White or cream
Application	The foam forms a blanket or film on the surface of a burning liquid. Conventional foam works well only on some liquids, so it's not good for use at home, but AFFF is very effective on most fires except electrical and chip-pan fires.
Dangers	'Jet' foam can conduct electricity back to you, though 'spray' foam is much less likely to do so. The foam could spread burning fat or oil around.
How to use	For solids, aim the jet at the base of the flames and move it over the area of the fire. For liquids, don't aim the foam straight at the fire – aim it at a vertical surface or, if the fire is in a container, at the inside edge of the container.

Table 2.3 Some fire extinguisher types

When considering using a fire extinguisher remember the following points:

- never use a fire extinguisher unless you have been trained to do so
- do not use water extinguishers on electrical fires due to the risk of electric shock and explosion
- do not use water extinguishers on oils and fats as this too can cause an explosion
- do not touch the horn on CO_2 extinguishers as this can freeze burn the hands
- do not use the CO_2 extinguisher in a small room as this could cause suffocation
- read the operating instructions on the extinguisher before use.

Sprinkler systems

To reduce as far as you possibly can the risk of dying in a fire, you should get fire sprinklers installed.

Sprinklers can be individually heat-activated, so the whole system doesn't go off at once, and they rarely get set off accidentally as they need high temperatures to trigger them. They operate automatically, whether you're in the building or not. Sprinklers should sound the alarm when they go off – so they alert you and also tackle the fire.

Sprinkler

Smoke alarms

Smoke alarms will alert you to slow-burning, smoke-generating fires that may not create enough heat to trigger a sprinkler.

General fire safety

Regular fire drills must be held and all personnel must be familiar with normal and alternative escape routes. Fire routes should be clearly marked and emergency lighting signs fitted above each exit, where applicable. Make sure you know where your assembly point is.

If you discover a fire:

- raise the alarm immediately
- leave by the nearest exit
- call the fire service out
- if in doubt stay out
- close, but don't lock the windows to help starve the fire of oxygen
- go to your assembly point and report to your supervisor
- *do not return to the building until you are authorised to do so.*

Some of the above duties may be allocated to a particular person.

Did you know?

In parts of the United States where sprinklers have become compulsory, almost no one dies from fire at home. Sprinklers are fitted in as many rooms as required, fed by small pipes and run off the mains water

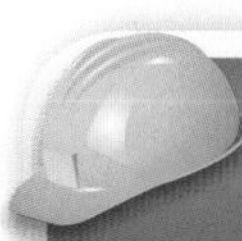

Safety tip

Test your smoke alarms regularly to make sure the battery is still working

Remember

Personal safety must always come before your efforts to contain a fire. Delay and you will have difficulty in finding your way through the smoke – and the fumes may choke you

Site safety

Working at height is classed as using access equipment. In this section we will start to look at more specific aspects of health and safety that affect your day-to-day work as an electrician.

Did you know?

According to the HSE, falls from height are the most common cause of fatal injury and the second most common cause of major injury to employees, accounting for 15 per cent of all such injuries

Working at height

All industry sectors are exposed to the risks presented by this hazard, although the level of incidence varies considerably.

Ladders

Ladders, stepladders and trestles are perhaps the most commonly used access equipment on sites and they are also perhaps the most misused. It is essential that safe working practices should be followed if accidents are to be avoided.

Before you use a ladder or stepladder, check it – make sure it is safe to use. Look for:

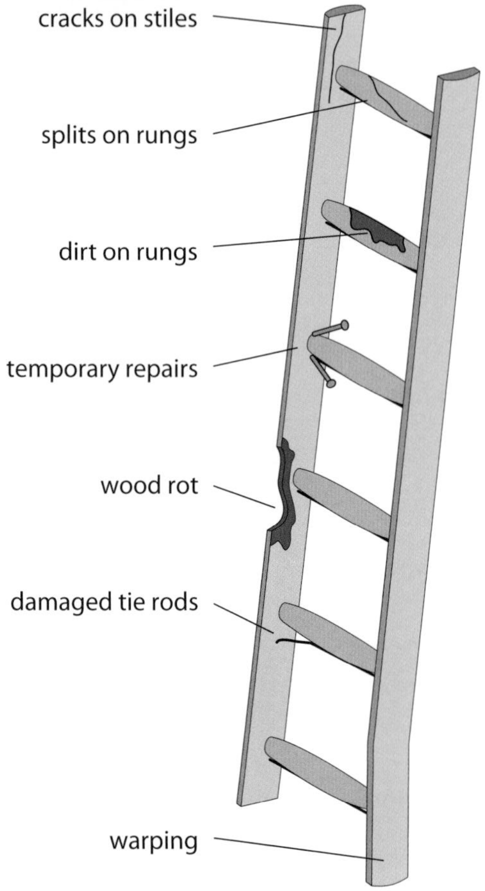

- cracks in the rungs or the stiles (the sides of the ladder)
- missing, broken or weakened rungs
- rungs depending solely on nails or spikes for support
- mud, grease or oil either on the rungs or the stiles
- obvious signs of permanent bending in the rungs or stiles
- items stuck in the feet such as swarf or stones, or grease or dirt, that prevent the feet from making a direct contact with the ground
- missing, damaged or worn anti-slip feet on metal and fibreglass ladders and stepladders; with ladders, check those at the top and bottom
- missing or loose screws or rivets
- cracked or damaged welds on metal ladders or stepladders
- rot, woodworm or tie rods that are either missing or damaged
- painted wooden ladders or stepladders: these should not be used as the paint can hide defects.

cracks on stiles
splits on rungs
dirt on rungs
temporary repairs
wood rot
damaged tie rods
warping

Figure 2.7 Ladder with defects

Short ladders can be carried by one person, on the shoulder, in either the horizontal or vertical position. Longer ladders should be carried by two people, horizontally on the shoulders, one at either end holding the upper stile. When carrying ladders take care in rounding corners or passing between or under obstacles.

There are certain rules for erecting ladders, which must be followed to ensure safe working. These are:

- The ladder should be placed on firm, level ground. Bricks or blocks should not be used to 'pack up' under the stiles to compensate for uneven ground.
- If using extension ladders they should be erected in the closed position and extended one section at a time. When extended there must be at least four or five rungs' overlap on each extension.
- Ladders should be placed clear of any excavation, and in such a position that they are not causing a hazard, or placed anywhere where they may be struck or dislodged. If the ladder is placed in an exposed position it should be guarded with barriers.
- The angle of the ladder to the building should be in the proportion of 4 up to 1 out, or 75°.

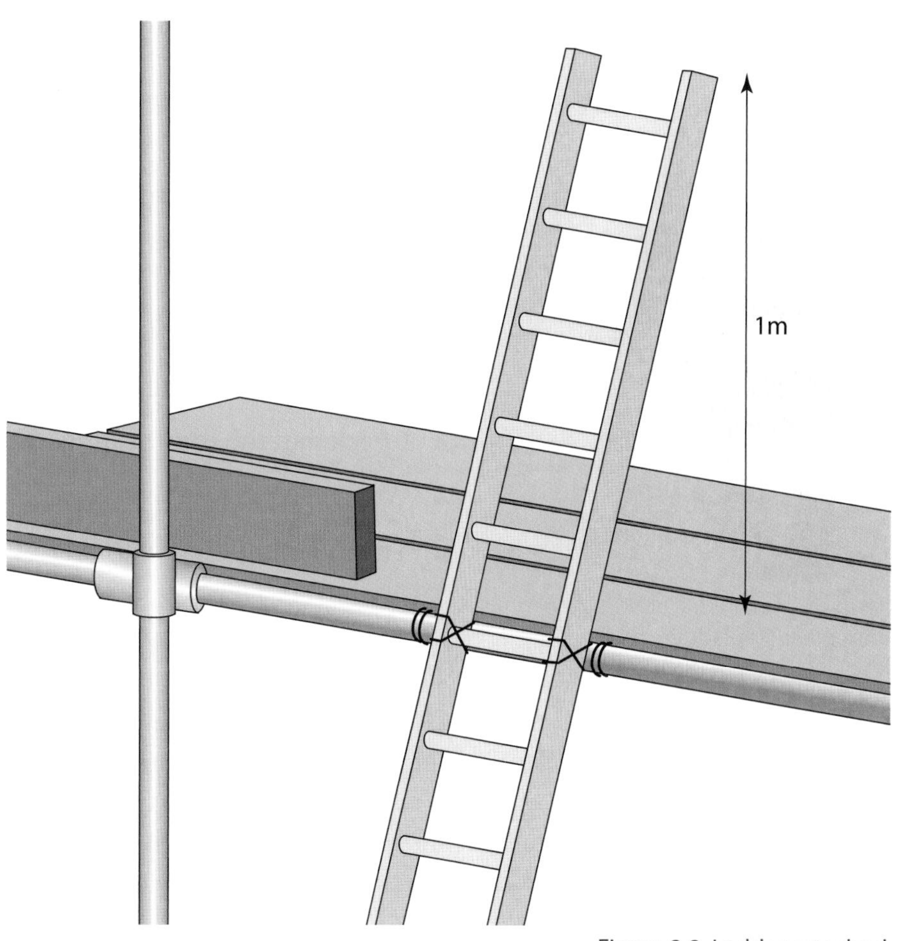

Figure 2.8 Ladder attached to scaffold platform

- The ladder should be secured at the top and as necessary at the bottom to prevent unwanted movement. Alternatively the ladder may be steadied by someone holding the stiles and placing one foot on the bottom rung. This is commonly known as 'footing' the ladder. This person must not under any circumstances move away from footing the ladder while someone is up it.
- When the ladder provides access to a roof or working platform, the ladder must extend at least 1 m, or 5 rungs, above the landing place.
- Make sure that the ladder is not resting against any fragile surface or against fittings such as gutters or drainpipes, as these may give way, resulting in an accident.
- If a ladder has to be secured, never secure it by the rungs; always use the stiles.
- When climbing up ladders you must use both hands to grip the rungs. This will give you better protection if you should slip.
- The working position should be not less than five rungs from the top of the ladder.

All ladders and stepladders should be tested and examined on an annual basis. The results of the tests should be recorded. Ideally the item tested should be marked to show it has been tested. A competent person must carry out this test.

Fibreglass stepladder

Stepladders

A lot of the rules that apply to ladders also apply to stepladders. However, specifically bear in mind the following:

- All four legs of a stepladder should rest firmly and squarely on the ground. They will do this provided that the floor or ground on which they stand is level and the steps themselves are not worn or damaged.
- Ensure the legs are fully opened.
- Check that the hinge is in good condition.
- Check that the rope or hinged bracket is of equal length and not frayed.
- When using the steps ensure that your knees are below the top of the steps.
- The top of the steps should not be used unless it is constructed as a platform.

Trestles

Some jobs cannot be done safely from a ladder or stepladder. In such cases, a working platform known as a 'trestle scaffold' should be used. This consists of two pairs of trestles or 'A' frames spanned by scaffolding boards, which then provide a simple working platform.

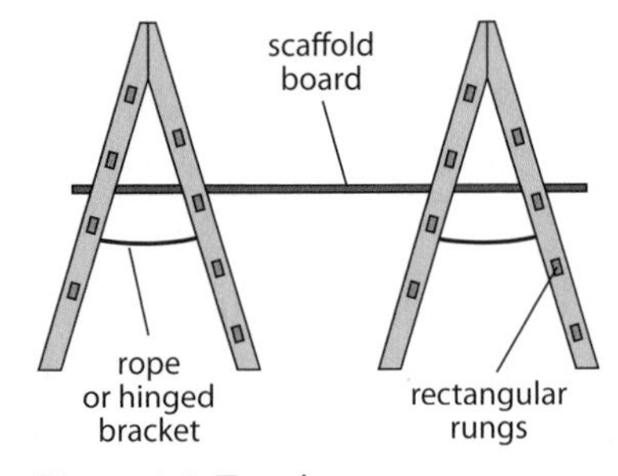

Figure 2.9 Trestles

When erecting trestle scaffolds the following rules should be observed:

- As with ladders they should be erected on a firm, level base with the trestles fully opened.
- The platform must be at least 600 mm wide.
- The platform should be no higher than two thirds of the way up the trestle; this ensures that there is at least one third of the trestle above the working surface.
- The scaffolding boards must be of equal length and thickness and should not overhang the trestle by more than four times their own thickness, e.g. a 40 mm board must not overhang by more than 160 mm.
- The trestles should be spaced at the following distances apart: 1 metre for 32 mm thick boards, 1.5 metres for 38 mm thick boards and 2.5 metres for 50 mm thick boards.
- If the platform is more than 2 m above the ground, toe boards and guardrails must be fitted and a separate ladder provided for access.
- Trestles must not be used where anyone can fall more than 4.5 m.
- Trestles over 3.5 m tall should ideally be 'tied' to the structure of the building.

Mobile scaffold towers

Tower scaffolds are widely used and sadly are involved in numerous accidents each year. These usually happen because the tower has not been properly erected or used. Aluminium towers are light and can easily overturn. Towers rely on all the parts being in place to ensure adequate strength and can easily collapse if sections are left out.

Normally made from light aluminium tube, the tower is built by slotting the sections together until the required height is reached. Towers may be either mobile ones fitted with wheels or static ones fitted with plates. Consequently, they are useful for installing long runs – in, say, a factory lighting installation.

When working with a mobile/static tower scaffold, the following points must be checked and followed:

- The person erecting the tower should be competent.
- Make sure the tower is resting on firm, level ground with the wheels or feet properly supported. Do not use bricks or building blocks to take the weight of any part of the tower.
- The taller the tower, the more likely it is to become unstable. As a guide, if towers are used in exposed conditions or outside, the height of the working platform should be no more than three times the minimum base dimension. If the tower is to be used inside, on firm level ground, the ratio may be extended to 3.5. Using this guide, if the base of the tower is 2 m by 3 m the maximum height would be 6 m for use outside and 7 m for inside.
- Before using the tower, always check that the scaffold is vertical and, in the case of a mobile tower, that all wheel brakes are on.
- There must be a safe way to get to and from the work platform. It is not safe to climb up the end frames of the tower except where the frame has an appropriately designed built-in ladder or a purpose-made ladder can be attached safely on the inside.
- Provide suitable edge protection on platforms where a person could fall more than 2 m. Guard rails should be at least 910 mm high and toe boards at least 150 mm high. An intermediate guard rail or suitable alternative should be provided so that the unprotected gap does not exceed 470 mm.
- When moving a tower, check that there are no power lines or other overhead obstructions; check that the ground is firm and level; push or pull only from the base; never move it while there are people or materials on the upper platforms; and never move it in windy conditions.

Safety tip

It is not safe to climb up the end frames of a mobile scaffold tower

- Outriggers can increase stability by effectively increasing the area of the base, but if used they must be fitted diagonally across all four corners of the tower and not on one side only. When outriggers are used they should be clearly marked (e.g. with hazard marking tape) to indicate that a trip hazard is present.
- When towers are used in public places, extra precautions may be needed such as minimising the storage of materials and equipment on the working platform, erecting barriers at ground level to prevent people from walking into the tower or work area, and removing or boarding over access ladders to prevent unauthorised access if the tower is to remain in position unattended.
- Before you use a tower on a pavement, check whether you need a licence from the local authority.
- Tower scaffolds must be inspected by a 'competent person' before first use and following substantial alteration or any event likely to have affected their stability. If a tower remains erected in the same place for more than seven days, it should also be inspected at regular intervals (not exceeding seven days) and a written report should be made. Any faults found should be put right.

Scissor and boom lifts

This category of access equipment is sometimes referred to as 'mobile elevating platforms'.

From a safety perspective as well as with regard to cost, scissor lifts can offer quick and efficient access solutions for a wide range of installation and maintenance tasks. Compact dimensions and tight turning circles give these machines great versatility.

There is a wide range of machines available for flat-slab and rough-terrain applications, thus enabling them to be used in circumstances ranging from installations within large factories to repairs of external lighting.

However, scissor lifts can only extend upwards. Sometimes it is necessary to 'reach' over objects to be able to carry out the work, for example repairing a street-lighting column. In these circumstances a telescopic boom platform is likely to be more appropriate.

In both types of lift, it is essential that the workers wear a safety harness. This must be attached to the lift and never to the structure being worked on.

Particular attention must also be paid to whether overhead power supplies are present.

Scissor lift

Boom lift

Working in excavations

Every year people are killed or seriously injured when working in excavations. Excavation work has to be properly planned, managed, supervised and carried out to prevent accidents.

Planning

Before digging any excavations, it is important to plan against the following:

- collapse of the sides
- materials falling onto people working in the excavation
- people and vehicles falling into the excavation
- people being struck by plant
- undermining nearby structures
- contact with underground services
- hazardous entry and exit points
- fumes
- accidents to members of the public.

Make sure the necessary equipment needed such as trench sheets, props, baulks etc. is available on site before work starts.

Precautions during excavation work

Once work starts:

- wear a hard hat when working in excavations
- prevent the sides and the ends from collapsing by battering them to a safe angle or supporting them with timber, sheeting or proprietary support systems
- never go into an unsupported excavation
- never work ahead of the supports
- never store spoil or other materials close to the sides of excavations. The spoil may fall in and the extra loading will make the sides more prone to collapse
- always make sure the edges of the excavation are protected against falling materials. Provide toe boards where necessary
- take steps to prevent people falling into excavations. If the excavation is 2 m or more deep, provide substantial barriers, e.g. guard rails and toe boards
- keep vehicles away from excavations wherever possible. Use barriers if necessary
- where vehicles have to tip materials into excavations, use stop blocks to prevent them over-running. Remember that the sides of the excavation may need extra support

Remember

Even shallow trenches can be dangerous. You may need to provide supports even if the work only involves bending or kneeling in the trench. You should never work in a trench alone without supervision

- keep workers separate from moving plant such as excavators. Where this is not possible, use safe systems of work to prevent people being struck
- make sure that any plant operators are competent
- make sure excavations do not affect the footings of scaffolds or the foundations of nearby structures. Walls may have very shallow foundations and these can be undermined by even small trenches
- look around for obvious signs of underground services, e.g. valve covers or patching of a road surface
- use locators to trace any services. Mark the ground accordingly
- make sure that the person supervising the excavation work has service plans and knows how to use them
- everyone carrying out the work should know about safe digging practices and emergency procedures
- provide good ladder access or other safe ways of getting into and out of the excavation
- fence off all excavations in public places to prevent people and vehicles falling in
- take precautions (e.g. securely covering excavations) where children might get onto a site out of hours, to reduce the chance of them being injured
- make sure that a competent person supervises the installation at all times.

Manual handling

The Manual Handling Operations Regulations 1992, as amended in 2002, apply to a wide range of manual handling activities, including lifting, lowering, pushing, pulling or carrying. The load may be either inanimate, such as a box or a trolley, or animate, such as a person or an animal.

Within the context of the electrical industry, manual handling can involve items such as scaffolding, tools, equipment, switchgear and motors.

Employer requirements	Employee responsibilities
Reduce the need for hazardous manual handling, so far as is reasonably practicable	Follow appropriate systems of work laid down for employee safety
Assess the risk of injury from any hazardous manual handling that cannot be avoided	Make proper use of equipment provided for employee safety
Reduce the risk of injury from hazardous manual handling, so far as is reasonably practicable	Co-operate with the employer on health and safety matters
	Inform the employer if you identify hazardous handling activities
	Ensure that your activities do not put others at risk

Table 2.4 Employer duty and employee responsibility regarding manual handling

Where possible, avoid manual handling. Check whether you need to move the item at all. For example, does a large work piece really need to be moved, or can the activity be done safely where the item already is? Think about using handling aids such as a conveyor, a pallet truck, an electric or hand-powered hoist or fork-lift truck.

However, beware of new hazards from automation or mechanisation. For example, automated plant still needs cleaning, maintaining etc., and fork-lift trucks must be suited to the work and have properly trained and certified operators.

The movement of loads requires careful planning in order to identify potential hazards before they cause injuries. This planning involves carrying out a risk assessment, which looks at a number of areas.

The task	*Does the task involve:* • holding loads away from the body? • twisting, stooping or reaching upwards? • large vertical movement? • long carrying distances? • strenuous pushing or pulling? • repetitive handling? • insufficient rest or recovery time? • a work rate imposed by a process?
The load	*Is the load:* • heavy? • bulky, unwieldy or difficult to grasp? • unstable or are the contents likely to shift? • sharp, hot or otherwise potentially damaging?
The working environment	*Does the working environment have:* • space constraints? • floors that are slippery or unstable? • poor lighting? • hot, cold or humid conditions?
Individual capability	*Does the individual:* • have a reach problem, that restricts their physical capability? • have knowledge of and training in manual handling?
Handling aids and equipment	• is the device the correct type for the job? • is it well maintained? • are the wheels on the device suited to the floor surface? • do the wheels run freely? • is the handle height between the waist and shoulders? • are the handle grips in good order and comfortable? • are there any brakes? If so, do they work?

Table 2.5 Risk assessment of a manual handling operation

Lifting and carrying

Here are some practical tips for safe manual handling.

Think before lifting/handling, plan the lift

- Can handling aids be used?
- Where is the load going to be placed?
- Will help be needed with the load?
- Remove obstructions such as discarded wrapping materials.
- For a long lift, consider resting the load midway on a table or bench to change grip.

Keep the load close to the waist

- Keep the load close to the body for as long as possible while lifting.
- Keep the heaviest side of the load next to the body.
- If a close approach to the load is not possible, try to slide it towards the body before attempting to lift it.

Adopt a stable position

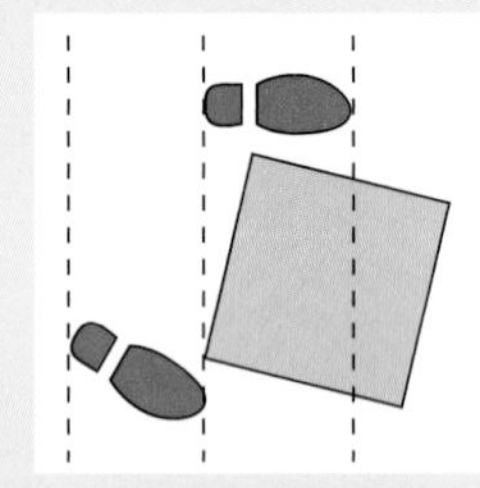

- The feet should be apart with one leg slightly forward to maintain balance (alongside the load, if it is on the ground).
- Test the weight of the load by pushing it with the foot.
- The worker should be prepared to move his or her feet during the lift to maintain stability.
- Avoid tight clothing or unsuitable footwear, which may make this difficult.

Get a good hold

Where possible the load should be hugged as close as possible to the body. This may be better than gripping it tightly with hands only.

Start in a good posture

At the start of the lift, slight bending of the back, hips and knees is preferable to fully flexing the back (stooping) or fully flexing the hips and knees (squatting).

Don't flex the back any further while lifting

- This can happen if the legs begin to straighten before starting to raise the load.
- Avoid twisting the back or leaning sideways, especially while the back is bent.
- Shoulders should be kept level and facing in the same direction as the hips.
- Turning by moving the feet is better than twisting and lifting at the same time.
- Keep the head up when handling.
- Look ahead, not down at the load, once it has been held securely.

Move smoothly

The load should not be jerked or snatched as this can make it harder to keep control and can increase the risk of injury.

Don't lift or handle more than can easily be managed

There is a difference between what people can lift and what they can safely lift. If in doubt, seek advice or get help.

Put down, then adjust

If precise positioning of the load is necessary, put it down first, then slide it into the desired position.

Pushing and pulling

Here are some practical points to remember when pushing and pulling loads.

Handling devices

- Aids such as trolleys should have handle heights that are between the shoulder and waist.
- Devices should be well-maintained with wheels that run smoothly (the law requires that equipment is maintained).
- When purchasing new trolleys etc. ensure they are of good quality with large-diameter wheels made of suitable material and with castors, bearings etc. that will last with minimum maintenance.
- Consultation with your employees and safety representatives will help, as they know what works and what doesn't.

Force

As a rough guide, the amount of force that needs to be applied to move a load over a flat, level surface using a well-maintained handling aid is at least 2 per cent of the load weight. You should try to push rather than pull when moving a load, provided you can see over it and control steering and stopping.

Did you know?

If a load weighs 400 kg, then the force needed to move the load is 8 kg. The force needed will be greater, perhaps far greater, if conditions are not perfect (e.g. wheels not in the right position or a device that is poorly maintained)

Slopes

You should enlist help whenever necessary if you have to negotiate a slope or ramp, as pushing and pulling forces can be very high. For example, if a load of 400 kg is moved up a slope of 1 in 12 (about 5°), the required force is over 30 kg even in ideal conditions – with good wheels and a smooth slope. This is above the guideline weight for men and well above the guideline weight for women.

Uneven surfaces

On an uneven surface, the force needed to start the load moving could increase to 10 per cent of the load weight, although this might be offset to some extent by using larger wheels. Soft ground may be even worse.

Remember

Moving an object over soft or uneven surfaces requires higher forces.

Stance and pace

To make it easier to push or pull, you should keep your feet well away from the load and go no faster than walking speed. This will stop you becoming too tired too quickly.

General risk assessment guidelines

There is no such thing as a completely 'safe' manual handling operation. But working within the guidelines shown in Figure 2.10 below will cut the risk.

Use the drawing to make a quick and easy assessment. Each box contains a guideline weight for lifting and lowering in that zone. (As you can see, the guideline weights are reduced if handling is done with arms extended, or at high or low levels, as that is when injuries are most likely to occur.)

The guideline weights are for infrequent operations – up to about 30 operations per hour – where the pace of work is not forced. Reduce the weights if the operation is repeated more often

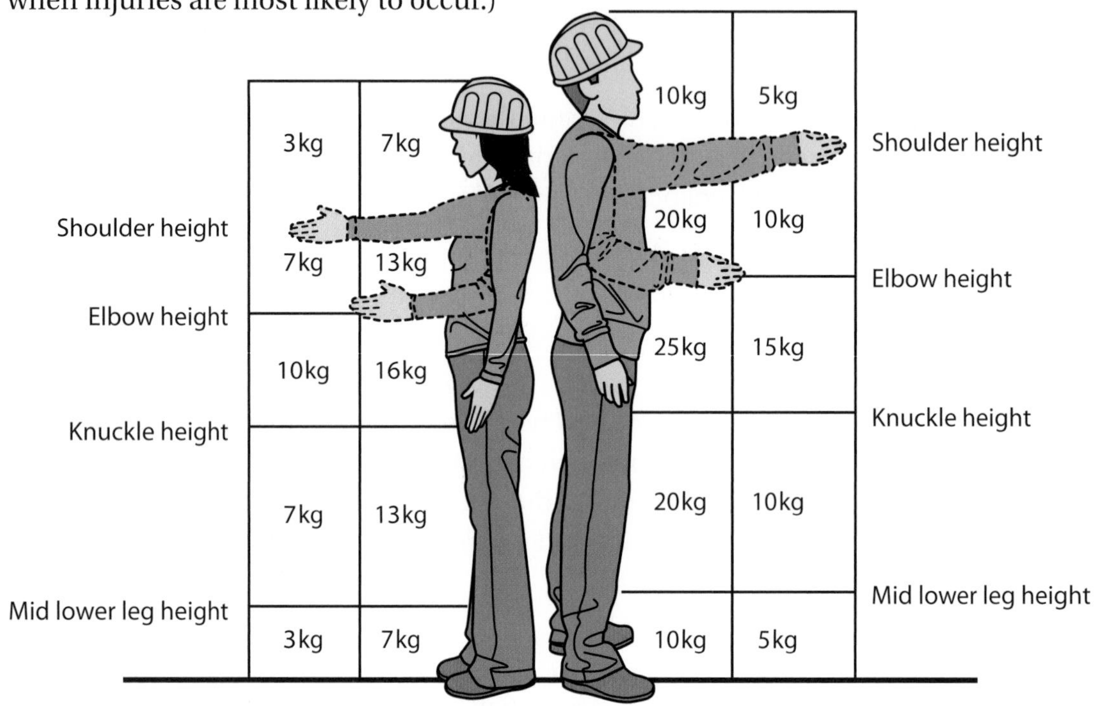

Figure 2.10 Lifting chart for men and women

Observe the work activity you are assessing and compare it to the diagram. First, decide which box or boxes the lifter's hands pass through when moving the load. Then, assess the maximum weight being handled. If it is less than the figure given in the box, the operation is within the guidelines.

If the lifter's hands enter more than one box during the operation, use the smallest weight. Use an in-between weight if the hands are close to a boundary between boxes.

The guideline weights assume that the load is readily grasped with both hands and that the operation takes place in reasonable working conditions, with the lifter in a stable body position.

Tools and equipment

You will use a variety of tools and pieces of equipment in your work as an electrician. All of them are potentially dangerous if misused or neglected. Instruction in the proper use of tools and equipment will form part of your training, and you should continue to follow safe working methods.

Hand tools and manually operated equipment are often misused. You should always use the right tool for the job, never just make do with whatever tool you may have to hand. For example, never use a hammer on a tool with a wooden handle as you may damage the wooden handle and create flying splinters. Here are some general guidelines.

- Keep cutting tools, saws, chisels, drills etc. sharp and in good condition.
- Ensure handles are properly fitted and secure, and free from splinters.
- Check that the plugs and cables of hand-held electrically powered tools are in good condition. Replace frayed cables and broken plugs.
- Electrically powered tools of 110 volts or 230 volts must be **P.A.T. tested** in accordance with your employer's procedures.

Other common items of equipment, e.g. barrows, trucks, buckets, ropes and tackle etc., are all likely to deteriorate with use. If they are damaged or broken they will eventually fail and may cause an accident.

Never use unserviceable tools and equipment. Either repair or replace them, and remove unsafe equipment from the site

Cartridge tools and high-pressure airlines

If you have to use cartridge-operated tools you will be given the necessary instruction on the safe and correct methods of use. They can be dangerous, especially if you have an accidental discharge. This can cause ricochets, which could lead to a serious injury.

You may also come into contact with high-pressure airlines. Used carelessly, compressed air can be dangerous. It can cause explosions or blow tools, equipment or debris about which may injure others. Never use an airline to blow dust away, never aim it at any part of your body and never point it at somebody else. If high-pressure air enters the body through a cut or abrasion or through one of the body's orifices it can cause an air embolism, which is very painful and can be fatal.

As an individual you may have no control over the general state of the workplace but you should see that your own work area is kept clear and tidy, as this is the mark of a skilled and conscientious electrician

Untidy working

Tools, equipment and materials left lying about, trailing cables and air hoses, spilt oil and so on cause people to trip, slip or fall. Clutter and debris, oily rags, paper etc. should be cleared away to prevent fire hazards.

Tools and equipment left in this state become obvious targets for thieves and make it difficult for an employer to maintain effective levels of insurance cover. Upon completion of any work activity, all tools etc. should be cleared away and the workplace left in a safe condition.

On the job: Site safety

Kelly, a 1st-year electrical apprentice, has been left in charge of a scissor boom while the other electricians take their lunch break in a local pub. Two carpenters climb onto the scissor boom and use it for working; while doing so, they crash into a low ceiling, damaging the boom. They tell Kelly not to worry because they all work for the same company. They finish the work and leave the site. A little later the electricians return.

1. What should Kelly have done in this situation?
2. Who was ultimately responsible for the safety of the scissor boom?
3. Who would have been responsible if the carpenters had hurt themselves?

Electrical safety

Electricity can kill. Even non-fatal shocks can cause severe and permanent injury. Shocks from faulty equipment may lead to falls from ladders, scaffolds or other work platforms. Those using electricity may not be the only ones at risk, as poor electrical installations and faulty electrical appliances can lead to fires. Yet most of these accidents can be avoided with careful planning and straightforward precautions.

Did you know?

Each year about 1,000 accidents at work involving electric shock or burns are reported to the Health and Safety Executive (HSE). Around 30 of these are fatal. Most of these fatalities arise from contact with overhead or underground power cables

Remember

Electricity is dangerous. Always take precautions

Hazards and precautions

Main hazards

- contact with live parts causing shock and burns (mains voltage at 230 volts a.c. can kill)
- faults which could cause fires
- fire or explosion where electricity could be the source of ignition in a potentially flammable or explosive atmosphere, e.g. in a spray-paint booth.

Hazards in harsh conditions

The risk of injury from electricity is also strongly linked to where and how it is used. The risks are greatest in harsh conditions.

- In wet surroundings – unsuitable equipment can easily become live and can make its surroundings live.
- Out of doors – equipment may not only become wet but may be at greater risk of damage.
- In cramped spaces with a lot of earthed metalwork, such as inside a tank. If an electrical fault were to develop here it could be very difficult to avoid a shock.
- Some items of equipment can also involve greater risk than others. Extension leads are particularly liable to damage – to their plugs and sockets, to their electrical connections, and to the cable itself. Other flexible leads, particularly those connected to equipment that is moved a great deal, can suffer from similar problems.

Safety precautions

There are precautions that can be taken to ensure that the electrical installation is safe.

- Install new electrical systems to a suitable standard, e.g. BS 7671 Requirements for Electrical Installations, and then maintain them in a safe condition.
- Existing installations should also be properly maintained.
- Provide enough socket outlets – overloading socket outlets by using adaptors can cause fires.

Safe and suitable equipment

- Choose and use equipment that is suitable for its working environment.
- Electrical risks can sometimes be eliminated by using air, hydraulic, hand or battery-powered tools. These are especially useful in harsh conditions.
- Ensure that equipment is safe when supplied and maintain it in a safe condition.
- Regularly inspect and test portable equipment (P.A.T.).
- Provide an accessible and clearly identified switch near each fixed machine to cut off power in an emergency.
- For portable equipment, use socket outlets that are close by so that equipment can easily be disconnected in an emergency.
- The ends of flexible cables should always have the outer sheath of the cable firmly clamped to stop the wires (particularly the earth) pulling out of the terminals.
- Replace damaged sections of cable completely.
- Use proper connectors or cable couplers to join lengths of cable. Do not use strip connector blocks covered in insulating tape.
- Some types of equipment are double insulated. These are often marked with a 'double-square' symbol (⧈). The supply leads have only two wires. Make sure they are properly connected if the plug is not a moulded-on type, noting the change in cable colours.
- Protect lamps and other equipment which could easily be damaged in use. There is a risk of electric shock if they are broken.
- Check electrical equipment used in flammable/explosive atmospheres. It should be designed to stop it from causing ignition. You may need specialist advice.

Reduce the voltage

One of the best ways of reducing the risk of injury when using electrical equipment is to limit the supply voltage to the lowest needed to get the job done. For example:

- temporary lighting can be run at lower voltages, e.g. 12, 25, 50 or 110 volt
- where electrically powered tools are used, battery-operated ones are safest
- portable tools are readily available that are designed to be run from a 110 volt centre-tapped-to-earth supply.

Provide a safety device

If equipment operating at 230 volts or higher is used, an **RCD** (**residual current device**) can provide additional safety. The best place for an RCD is built into the main switchboard or the socket outlet, as this means that the supply cables are permanently protected. If this is not possible, a plug incorporating an RCD, or a plug-in RCD adaptor, can also provide additional safety.

RCDs for protecting people have a rated tripping current (sensitivity) of usually not more than 30 milliamps (mA). However, please remember:

- an RCD is a valuable safety device: never bypass it
- if an RCD trips, this is a sign that there is a fault
- check the system before using it again
- if the RCD trips frequently and no fault can be found in the system, consult the manufacturer of the RCD
- the RCD has a test button to check that its mechanism is free and functioning. Use this regularly.

Carry out preventive maintenance

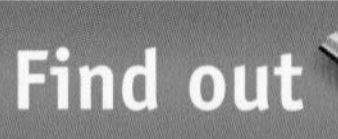

Find out

How often does equipment in a car repair garage need inspecting?

All electrical equipment and installations should be maintained to prevent danger. It is strongly recommended that this includes an appropriate system of visual inspection and, where necessary, testing.

By concentrating on a simple, inexpensive system of looking for visible signs of damage or faults, most of the electrical risks can be controlled. This will need to be backed up by testing as necessary. It is always recommended that fixed installations are inspected and tested periodically by a competent person.

The frequency of inspections and any necessary testing will depend on the type of equipment, how often it is used, and the environment in which it is used (see BS 7671).

Records of the results of inspection and testing can be useful in assessing the effectiveness of the system.

Work safely

Remember

Label and report faulty equipment

Make sure that people who are working with electricity are competent to do the job. Even simple tasks such as wiring a plug can lead to danger – ensure that people know what they are doing before they start. Therefore check that:

- suspect or faulty equipment is taken out of use, labelled 'DO NOT USE' and kept secure until examined by a competent person
- where possible, tools and power socket outlets are switched off before plugging in or unplugging
- equipment is switched off and/or unplugged before cleaning or making adjustments
- more complicated tasks, such as equipment repairs or alterations to an electrical installation, should only be tackled by people with a knowledge of the risks and the precautions needed
- you must not allow work on or near exposed live parts of equipment unless it is absolutely unavoidable and suitable precautions have been taken to prevent injury, both to the workers and to anyone else who may be in the area.

Other situations

Underground power cables

Although this will be a rare occurrence for an electrician, always assume cables will be present when digging in the street, pavement or near buildings. Use up-to-date service plans, cable-avoidance tools and safe digging practice to avoid danger. Service plans should be available from regional electricity companies, local authorities, highways authorities etc.

Remember

Electricity can flash across from overhead lines, even though plant and equipment do not touch them

Overhead power lines

When working near overhead lines, it may be possible to have them switched off if the owners are given enough notice. If this cannot be done, consult the owners about the safe working distance from the cables. Over half of the fatal electrical accidents each year are caused by contact with overhead lines.

Electrified railways and tramways

If working near electrified railways or tramways, consult the line or track operating company. Not all trains are diesel powered, and some railways and tramways use electrified rails rather than overhead cables.

Safe isolation of electrical supplies

Many fatal electrical accidents occur during the proving of isolation. You must be on guard, as you may have no idea whatsoever of the type of supply you are dealing with. Before beginning work on any electrical circuit you should make sure that it is completely isolated from the supply by following recognised procedures drawn up by the Joint Industry Board for the Electrical Contracting Industry:

- identify sources of supply
- isolate
- secure the isolation
- test the equipment/system is dead
- begin work.

Identify sources of supply

Not only is it important to identify the source of supply, but you also need to know what type it is. You need to be aware that a Band I supply covers extra-low voltage, that is a voltage normally not exceeding 50 volts a.c.

Fifty volts should not be enough to kill you by passing/forcing a current through your body, but a shock could cause you to lose balance and fall from height if you were working on ladders or scaffold, and that could result in a personal injury.

Did you know?

Emergency switches and associated alarms and equipment may have essential or dedicated supplies

Band I supplies also cover telecommunications, signalling, bell, and control and alarm installations where the voltage is limited for operational reasons.

Band II supplies contain low voltage, which is a voltage normally exceeding extra-low voltage but not exceeding 1000V a.c. between conductors or 600V a.c. between conductors and earth. Band II also includes the voltages for supplies to households and most commercial and industrial installations. Band II will therefore contain both single-phase and three-phase supplies. The actual voltage of the supply may differ from the nominal value by a quantity within normal tolerances, which are +10 per cent to –6 per cent.

After identifying the type of supply, you now need to identify the source of supply. It is not unusual for rooms or buildings to be fed from more than one source.

The lighting circuit may be fed from a different distribution board than the ring circuit. You may also find that a radial circuit has been added to the installation and is fed from a completely different source.

All these sources need to be located and isolated before safe work can proceed. Schematic drawings will help you to identify the source of supply for isolation.

Isolate

Definition

Isolators – mechanical switching devices that separate an installation from every source of electrical energy

Now that the source and type of supply have been identified they need to be isolated. Regulations require that a means of isolation must be provided to enable electrically skilled persons to carry out work on or near parts that would otherwise normally be energised.

Isolating devices must comply with British Standards, and the isolating distance between the contacts must comply with the requirements of BS EN 60947-3 for an **isolator**. The position of the contacts must either be externally visible or be clearly, positively and reliably indicated.

If it is installed remotely from the equipment to be isolated, the device must be capable of being secured in the open position using a lock and key that are unique to that isolator.

You must appreciate the basic difference between switching off and isolating:

- Switching off may involve the breaking of normal load current (or even higher current due to overload or short circuit).
- Isolation is concerned with cutting the already dead circuit so that re-closing the switch will not make it live again.

A semi-conductor device, e.g. a dimmer, cannot be used for isolation. While a switch must be able to break currents, an isolator need not be designed to do so because it should only be operated after the current has been broken. Typically a single-phase installation would be isolated using a double-pole device and an installation supplied with a three-phase four-wire system would be isolated using a three-pole switch.

Secure isolation

To prevent any unauthorised or unexpected re-closing of the contacts the isolator must be provided with a means of locking in the 'off' position, with a padlock and a key that is unique to that padlock and can be kept by the person at work on the circuit or equipment (Reg. 537.02.07).

Remember

All live conductors must be isolated before work can be carried out. As the neutral conductor is classified as a live conductor this should also be disconnected. This may mean removing the conductor from the neutral block in the distribution board. Not all distribution boards are fitted with double pole isolators so the connecting sequence for neutral conductors needs to be verified and maintained

Test that the equipment and system is dead

All circuits to be worked on must be tested to ensure they are dead. First, test the voltage indicator on a known supply or proving unit before use. Then test between phase and neutral, phase and earth, neutral and earth, for single-phase supplies. For three-phase supplies, test between brown and black, brown and grey, grey and black, brown and blue, black and blue, grey and blue, and brown and green/yellow, grey and green/yellow, black and green/yellow, blue and green/yellow.

On 1 April 2004 the insulation colours changed. However, for many years to come you will find existing installations with the old colours. These were:

- Single-phase – red, black
- Three-phase – red, yellow and blue.

Retest the voltage indicator on a known supply or proving unit after use. You must also erect warning notices stating 'Danger: Electrician at Work'.

Begin work

Some companies make use of a paper-based 'Permit to Work' system for all isolation procedures. Make use of the warning notices 'Danger: Electrician at Work' when you are working, and you should ensure that your name and the time and date you started work are included on the notice. This will help anyone to establish why power has been turned off. The charts on pages 112 and 113 show you standard isolation procedures.

Test equipment

All test equipment must be regularly checked to make sure it is in good and safe working order. You must ensure that your test equipment has a current calibration certificate indicating that the instrument is working properly and providing accurate readings. If you do not do this test, results could be void.

When checking the equipment yourself the following points should be noted:

- Check equipment for any damage, see if the case is cracked or broken. This could indicate a recent impact, which could result in false readings.
- Check that the batteries are in good condition and have not leaked (local action) and check that they are all of the same type.
- Check that the insulation on the leads and probes is not damaged but complete and secure.
- Check the operation of the meter with the leads both open and short-circuited.
- Then zero your instrument on the ohm scale. If you have any doubt about an instrument or its accuracy ask for assistance.

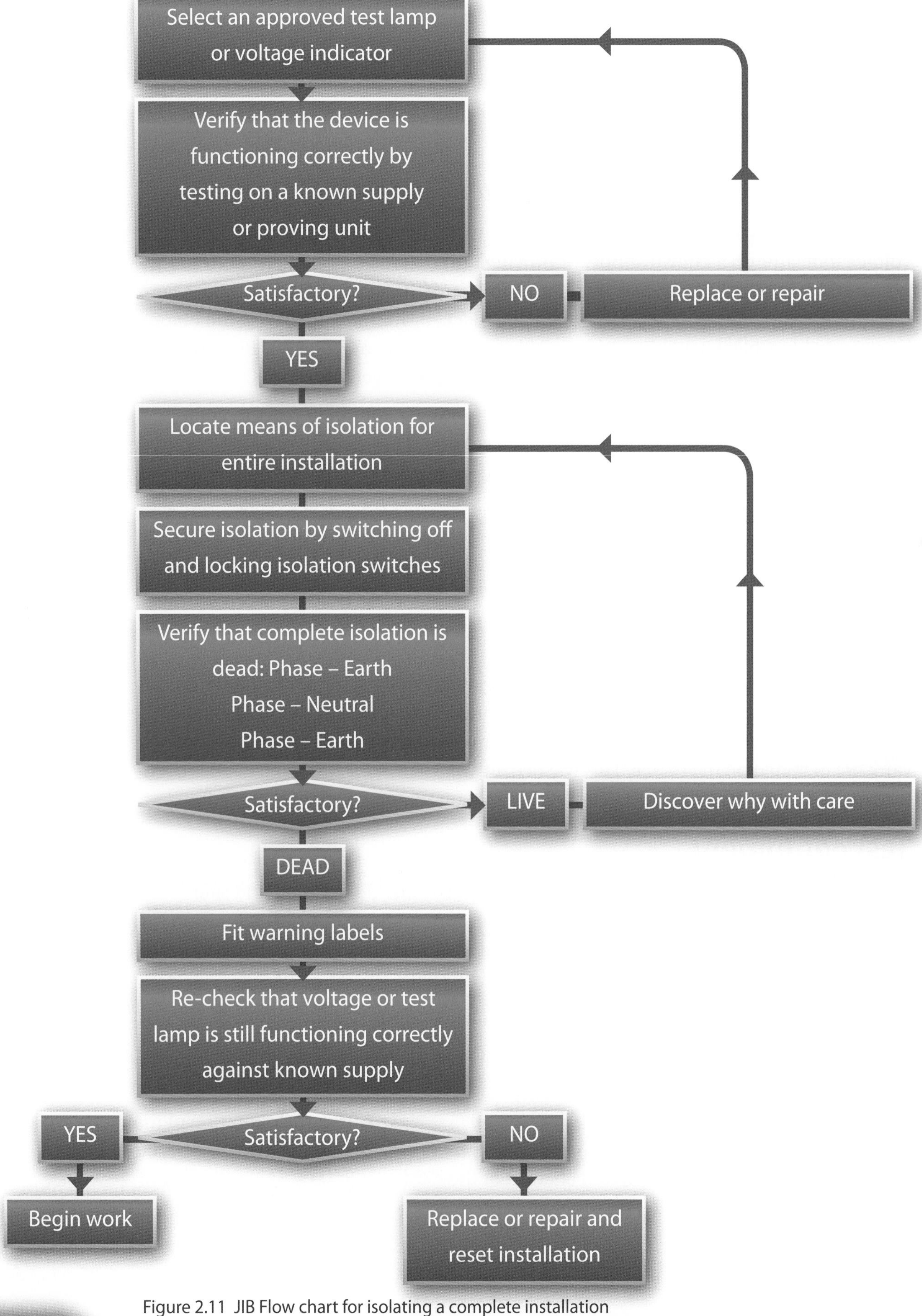

Figure 2.11 JIB Flow chart for isolating a complete installation

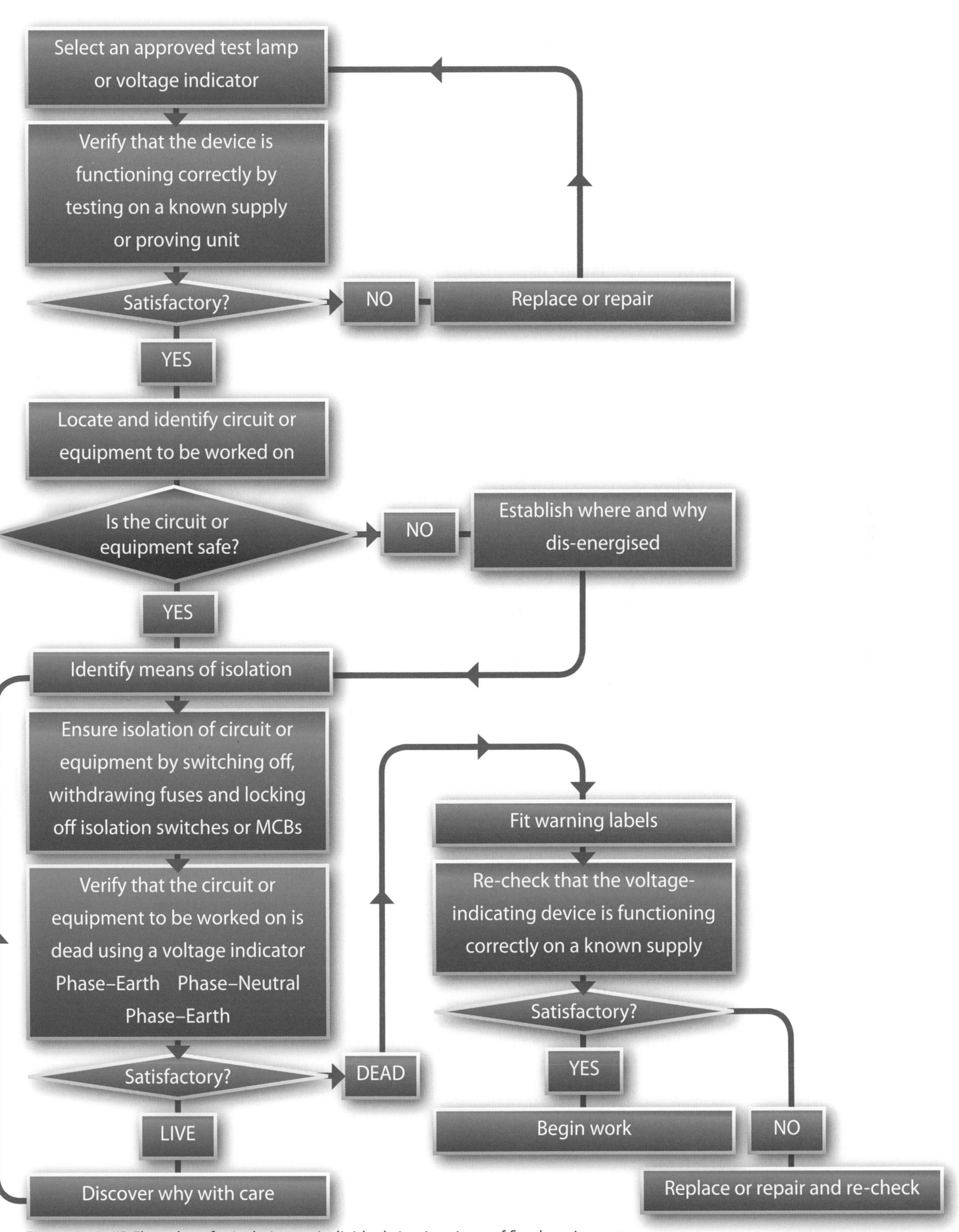

Figure 2.12 JIB Flow chart for isolating an individual circuit or item of fixed equipment

Guidance Note GS 38

Published by the HSE, this document covers electrical test equipment used by electricians and gives guidance to electrically competent people involved in electrical testing, diagnosis and repair. Electrically competent people may include electricians, electrical contractors, test supervisors, technicians, managers or appliance repairers. Advice is given in the following areas.

Test probes

Test probes and leads used in conjunction with a voltmeter, multimeter, electrician's test lamp or voltage indicator should be selected to prevent danger. Good test probes and leads will have the following:

Probes:

- have finger barriers or are shaped to guard against inadvertent hand contact with the live conductors under test
- are insulated to leave an exposed metal tip not exceeding 2 mm measured across any surface of the tip. Where practicable it is strongly recommended that this is reduced to 1 mm or less or that spring-loaded retractable-screened probes are used
- should have suitable high-breaking capacity fuse or fuses with a low current rating usually not exceeding 500 mA or a current limiting resistor and a fuse.

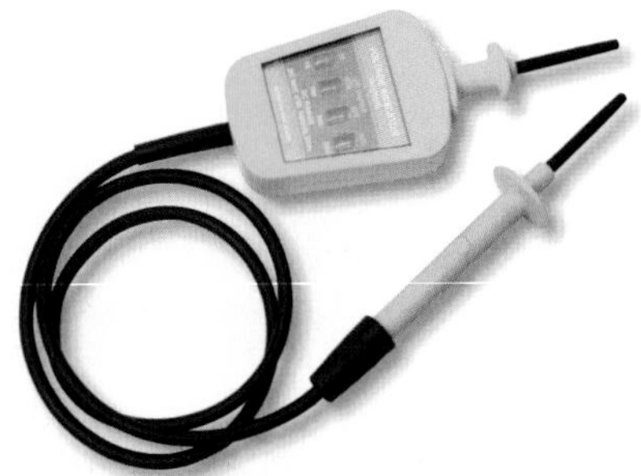

Recommended test probe

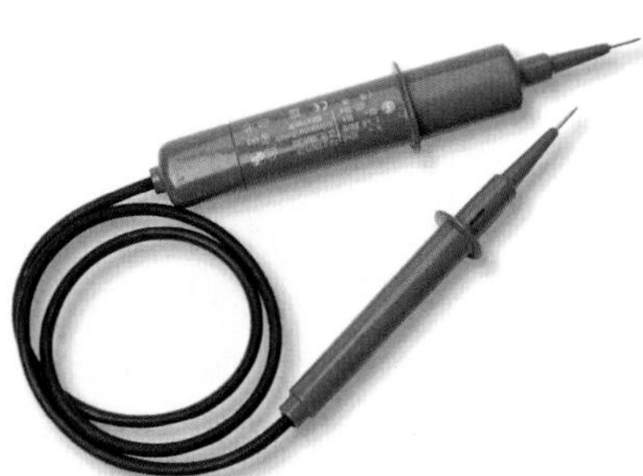

Non-recommended test probe (non-retractable tips)

Leads:

- are adequately insulated
- are coloured so that one lead can be easily distinguished from the other
- are flexible and of sufficient capacity for the duty expected of them
- are sheathed to protect against mechanical damage
- are long enough for the purpose while not being so long that they are clumsy or unwieldy
- do not have accessible exposed conductors other than the probes or tips or have live conductors accessible to the person's finger should a lead become detached from a probe. The test lead or leads are held captive and sealed into the body of the voltage detector.

Voltage-indicating devices

Instruments that are used solely for detecting a voltage fall into two categories as described below.

(a) Detectors that rely on an illuminated lamp (test lamp) or a meter scale (test meter)

Test lamps are fitted with a 15 W lamp and should not give rise to danger if the lamp is broken. They should also be protected by a guard.

These detectors require protection against excess current. This may be provided by a suitable high breaking capacity fuse with a low current rating usually not exceeding 500 mA or by means of a current-limiting resistor and a fuse. These protective devices are housed in the probes themselves. The test lead or leads are held captive and sealed into the body of the voltage detector.

(b) Detectors that use two or more independent indicating systems (one of which may be audible) and limit energy input to the detector by the circuitry used

An example is a 2-pole voltage detector, i.e. a detector unit with an integral test probe, an interconnecting lead and a second test probe.

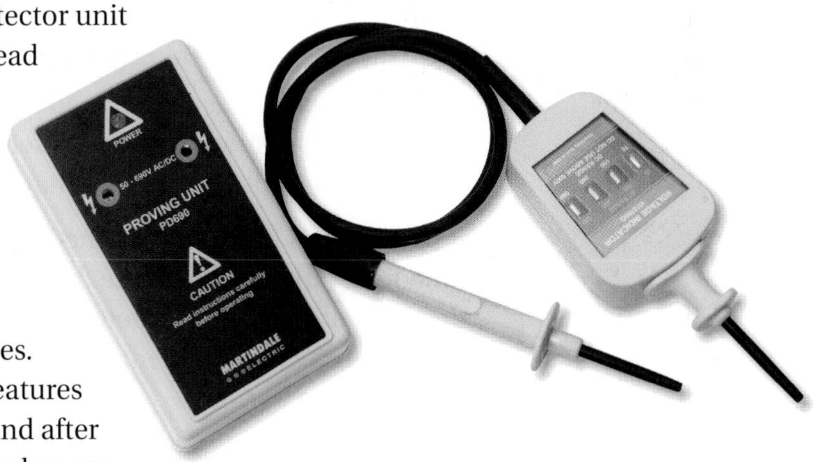

These detectors are designed and constructed to limit the current and energy that can flow into the detector. The limitation is usually provided by a combination of the circuit design using the concept of protective impedance and current-limiting resistors built into the test probes. These detectors are provided with in-built test features to check the functioning of the detector before and after use. The interconnecting lead and second test probes are not detachable components.

These types of detector do not require additional current-limiting resistors or fuses to be fitted provided that they are made to an acceptable standard and the contact electrodes are shrouded as previously mentioned.

It is recommended that test lamps and voltage indicators are clearly marked with the maximum voltage that may be tested by the device and any short-time rating for the device if applicable. This rating is the recommended maximum current that should pass through the device for a few seconds; these devices are generally not designed to be connected for more than a few seconds.

Systems of work

The use of test equipment by electricians falls into three main categories:

1. testing for voltage (voltage detection), **2.** measuring voltage, **3.** measuring current, resistance and, occasionally, inductance and capacitance.

Item 1 forms an essential part of the procedure for proving a system dead before starting work, but it can be associated with simple tests to prove presence of voltage. Items 2 and 3 are more concerned with commissioning procedures and fault-finding.

Before testing begins it is essential to establish that the device, including all leads, probes and connectors, is suitably rated for the voltage and currents that may be present on the system under test. Where a test is being made simply to establish the presence or absence of voltage, the proprietary test lamps or 2-pole voltage detectors suitable for the working voltage of the system, are preferred to a multimeter.

For voltage detection or measurement, fuse-protected test leads are recommended when voltmeters and in particular multimeters are used. Some multimeters have electromechanical overload devices, but these are often inadequately rated to deal with short-circuit energy present on electrical power systems.

It is usually necessary to use leads that incorporate high breaking capacity fuses even if the multimeter has an overload trip. If thermal clips are provided for connection to test points they should be adequately insulated and arranged to be suitable for use with the test leads as a safe alternative to the use of test probes.

It is important that a multi-function or multi-range meter is set to the correct function and/or range before the connections are made. Where there is doubt about the value of voltage to be detected or measured, the highest range should be selected at first provided that the maximum voltage possible is known to fall within the range of the instrument.

Progressive voltage detection or measurement is often used to prove circuit continuity. The dangers from exposed live conductors should be borne in mind when using this method. In many cases continuity testing can be carried out safely with the apparatus dead using a self-contained low voltage d.c. source and indicator.

Keep records of inspection and testing of the equipment, particularly when faults are found. These will help decide how often visual inspection or testing will need to be carried out. Test instruments should be re-calibrated at regular intervals and the calibration certificate retained for reference purposes.

Did you know?

Accident history shows that the use of incorrectly set multimeters or makeshift devices for voltage detection has often caused accidents

Safety tip

All items of test equipment, including those items issued on a personal basis, should receive a regular inspection and where necessary a test by a competent person

FAQ

Q **Is it possible to use any multimeter to measure voltage on site?**

A No, as there is a risk that you could set the instrument range incorrectly, causing damage to the instrument and yourself. You should use an approved GS 38 voltage-indicating device.

First aid

First aid for electric shock

Electric shock occurs when a person becomes part of the electrical circuit. The severity of the shock will depend on the level of current and the length of time it is in contact with the body.

The lethal level is approximately 50 mA, above which muscles contract, the heart fibrillates and breathing stops. A shock current above 50 mA could be fatal unless the person is quickly separated from the supply. Below 50 mA only an unpleasant tingling sensation may be experienced. However, this may cause you to fall from a roof or ladder, which could lead to a serious injury.

In the event of an electric shock

- ✔ First of all check for your safety to ensure that you would not put yourself at risk by helping the casualty.
- ✔ Then break the electrical contact to the casualty by switching off the supply, removing the plug (if it is undamaged) or wrenching the cable free (only attempt this if the cable and plug etc. are undamaged).
  If this is not possible break the contact by pushing or pulling the casualty free using a piece of non-conductive material, e.g. a piece of wood.
- ✔ If the casualty is conscious, guide them to the ground, making sure that further injuries are not sustained, e.g. banging their head on the way down.
  If the casualty is unconscious, get help straight away, then check the casualty's response. Talk to and gently shake the casualty to gauge their level of response. If the casualty appears unharmed they should be advised to rest and see a GP.
- ✔ If there is no movement or any sign of breathing summon help immediately. If there is someone with you, tell them to get help, i.e. ring 999; if you are on your own with the casualty you will have to leave them for a moment while you get help yourself.
- ✔ As soon as you return to the casualty you need to begin CPR (cardio-pulmonary resuscitation).

Airways, Breathing, Circulation

Remember

This is best remembered as the ABC of resuscitation: Airways, Breathing, Circulation

When a casualty is unconscious, you need to know to what degree they are unconscious. Ask a question or give a command – for example, 'What happened?' or 'Open your eyes'. Speak loudly and clearly, close to the casualty's ear. Carefully shake their shoulders. A casualty in a serious state of 'altered consciousness' may mumble, groan or make slight movements. A fully unconscious casualty will not respond.

Open the airway. An unconscious casualty's airway may be narrowed or blocked, making breathing difficult and noisy, or impossible. The main reason for this is that muscular control in the throat is lost, which allows the tongue to sag back and block the throat. Lifting the chin and tilting the head back lifts the tongue away from the entrance to the air passage.

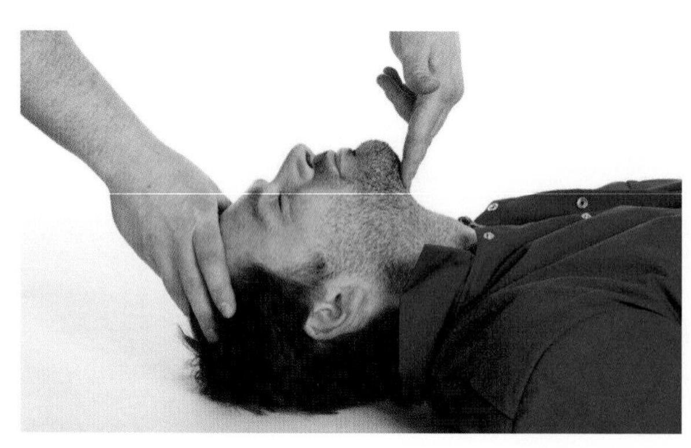

Opening the airway

To open the airway, remove any obvious obstruction from the mouth. Place two fingers under the point of the casualty's chin as shown here and lift the jaw. At the same time, place your other hand on the casualty's forehead and tilt the head well back.

To check for breathing put your face close to the casualty's mouth. Look and feel for breathing as well as looking for chest movements. Listen for the sound of breathing. Feel for breath on your cheek. Look, listen and feel for 10 seconds before deciding that breathing is absent.

Check for signs of life and a pulse. Because of the time being lost trying to find the carotid pulse in the neck, the National Institute of Clinical Excellence currently does not advise this method of checking for a pulse. Current thinking is to listen to the patient's chest instead. Check for about 10 seconds to see whether a pulse is present. If the casualty is breathing, place them in the recovery position (see page 118). If not, deliver CPR (cardiopulmonary resuscitation) in the following manner.

Mouth-to-mouth ventilation is given with the casualty lying flat on their back:

- first, remove any obvious obstruction including broken or displaced dentures from the mouth
- open the airway by tilting the head and lifting the chin
- close the casualty's nose by pinching it with your index finger and thumb
- take a full breath, and place your lips around their mouth, making a good seal
- blow into the casualty's mouth until you see the chest rise. Take about two seconds for full inflation
- remove your lips and allow the chest to fall fully
- repeat
- if there is a response from the casualty, place him or her in the recovery position. If not, begin chest compressions.

Chest compressions

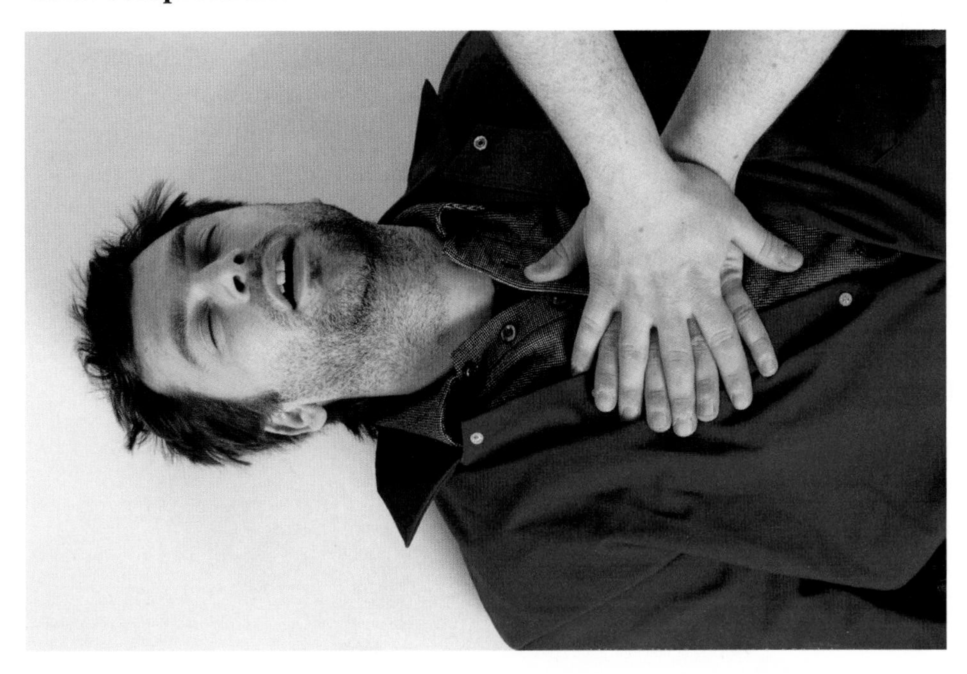

Chest compression

- with the casualty still lying flat on their back on a firm surface, kneel beside them, and find one of the lowest ribs using your index and middle fingers
- slide your fingers upwards to the point in the middle where the rib margins meet the breastbone
- place your middle finger over this point and your index finger on the breastbone above
- place the heel of your hand on the breastbone, and slide it down until it reaches your index finger. This is the point at which you will apply pressure
- place the heel of your first hand on top of the other hand, and interlock fingers
- leaning well over the casualty, with your arms straight, press down vertically on the breastbone to depress it approximately 4–5 cm, then release the pressure without removing your hands
- repeat the compressions 15 times, aiming for a rate of approximately 100 compressions per minute
- return to the head and give two more ventilations, then 15 further compressions
- continue to give two ventilations to every 15 compressions until help arrives or you are unable to continue. Do not interrupt CPR to make pulse checks unless there is any sign of a returning circulation. With a pulse confirmed, check breathing and if it is still absent continue with artificial ventilation. Check the signs of life/pulse after every 10 breaths, and be prepared to re-start chest compressions if it disappears. If the casualty starts to breathe unaided, place them in the recovery position. Re-check breathing and pulse every three minutes.

The recovery position

Any unconscious casualty whose breathing is not impaired (see Opening the airway page 118) should be placed in the recovery position. This position prevents the tongue from blocking the throat, and, because the head is slightly lower than the rest of the body, it allows liquids to drain from the mouth, reducing the risk of the casualty inhaling stomach contents.

The head, neck and back are kept in a straight line while the bent limbs keep the body propped in a secure and comfortable position. If you must leave an unconscious casualty unattended, they can be left safely in the recovery position while you get help.

The recovery position

The technique for turning is described below and assumes that the casualty is found lying on their back from the start:

- Kneeling beside the casualty, open the airway by tilting the head and lifting the chin; straighten the legs. Place the arm nearest you out at right angles to the body, elbow bent, and with the palm uppermost.
- Bring the arm furthest from you across the chest, and hold the hand, palm outwards, against the casualty's nearer cheek. (Remember to turn any rings with stones in them to face the palm so that there is no contact between the ring and the casualty's face. Also remember to check pockets for keys etc. that could cause more injury when the casualty is turned on to his or her side.)
- With your other hand, grasp the lower thigh just above the knee furthest from you and pull the knee up, keeping the foot flat on the ground.
- Keeping the casualty's hand pressed against his or her cheek, pull at the knee to roll the casualty towards you and onto his or her side.

- Tilt the head back to make sure the airway remains open. Adjust the hand under the cheek, if necessary so that the head stays in this tilted position and circulation to the hand is not restricted.
- Adjust the upper leg, if necessary, so that both the hip and the knee are bent at right angles.
- The casualty should then be sent to hospital in an ambulance.

Treatment for burns, shock and breaks

An electric shock may result in other injuries as well as unconsciousness. There may be burns at both the entry point and exit point of the current. The treatment for these burns is to flood the site of the injury with cold water for at least 10 minutes. This will halt the burning process, relieve pain and minimise the risk of infection.

You would then need to treat for shock, which is the medical condition where the circulatory system fails and insufficient oxygen reaches the tissues. If this shock is not treated quickly the vital organs can fail, leading ultimately to death. To treat shock you need to:

- stop external bleeding if there is any
- lay the casualty down, keeping the head low
- raise and support the legs but be careful if you suspect a fracture
- loosen tight clothing, braces, straps or belts to reduce constriction at the neck, chest and waist
- keep the casualty warm by wrapping him or her in a blanket or coat
- continue to check and record breathing, pulse and level of response
- be prepared to resuscitate if necessary.

There are situations in construction and maintenance work where the nature of the task involves risks that are so high that you should not undertake those tasks alone

Once you are satisfied that the casualty is in a stable condition, you should cover any burned skin with a loose, lint-free dressing, or even with sheets of cling film (do not wind any dressing around the injured area). If the casualty had sustained a broken bone then your first aim is to prevent movement at the site of the injury.

Do not move the casualty until the injured part is secured and supported, unless he or she is in danger, i.e. from further electric shock. You must arrange for the casualty's immediate removal to hospital, maintaining comfortable support during transport.

Treatment for smoke and fume inhalation

If you find someone suffering from the effects of fume inhalation or asphyxiation, provided that it is safe to do so, get them outside into the fresh air as soon as possible. Loosen any clothes around their neck or chest that may impair their breathing. Call the emergency services and refer to the First Aid section of this book regarding the ABC procedure (page 118).

Safety tip

Hazards that cannot be removed or reduced should be guarded against: safety guards and fences can be put on or around machines, safe systems of work introduced, safety goggles, helmets and shoes issued. Your employer should provide any additional PPE required, such as ear-defenders, respirators, eye protection and overalls

General hazards

This section covers:

- accident prevention
- personal hygiene
- general electrical hazards
- liquid petroleum gas
- chemical hazards.

Accident prevention

The best way to reduce the risk of accidents is to try to remove the cause. If substances such as oil, grease, cutting compounds, paints and solvents are spilled on the floor and not dealt with immediately, they can cause hazards. Even things like food, off-cuts from conduit, cables and tools are dangerous if left underfoot. Clearly employees have some responsibility both to themselves and others to keep the workplace hazard free.

Where necessary the workplace should be tidy, with clearly defined passageways, good lighting, ventilation, reduced noise levels and non-slip floorings. Locking dangerous substances away into approved areas will remove another potential hazard, so that people will not be tempted to interfere with them, causing danger to themselves and others. Conduit, metal trunking and cable tray should be stored horizontally to minimise the hazard of their falling.

General electrical hazards

Before using any of this equipment you are strongly advised to carry out a number of visual checks:

- ✓ Check that the cable is not damaged or frayed.
- ✓ Check that the cable is properly secured at both ends and that none of the conductors are visible.
- ✓ Check that the plug is in good condition.
- ✓ If using an extension lead of the coiled type, uncoil all the cable before use, check it for serviceability and leave uncoiled during use. Recoil carefully after use.
- ✓ Check that any portable power tools are suitable for working from a reduced voltage (110V or 55V) and that a suitable transformer is supplied.
- ✓ Never carry power tools by their cables.
- ✓ Keep plugs and sockets clean and in good order.

When working with electricity, always make certain that the circuit is isolated from the supply by switching it off and removing the fuses before touching any of the conductors. Make sure that the circuit cannot be turned on while you are working on it. Use the company's Permit to Work or Safe Isolation Procedures.

Personal hygiene

Use barrier cream before starting work. This fills the pores of the skin with a water-soluble antiseptic cream, so that when you wash your hands the dirt and germs are removed with the cream. Always wash at the end of the work period, before and after using the toilet, and before handling food. Re-apply barrier cream after washing. Do not use solvents to clean your hands. They remove protective oils from the skin and can cause serious problems such as dermatitis. Change overalls regularly before they become too dirty and a health hazard.

LPG (liquid petroleum gas)

Even small quantities of LPG when mixed with air create an explosive mixture (LPG is a gas above –42 °C).

LPG is widely used in construction and building work as a fuel for burners, heaters and gas torches. The liquid, which comes in cylinders and containers, is highly flammable and needs careful handling and storage. Here are some guidelines:

- Everyone using LPG should understand the procedures to be adopted in case of an emergency.
- Appropriate fire-fighting extinguishers (dry powder) should always be available.
- Cylinders must be kept upright whether in use or in storage.
- When not in use the valve should be closed and the protective dust cap should be in place.
- When handling cylinders do not drop them or allow them to come into violent contact with other cylinders.
- When using a cylinder with an appliance ensure it is connected properly, in accordance with the instructions you have been given, and that it is at a safe distance from the appliance or equipment that it is feeding.

It is essential to make sure the gas does not leak. LPG is heavier than air: if it leaks, it will not disperse in the air but will sink to the lowest point and form an explosive concentration that could be ignited by a spark. Leakages are especially dangerous in trenches and excavations because the gas cannot flow out of these areas. For this reason LPG cylinders should not be taken into them.

Did you know?

One litre of LPG when boiled or evaporated becomes 250 litres of gas. This is enough to make an explosive mixture in a large shed, room, store or office

Safety tip

On no account should you smoke in areas where LPG is in use

Chemical hazards

Certain chemicals can be harmful if they come into contact with your body by accident. When not contained or handled properly chemicals can be:

- inhaled as a dust or gas
- swallowed in small doses over a long time
- absorbed through skin or clothing
- touched or spilled onto unprotected skin.

Did you know?

The HASAWA defines risk assessments for substances

Some chemicals can cause:

- injury to eyes, skin, organs, injury from fires and burns etc.
- illness that can sometimes leave you feeling fine after the exposure but causes medical problems after many months or years of exposure
- allergy to the skin as a rash; coughing and breathing problems
- death – some poisonous chemicals can kill outright.

Chemical accidents may be caused by:

- hurrying, overconfidence, fooling around or not following instructions
- spills and leaks that aren't noticed or wiped up
- unsafe working conditions – vapours may build up in poor ventilation
- exposure of some chemicals to heat or sunlight, causing explosion, fire and poisonous reactions
- contact between a chemical and the wrong material, causing harmful reactions
- neglect or failure to dispose safely of certain old chemicals – chemical changes can happen with time, leading to a dangerous situation
- chemicals not stored in a safe manner falling and rupturing, causing spillage.

There are four main types of chemicals. You need to know them and how to guard against their hazards.

- toxic agents
- corrosives
- flammables
- reactives

Toxic agents

Poisons such as hydrogen sulphide and cyanide can cause injury, disease and death. To protect yourself:

- close containers tightly when not in use
- be sure the work area is well ventilated
- wear personal protective equipment
- wash hands often
- carry cigarettes in a protective packet
- safely dispose of contaminated clothing
- keep any proper antidotes handy.

Corrosives

These are irritants such as acids and alkalis, which are especially dangerous to the eyes and respiratory tract. To protect yourself:

- wear personal protective equipment, goggles, breathing devices, protective gloves
- make sure ventilation is good
- run for water if corrosives come into contact with you; use a safety shower
- if your eyes are affected, flush with water for 15 minutes and get medical aid. Most first-aid kits contain eye-irrigation kits.

Flammables

These are liquids and gases that burn readily such as ethyl, ether and petrol. For them to burn they require just the right amount of flammable (fuel), oxygen and a spark or other source of ignition.

To protect yourself:

- make sure no flames, sparks or cigarette lighters are near flammables
- keep only a small amount of flammables in the work area
- store and dispose of flammables safely
- in an emergency you may have to evacuate the area; if safe to do so turn off all flames and call the emergency services.

Reactives

These are substances that can explode, such as nitro compounds. To protect yourself:

- know your chemicals: read about them and test them for stability
- handle reactives with great care. For example, Toilet Duck mixed with bleach can give off hydrogen peroxide
- at the first sign of trouble close the doors and evacuate the room through doors that do not lead to the danger area.

Safety tip

When using chemicals of any sort you must understand the dangers involved, you must follow all safety rules and procedures recommended and you must know what to do in an emergency

Asbestos dust safety

Asbestos is a mineral found naturally in certain rock types. When separated from rock it becomes a fluffy, fibrous material that has many uses in the construction industry. It is used in cement production, roofing, plastics, insulation, floor and ceiling tiling, and fire-resistant board for doors and partitions.

Asbestos becomes a health hazard if the dust is inhaled; some of the fine rod-like fibres may work their way into the lung tissue and remain embedded for life. This will become a constant source of irritation.

To protect yourself:

- know the hazards, avoid exposure and always follow recommended controls
- wear and maintain any personal protective equipment provided
- practise good housekeeping. Use special vacuums and dust-collecting equipment
- report any hazardous conditions, e.g. unusually high dust levels, to your supervisor.

Remember

The removal of asbestos must only be carried out by specialist contractors. Consider your health at all times. Don't work with asbestos under any circumstances. After all it's your health and safety, and nothing is more important

Other safety topics

Permit to Work

Construction sites can be dangerous places, but with a little thought no one needs to come to any harm. One area that continues to be of concern, however, is that people remain unaware of the dangers of working in isolation.

Remember

There are situations in construction and maintenance work where the nature of the task involves risks that are so high that you should not undertake these tasks alone

On the job: Dangers of lone working

Roger, a painter, had gone off to paint some rooms one Saturday morning. Arriving on site he said hello to a few people and went off to start his work. Once there he realised he also had a fairly large storage tank to paint. Many hours later, his body was found inside the storage tank. He was killed by the fumes from the specialist paint he was using.

1. What precautions should Roger have taken before starting this work?
2. Who would have been involved in the investigation of this accident? And what recommendations do you think they would make?

Find out

Are there any other circumstances where you should not work alone?

This is one example, but there are many circumstances where you should not work by yourself, such as:

- in confined spaces
- in trenches
- near, or on, live sources or equipment
- at height
- near to unguarded machinery
- where there is a risk of fire or in hazardous atmospheres
- with toxic or corrosive substances.

There will always be circumstances where you will be involved in potentially dangerous tasks. For many situations, the answer is the use of the **Permit to Work** system. One contributing factor to the painter's death in the case study above was that no one knew exactly where he was or what he was doing. A Permit to Work is a document that specifies the work to be done, the person(s) involved, when it is going to be done, the hazards involved and the precautions to be taken. The permit is then only 'active' for a set period, and if you haven't returned by the allocated time, someone will investigate.

As the Permit to Work has to be authorised, it is essential that the authoriser is competent and fully able to understand the work to be done, the hazards and the proposed system of work and precautions needed. A sound knowledge of Regulations such as the Confined Spaces Regulations 1997 and COSHH is also essential.

The Permit to Work form aids communication between everyone involved, and employers must train staff in its use. It should ideally be designed by the company issuing the permit, taking into account individual site conditions and requirements. On certain sites, separate permit forms may be required for different tasks, such as hot work and entry into confined spaces, so that sufficient emphasis can be given to the particular hazards present and precautions required.

Portable appliance testing (P.A.T.)

As we have previously discussed, the Health and Safety at Work Act puts a duty of care on both employer and employee to ensure the safety of all persons using the work premises. However, the specific legal requirements regarding the use and maintenance of electrical equipment are contained in the Electricity at Work Regulations 1989 (EAWR).

The EAWR require all electrical systems to be maintained to prevent danger. This requirement covers all electrical equipment, including fixed and portable equipment. Employers must therefore maintain their electrical equipment in order to prevent accidents.

The majority of equipment defects can be found by a detailed visual inspection (after disconnecting the appliance). This is likely to eliminate hazards caused by damaged plugs, cuts in cable or equipment being used in the wrong circumstances. Other signs such as burn marks may show that the equipment's condition could create faults or render it a danger to the user.

However, a visual inspection alone is insufficient and will not identify all dangerous faults. Therefore a visual inspection needs to be linked to a testing programme carried out with an approved test instrument, to reveal less obvious electrical faults with earth continuity, insulation resistance and earth leakage.

There is no legal requirement to keep records regarding the frequency of testing. It would seem sensible to keep records, though, as this would allow employers to monitor P.A.T. within their organisation. Guidance given by the Institute of Electrical Engineers (IEE) suggests the following frequencies for industry applications:

Equipment type	When	Class I		Class II	
		Visual only	*Visual + test*	*Visual only*	*Visual + test*
Stationary	Weekly	None	1 year	None	1 year
IT equipment	Weekly	None	1 year	None	1 year
Moveable	Before use	1 month	1 year	3 months	1 year
Portable	Before use	1 month	6 months	3 months	6 months
Hand-held	Before use	1 month	6 months	3 months	6 months

Table 2.6 Recommended frequency of equipment testing

Insurance

Did you know?

Employers must ensure that they use an authorised insurer. If they do not, they may be breaking the law. They should check that their insurer is authorised before taking out employers' liability insurance

Employers are responsible for the health and safety of their employees while they are at work. An employee may be injured at work, or a former employee may become ill as a result of their work while in employment. The employee might try to claim compensation from the employer if they believe the employer is responsible.

The Employers' Liability (Compulsory Insurance) Act 1969 ensures that employers have at least a minimum level of insurance cover against any such claims. Employers' liability insurance will enable the employer to meet the cost of compensation for an employee's injuries or illness whether they are caused on or off site; any injuries or illness relating to motor accidents that occur while employees are working for an employer may be covered separately by the employer's motor insurance.

Public liability insurance is different. It covers an employer for claims made against it by members of the public or other businesses, but not for claims by employees. While public liability insurance is generally voluntary, employers' liability insurance is compulsory. Employers can be fined if they do not hold a current employers' liability insurance policy that complies with the law.

When an employer takes out employers' liability insurance, it will have an agreement with the insurer about the circumstances in which the latter will pay compensation. For example, the policy will cover the specific activities that relate to the business. There are certain conditions that could restrict the amount of money an insurer might have to pay. Employers must make sure that their contract with the insurer does not contain any of these conditions. The insurer cannot refuse to pay compensation purely because the employer:

- has not provided reasonable protection for employees against injury or disease
- cannot provide certain information to the insurer
- did something the insurer told it not to do (for example, said it was at fault)
- has not done something the insurer told it to do (for example, report the incident)
- has not met any legal requirement connected with the protection of the employees.

Remember

Employers must carry out a risk assessment, take practical measures to protect the employees and report incidents

However, this does not mean employers can forget about their legal responsibilities to protect the health and safety of their employees. If an insurer believes that an employer has failed to meet its legal responsibilities for the health and safety of its employees and that this failure has led to the claim, the policy may enable the insurer to sue the employer to reclaim the cost of the compensation.

Employers must be insured for at least £5 million. However, they should look carefully at their risks and liabilities and consider whether insurance cover of more than £5 million is needed. In practice, most insurers offer cover of at least £10 million. If the business is part of a group, a policy for employers' liability insurance can be taken out for the group as a whole. In this case, the group as a whole, including subsidiary companies, must have cover of at least £5 million. Employers can have more than one policy for employers' liability insurance, but the total value of the cover provided by the policies must be at least £5 million.

Access and exit

During normal on-site work activities it is inevitable that the site will receive visitors, from delivery drivers to architects and clients. This would not be a problem in areas such as factories, office blocks or hospitals where a construction site may be located within the boundary of these organisations, as they probably already have a visitor facility.

However, at sites without designated facilities, the motivation and co-operation of site personnel is relied on to display the correct actions to any expected visitors. Irrespective of the quality of the facilities available for receiving visitors, the reasons for putting into practice a 'visitors procedure' remain the same:

- to meet with health and safety requirements
- to maintain site security
- to project a professional approach for the company
- to establish and maintain good client relationships.

Generally speaking, the following good practice is recommended:

- Receive visitors in a manner that fosters goodwill.
- Check the validity of the visitor.
- Check the reason for their visit.
- Establish whom they wish to see.
- Log their arrival time (and departure time when they leave).
- Brief them on site safety.
- Issue an identification badge if necessary.
- Issue them with a hard hat and high visibility jacket if necessary.
- If a request is made that is beyond your authority, contact an authorised person.

Date	Visitor's name	Company	To see	Time in	Time out	I.D. checked	Badge number	H&S briefed	Visitor's signature
Enter date	PRINT visitor's name	Enter name of company or organisation visitor is representing	Name of person to be visited	Time visitor is booked into the workshop	Time visitor is booked out of workshop	Type of ID used – e.g. student card, letterhead, etc.	Number of visitor pass or badge issued	Enter Yes when briefed and PPE issued, if required	Ask visitor to sign here

Table 2.7 Site visitors' book

FAQ

Q Do I always have to use stepladders in the open position? Can't I just lean them against the wall?

A Not if you want to use them safely. The bottoms of the stiles are cut at an angle, so when the steps are open the base of the stile sits squarely on the floor. If you rest the ladder against the wall, it becomes impossible to get the ladder at this angle. Invariably the ladder sits only on a small part of the base, making it liable to slip away while you are working on it.

Q What do I do if the boss asks me to lift and carry something that I think is too heavy?

A Look around for things and people to assist you to move the object – a sack truck, a small hand cart, or rollers to put under it. Whatever method you decide on, do not be afraid to tell the boss or to ask for assistance, even if its just to lift the load on to the trolley.

Q Why do I need to know about mouth-to-mouth ventilation – I won't ever use it?

A I hope you won't! But working with electricity means that there is always a risk that someone you are working with may receive an electric shock that causes them to stop breathing. Just think, if you were in such a predicament, wouldn't you want someone to be able to help you!

Work and the environment

In 1988, the United Nations set up the Intergovernmental Panel on Climate Change (IPCC) – a body of scientists from all parts of the world who assess the best available scientific and technical information on climate change.

Climate change is a serious risk to the planet

Their 2007 report warned of an increase in average global temperatures of up to 6.4 degrees Celsius by the end of this century, depending on future levels of emissions. It also said that such changes to the climate were 90 per cent likely to be the result of human activity. Tackling climate change needn't damage the economy, but industry will have to adapt and jobs may change. There may even be more jobs created overall.

Reducing the use of energy

Using less energy can also save companies and households money and 'greener' living and working are going to be key factors. As an overview of practical measures for householders, the UK government currently suggests the following options, but also bear in mind how these could then be applied to the world of work.

Save energy and water

Burning fossil fuels to either heat our homes or produce electricity releases carbon emissions, which cause climate change. The energy you use at home is likely to be your biggest contribution to climate change. Approximately 80 per cent of it goes on heating and hot water, so this is a good place to look for savings.

- **Turn down the thermostat** – turning your thermostat down by one degree could reduce carbon emissions and cut fuel bills by up to 10 per cent.
- **Look for the labels** – when buying products that use energy (this could be anything from light bulbs to fridge-freezers) help tackle climate change by looking for the Energy Saving Recommended label or European energy label rating of A or higher. The European energy label also tells you how much water appliances use, so you can choose a more efficient model.
- **Improve insulation** – more than half the heat lost in your home escapes through the walls and roof. Cavity wall insulation costs about £450, can take a couple of hours to install, and could save about £100 a year on fuel bills, as well as reducing your carbon footprint.
- **Install water efficient products** – low flush volume toilets, water efficient showerheads and aerating heads on washbasin taps help to reduce your water use significantly. Also, fixing dripping taps and fitting a 'hippo' in toilet cisterns are cheap ways of saving water. You can also collect rainwater in water butts and use it for watering your garden instead of a hose.

Getting around

Travelling accounts for around a quarter of all the damage individuals do to the environment, including climate change effects. Individual car travel is responsible for the majority of these impacts. If you're buying a new car, look for the fuel efficiency label to choose a more efficient model.

- **Try to reduce car use** – reduce the number of short trips you make in the car. Walking, cycling, or taking the bus or train will help reduce local air pollution and climate change effects.
- **Tackling the environmental impact of flying** – consider the need for a flight and the alternatives to taking a plane. If you do fly, you can offset your CO_2. You could consider options for reducing your travel, for example taking fewer, longer breaks if possible instead of several short ones. Maybe you can find what you want closer to home, by taking a holiday in the UK or travelling to nearby countries by rail or sea.

Eating and drinking

Producing, transporting and consuming food is responsible for nearly a fifth of our climate change effects. Some foods have a much bigger impact on the environment than others.

- **Look for the labels** – look for the labels to help you choose food that has been produced with the aim of reducing the negative impact on wildlife and the environment.
- **Buy fresh and in season** – buying food and drink when locally in season, and unprocessed or lightly processed food, is likely to mean that less energy has been used in its production. Providing it has been produced and stored under similar conditions, choosing food that has travelled a shorter distance will help to reduce any congestion and transport emissions that contribute to climate change.
- **Reduce your food waste** – the average UK household spends £424 a year on food that goes in the bin – if this ends up in landfill it produces methane, a greenhouse gas judged to be more than 20 times as powerful as carbon dioxide in causing climate change. Throwing less food away produces less methane and reduces other harmful environmental impacts from producing, packaging and transporting food.

Recycling and cutting waste

Reducing, reusing and recycling waste saves on the raw materials and energy which are needed to make new paper, metal, glass and other items.

- **Re-use and repair** – avoiding waste in the first place, by re-using and repairing items, is the most efficient way to reduce waste. For example, buy items that can be re-used rather than disposables, and pass things on when you've finished with them.
- **Recycle more** – nearly two thirds of all household rubbish can be recycled. Most councils run doorstep recycling collections for paper, glass and plastics, often more. But local civic amenity sites often accept many other things – from wood and shoes, to textiles and TVs.
- **Take a bag** – hang on to your shopping bags and take some with you when you next go to the supermarket.

The above list applies to households and it may be fair to say that in general the work of an electrician has little direct environmental impact. It may also be fair to say that some of these issues are the responsibility of designers or of other occupations. However, this logic can be applied to the way electricians work.

Reducing energy use in the workplace

Buildings produce nearly half of the UK's carbon emissions and the way a building is constructed, insulated, heated, ventilated and the type of fuel used, all contribute to its carbon emissions. New measures to improve the energy performance of our buildings are being introduced by the Government and these include:

- the introduction by the end of 2008 of Energy Performance Certificates (EPCs) for ALL buildings, whether they are built, sold or rented out. The EPC provides 'A' to 'G' ratings for the building, with 'A' being the most energy efficient and 'G' being the least, with the average to date being 'D'

- requiring public buildings to display energy performance certificates
- requiring inspections for air conditioning systems
- giving advice and guidance for boiler users.

Obviously the selection of construction materials, air conditioning systems and boilers is the work of others, but it is clear that the selection of efficient materials and systems prior to installation and the maintenance of any existing systems will be key factors.

Remember that although an electrician may not design such systems, they may be involved in a systems installation or maintenance, and every visit to site, along with the electrical systems used, has an environmental effect. Electricians therefore have a responsibility to work in an environmentally friendly way.

As can be seen from the householder's list, planning can make a big difference. A carefully planned installation can reduce the environmental impact in terms of the number of visits to site, consequent travel and welfare arrangements, material and equipment delivery arrangements, selection of material and equipment both in terms of installation and maintenance factors and the removal or disposal of waste.

Consequently, one trend in construction that is gathering momentum is the move away from traditional on-site construction to off-site (modular) construction. Pre-fabricated construction has been around for some time, but with modern technology at our disposal, the modular approach is felt to offer many advantages. Although definitive statistics are difficult to find, some well respected architects have said that some of the advantages are:

- Up to 60 per cent less energy is required to produce a modular building compared to an equivalent traditionally-built project.
- Up to 90 per cent fewer vehicle movements to site, thus reducing carbon emissions, congestion and disruption.
- A reduction in on-site waste by up to 80 per cent.

Environmental technologies

As well as using technology within the building, we need to consider its application elsewhere. We know that conventional power generation can be very damaging to the environment, and there is a range of alternatives that come under the heading of renewable energy sources.

Solar photo-voltaic

These systems work by converting solar radiation into electricity that can be used immediately, stored or even sold back to the electricity provider. Correctly installed systems, even in the UK, could allow efficiently managed domestic properties to generate about 50% of their own electricity.

Wind

Wind power is one of many alternative power sources whose use may increase over the next few years

These systems rely upon wind to turn a shaft linked to a generator that in turn produces electricity. The greater the wind speed, the greater the power produced, and as wind speed increases with height, this is why the large rotating blades are normally seen at the top of a tall supporting mast or tower. With wind speed and direction being variable, this may not be the best option for domestic properties unless they are in remote locations.

Biomass

As this method can be replaced at the same rate as it used it is therefore considered a renewable energy source. Generally there are two acknowledged sources: wood (e.g. forests) and non-wood (e.g. high energy crops such as rape). One example of this method would be the wood burning stove that heats a room and powers a boiler that heats the rest of the house.

Ground-source heat pump

This system makes use of the fact that a few metres under the ground surface, the temperature remains at a fairly constant 12°C. This system does not require any external fuel and is designed to heat a whole building. Working in a similar way to a fridge, a buried closed-loop of pipe is filled with a mixture of water and anti-freeze which is then pumped around the loop, absorbing heat from the ground and fed via a condenser to a hot water tank before being distributed to provide heating.

Small scale hydro

This system uses running or falling water to turn a turbine to produce electricity. Useful power may be produced from even a small stream therefore for houses with no mains connection but with access to a micro hydro site, a good hydro system can generate a steady, more reliable electricity supply than other renewable technologies, at a lower cost. Total system costs can be initially high, but often less than the cost of a grid connection, with no monthly bills to follow and low maintenance costs.

Solar water heating

Solar water heating systems use heat from the sun to work alongside a conventional water heater. They collect heat from the sun's radiation via a flat plate system or evacuated glass tube system before being stored in a hot water cylinder. Such systems can supply about one third of domestic hot water needs and tend to require little maintenance.

Micro Combined Heat and Power (Micro-CHP) unit

Although not renewable, this system, sometimes called cogeneration, is essentially a domestic boiler that contains a condensing boiler to heat the home and provide hot water, but also include a Stirling engine to generate electricity.

Invented by Scotsman Dr. Robert Stirling in 1816, these engines generate motion from heat without combustion, but as you only generate electricity when the boilier is generating heat and they only generate approximately 1.5 kW electricity, the idea is to reduce your electricity bills, not replace your electricity supply. However, the concept can be operated on a larger scale and there are currently trials under way at community level.

Energy saving

The current Building Regulations set standards for the design and construction of most buildings, primarily to ensure the safety and health for people in or around those buildings, but also for energy conservation and access to and about buildings.

There are 14 parts to the Regulations and as electricians we may already be aware of Part P, which looks at electrical safety. However, as regards this book, 'Part L – Conservation of fuel and power' is relevant and particularly Approved Document L2A.

Part L states that *"Responsible provision shall be made for the conservation of fuel and power in buildings by ... providing and commissioning energy efficient fixed building services with effective controls"*. Document L2A clarifies 'fixed building services' as including "any part of or control associated with, fixed internal or external lighting, excluding emergency lighting".

Document L2A also states that *"Lighting controls should be provided so as to avoid unnecessary lighting during times when daylight levels are adequate or when spaces are unoccupied ... with automatically switched systems being subject to risk assessment"*.

Part L also makes requirements for the type of lamps used in luminaires and these are covered in more detail in Chapter 3 (Lighting) of *Electrical Installations NVQ and Technical Certificate Book 2*. But in essence we still rely too heavily on the incandescent lamp and older low efficiency fluorescent tubes (the incandescent lamp is a very inefficient lamp converting only about 5 per cent of the energy received into light). However, there are advancements with modern compact fluorescent tubes and LEDs that should be considered.

"Knowledge is power" is an old expression, but in this case it's true and legislated for, as Part L states that the owner of a building must be provided with sufficient information about the building, its fixed building services and any maintenance requirements, to enable the building to be operated so as to use no more fuel and power than is reasonable under the circumstances.

Saving energy in an eco-friendly building

Many things interact to reduce our impact on the environment and the phrase 'eco-friendly' is often used without its understanding being defined. Certainly in construction there would be many definitions suggested. However, in addition to those things mentioned previously, some of the following logic applies to energy saving within an eco-friendly building:

Materials

- From natural, sustainably managed renewable sources (e.g.these are timber building companies in Norway that have planting policies which mean they are growing more trees than they cut down to make their products).
- Should be sourced materials near to the point of use, therefore reducing transportation effects.
- Should be made using minimal processing or with added content such as chemicals.
- Make use of their natural insulation properties.
- Should be non-toxic and not hazardous to users or building occupants (e.g. paint fumes)
- Should be durable with low maintenance.
- Should have the capability to be recycled.

Energy

- Use natural light wherever possible.
- Use low energy appliances wherever possible (e.g. 'A' rated washing machines).
- Use local expertise and labour wherever possible, thus reducing the effects of transportation.
- Use renewable sources where possible.

Water

- Use rainwater harvesting (using rainwater for irrigation, vehicle washing or toilet flushing).
- Use greywater recycling (using water from baths, showers etc. for toilet flushing).
- Use low volume flush toilets.
- Use aerated taps (these make the water spray).
- Use instantaneous water heaters over sinks instead of heating large volumes of water.
- Lag hot water pipes.

Waste reduction

The Environmental Protection Act 1990 defined waste as including any substance which constitutes a scrap material, an effluent or other unwanted surplus arising from the application of any process or any substance or article which requires to be disposed of which has been broken, worn out, contaminated or otherwise spoiled.

This definition was amended by the Waste Management Licensing Regulations 1994. It re-defined waste as *"any substance or object which the producer or the person in possession of it, discards it, or intends/is required to discard it, but with exception of anything excluded from the scope of the Waste Directive."*

Did you know?

If you leave materials on site when your work is complete, you may be discarding them. If they are discarded they will be 'waste' and, as the producer of that waste, you will be responsible for it

Therefore, the producer of a waste must ask themselves certain questions that will help identify it as such, e.g. Is it a scrap material? Is it an unwanted surplus substance? Or is it broken or worn out?

Any substance or object that you discard, intend to discard or are required to discard is waste and as such is subject to a number of regulatory requirements. The term 'discard' has a special meaning. Even if material is sent for recycling or undergoes treatment in-house, it can still be waste.

Various legislation including the Environmental Protection Act 1990, the Controlled Waste Regulations 1992 and the Waste Management Licensing Regulations 1994 legally defines the types of wastes as follows.

Controlled waste

This includes household, industrial and commercial waste:

- **Household waste** – from dwellings including houses, caravans, houseboats, campsites, prisons and also from schools, colleges and universities.
- **Commercial waste** – from premises used for business, sport, recreation & entertainment.
- **Industrial waste** – from factories or industrial processes.

Because of their nature, some controlled wastes are subject to further regulation and need to be handled differently, namely:

- **Clinical waste** – from hospitals, nursing homes, doctors surgeries and dentists.

Did you know?

If you transport your own waste, which is either building or demolition waste, you will need to be registered as a waste carrier with your Environmental Register

Under the Controlled Waste Regulations 1992, the Duty of Care requires that you ensure all waste is handled, recovered or disposed of responsibly, that it is only handled, recovered or disposed of by individuals or businesses that are authorised to do so, and that a record is kept of all wastes received or transferred through a system of signed Waste Transfer Notes.

If you work as a subcontractor and the main contractor arranges for the recovery or disposal of waste that you produce, you are still responsible for those wastes under this Duty of Care.

Be aware that materials requiring recycling (such as copper cable), either on your premises or elsewhere, are likely to be waste and will be subject to the waste management regime and the Duty of Care. If in any doubt, seek advice from the Environmental Regulator.

Hazardous waste

On 16 July 2005 the Hazardous Waste (England and Wales) Regulations and the List of Wastes (England) Regulations came into force replacing the Special Waste Regulations. Therefore, if the material you are handling has hazardous properties (e.g. toxic or explosive), it may need to be dealt with as waste that is potentially hazardous or dangerous and which may therefore require extra precautions during handling, storage, treatment or disposal.

The Hazardous Waste Regulations therefore try to ensure safe management of hazardous wastes and provide cradle-to-grave documentation for the movement of all hazardous waste.

Form HWCN01v051

The Hazardous Waste Regulations 2005: Consignment Note

Environment Agency

PRODUCER'S/HOLDER'S/CONSIGNOR'S COPY (Delete as appropriate)

PART A Notification details

1 Consignment note code:

2 The waste described below is to be removed from (name, address, postcode, telephone, e-mail, facsimile):

3 Premises code (where applicable):

4 The waste will be taken to (name, address and postcode):

5 The waste producer was (if different from 2) (name, address, postcode, telephone, e-mail, facsimile):

PART B Description of the waste — If continuation sheet used, tick here

1 The process giving rise to the waste(s) was:

2 SIC for the process giving rise to the waste:

3 WASTE DETAILS (where more than one waste type is collected all of the information given below must be completed for each EWC identified)

Description of waste	List of wastes (EWC code)(6 digits)	Quantity (kg)	The chemical/biological components of the waste and their concentrations are:		Physical form (gas, liquid, solid, powder, sludge or mixed)	Hazard code(s)	Container type, number and size
			Component	Concentration (% or mg/kg)			

The information given below is to be completed for each EWC identified

EWC code	Packing group(s)	UN identification number(s)	Proper shipping name(s)	UN class(es)	Special handling requirements

PART C Carrier's certificate

(If more than one carrier is used, please attach schedule for subsequent carriers. If a schedule of carriers is attached tick here.)

I certify that I today collected the consignment and that the details in A2, A4 and B3 are correct and I have been advised of any specific handling requirements.

1 Carrier name:

On behalf of (name, address, postcode, telephone, e-mail, facsimile):

2 Carrier registration no./reason for exemption:

3 Vehicle registration no. (or mode of transport, if not road):

Signature

Date D D M M Y Y Y Y Time H H M M

PART D Consignor's certificate

I certify that the information in A, B and C above is correct, that the carrier is registered or exempt and was advised of the appropriate precautionary measures. All of the waste is packaged and labelled correctly and the carrier has been advised of any special handling requirements.

1 Consignor name:

On behalf of (name, address, postcode, telephone, e-mail, facsimile):

Signature

Date D D M M Y Y Y Y Time H H M M

PART E Consignee's certificate (where more than one waste type is collected all of the information given below must be completed for each EWC)

Individual EWC code(s) received	Quantity of each EWC code received (kg)	EWC code accepted/rejected	Waste management operation (R or D code)

1 I received this waste at the address given in A4 on: Date D D M M Y Y Y Y Time H H M M

2 Vehicle registration no. (or mode of transport if not road):

3 Where waste is rejected please provide details:

I certify that waste management licence/permit/authorised exemption no(s).

authorises the management of the waste described in B at the address given in A4.

Name:

On behalf of (name, address, postcode, telephone, e-mail, facsimile):

Signature

Date D D M M Y Y Y Y Time H H M M

HWCN01v051

Figure 2.13 Waste consignment note

In reality most businesses are likely to produce some 'special wastes'. The following are examples:

- asbestos
- used engine oils and oil filters
- solvents and/or solvent-based paint and ink
- pesticides
- lead-acid batteries
- oily sludges
- chemical wastes
- fluorescent tubes.

Disposal of these requires technical competence. If you intend to discard containers, an assessment must be made as to whether they are hazardous waste. With very few exceptions, a Consignment Note (see Figure 2.13) must be completed to accompany hazardous waste when it's moved from any premises, including premises that are exempt from registration.

Containers may be hazardous waste if they contain residues of hazardous or dangerous substances/materials. If the residue is 'hazardous', then the whole container is hazardous waste.

Here are some guidelines for good waste disposal practice.

- Do not burn cable or cable drums on site. Find another method of disposal.
- Before allowing any waste haulier or contractor to remove a waste material from your site, ask where the material will be taken and ask for a copy of the waste management licence or evidence of exemption for that facility.
- Segregate the different types of waste that arise from your works. This will make it easier to supply an accurate description of the waste for waste transfer purposes.
- Minimise the quantity of waste you produce to save you money on raw materials and disposal costs.
- Label all waste skips –make it clear to everyone which waste types should be disposed of in that skip.

Remember

In general the burning of waste in the open is an environmentally unsound practice and you should use less damaging options of waste disposal.

One further piece of legislation applicable to waste is The Waste Electrical and Electronic Equipment Regulations 2006 (WEEE).

These Regulations apply to electrical and electronic equipment (EEE) in the categories listed below with a voltage of up to 1000 V for a.c. or up to 1500 V for d.c. You need to comply with the WEEE Regulations if you generate, handle or dispose of waste that falls under one of the following ten categories:

- Large household appliances
- Small household appliances
- IT and telecommunications equipment
- Consumer equipment
- Lighting equipment
- Electrical and electronic tools
- Toys, leisure and sports equipment
- Medical devices
- Monitoring and control equipment
- Automatic dispensers.

The regulations aim to:

- reduce waste from electrical and electronic equipment
- encourage the separate collection of waste EEE
- encourage treatment, reuse, recovery, recycling and sound environmental disposal of waste EEE
- make producers of EEE responsible for the environmental impact of their products
- improve the environmental performance of all those involved during the lifecycle of EEE.

All businesses that use electrical and electronic equipment (EEE) must comply with the Waste Electrical and Electronic Equipment (WEEE) Regulations. This includes all domestic or household EEE that you may use on your premises. For all non household EEE either the producer or end user is responsible for the disposal of the products.

You must obtain and keep proof that your WEEE was given to a waste management business, and was subsequently treated and disposed of in an environmentally sound way.

The Packaging (Essential Requirements) Regulations 2003 (Amended 2004 and 2007)

One part of the waste process that is easy to forget, is the packaging that helped deliver the items wanted, in a sound condition.

The official definition is that 'packaging' means all products, made of any materials of any nature, used for the containment, protection, handling, delivery and presentation of goods, from raw materials to processed goods, from the producer to the user or the consumer, including non-returnable items used for the same purposes.

However, there can be much waste generated by packaging and therefore these Regulations state packaging shall be:

- manufactured so that the packaging volume and weight be limited to the minimum adequate amount to maintain the necessary level of safety, hygiene and acceptance for the packed product and for the consumer
- designed, produced and commercialised in such a way as to permit its reuse or recovery, including recycling, and to minimise its impact on the environment when packaging waste or residues from packaging waste management operations are disposed of
- manufactured so that the presence of noxious and other hazardous substances and materials as constituents of the packaging material or of any of the packaging components is minimised with regard to their presence in emissions, ash or leachate when packaging or residues from management operations or packaging waste are incinerated or landfilled.

Other environmental issues

Hired plant or equipment

Should hired plant and equipment cause an environmental incident while on hire, the responsibility for that incident rests with the person or business that has control of that plant or equipment. Therefore check that the hire company supplying you with plant and equipment operates a system of Preventive Plant Maintenance – well-maintained plant and equipment is less likely to break down, will emit fewer pollutants to the air and can help prevent spillage of oil and fuel to the environment.

Remember

A statutory nuisance could arise from the poor state of your premises or any noise, smoke, fumes, gases, dust, steam, smell, effluvia, the keeping of animals, deposits and accumulations of refuse and/or other material

Noise and statutory nuisance

Part III of the Environmental Protection Act 1990 (as amended) contains the main legislation relating to statutory nuisance. It applies in England, Wales and Scotland and is enforced by the local authorities; under Part III of the Pollution Control and Local Government (Northern Ireland) Order 1978, district councils have powers to deal with noise nuisance. Your local authority also operates a system of Consents and Prohibition Notices, which could affect your site.

In England, Scotland and Wales this system is set out in the Control of Pollution Act 1974: s.60 on Prohibition Notices and s.61 on Consents – A Code of Practice. The Noise Act 1996 also applies to domestic dwellings and therefore has an impact on installations, as does the Noise and Statutory Nuisance Act 1993.

Although there is no legal definition of a statutory nuisance, for action to be taken the nuisance complained of must be, or be likely to be, prejudicial to people's health or interfere with a person's legitimate use and enjoyment of land. This particularly applies to nuisance to neighbours in their homes and gardens.

Good practices would be to:

- Establish whether your business might cause a nuisance to neighbours by checking noise, odours and other emissions near the boundary of your site during different operating conditions and at different times of the day. Take all reasonable steps to prevent or minimise a nuisance or a potential nuisance.
- Make sure there is a good level of 'housekeeping' on your site and that your site manager and staff are aware of the need to avoid nuisances. Regularly check your site for any waste accumulations, evidence of vermin, noise or smell as applicable.

Remember

Where appropriate use mains-generated electricity in preference to diesel generators. This will help you to reduce noise levels and to reduce the risk of pollution through fuel spillage

- Even if a complaint does not amount to a statutory nuisance you should consider if there are simple, practical things that you can do to keep the peace.
- Try to establish a good relationship with your neighbours, particularly in relation to transient events likely to affect them. Advise neighbours in advance if you believe that a particular operation such as building work or an installation process for new plant could cause an adverse effect. If neighbours are kept informed they will perceive the business as more considerate and are less likely to make a complaint.
- Use good practice to minimise noise escaping from your buildings by, for example, keeping doors and windows closed.
- At noise-sensitive locations, undertake monitoring of background noise before your works begin.
- At noise-sensitive locations, include in your **method statement** any actions you can take to reduce noise levels.
- Reduce noise levels outside your buildings by increasing insulation to the building fabric and keeping doors and windows closed.
- Ensure that any burglar alarms on your premises have a maintenance contract and a callout agreement.
- Avoid or minimise noisy activities, especially at night; pay particular attention to traffic movements, reversing sirens, deliveries, external public address systems and radios.
- Where practical, schedule or restrict noisy activities to the normal working day (for example 0800 to 1800, Monday to Friday and 0800 to 1300 on Saturday).
- Consider where noisy operations are undertaken in relation to site boundaries and relocate them if you can, perhaps further away, or make use of existing buildings/stockpiles/topography as noise barriers.
- Consider replacing any noisy equipment and take account of noise emissions when buying new or replacement equipment. Maintain fans and refrigeration equipment.
- Do not have any bonfires; find other ways to re-use or recover wastes (see the Clean Air and Waste Management Guidelines).
- Keep abatement equipment, such as filters and cyclones, in good working order. Ensure boilers, especially oil or solid-fuel units, are operating efficiently and do not emit dark smoke during start up.

Pollution prevention and control

Within the UK, the new Pollution Prevention and Control (PPC) regime implements the EU's Integrated Pollution Prevention and Control (IPPC) Directive. The UK implementing legislation for IPPC (collectively referred to as the PPC Regulations) is the Pollution Prevention and Control (England and Wales) Regulations 2000.

PPC Part A (A(1) and A(2) in England and Wales) includes new issues not previously covered by IPPC such as:

- energy efficiency
- waste minimisation
- vibration
- noise.

The new regime also requires an effective system of management to be implemented to ensure that all appropriate pollution prevention and control measures are taken. Special emphasis is placed on the application of Best Available Techniques (BAT) to reduce the environmental impact of the process.

The Dangerous Substances and Preparations (Safety) and Chemicals (Hazard Information and Packaging for Supply) Regulations 2002 concern the marketing and use of substances and preparations that contain Carcinogens, **Mutagens** and substances toxic to Reproduction (CMRs), and the subsequent restriction of their supply to the general public.

Definition

Mutagens – agents, such as radiation or chemical substances, which cause genetic mutation

Radioactive substances

In the course of exploration, exploitation and construction programmes, specialist contractors use a variety of sealed and unsealed radioactive sources. Unsealed sources may be used during grouting or cementing operations or for reservoir or installation equipment tracing studies.

Sealed radioactive sources may be used for radiography, well logging or in liquid level and density gauges. Such substances are covered by the Radioactive Substances Act 1993 (RSA 1993), which is enforced by the Environment Agency.

Did you know?

Organisations are increasingly adopting environmental management systems that comply with ISO 14001 or the European Eco-Management and Audit Scheme (EMAS) Regulation

Environmental Management Systems (EMS)

Concern for the environment and awareness of the need to improve management of resources is on the increase. The issue has found its way into the political arena and our everyday lives and is creating pressures on businesses to demonstrate a commitment to minimising their impact on the environment.

An effective EMS certified to ISO 14001 or registered to EMAS (the European Eco-Management and Audit Scheme) can help an organisation to operate in a more cost-efficient and environmentally responsible manner by managing its activities while also complying with relevant environmental legislation and its own environmental policy.

There are a number of key benefits associated with the implementation of a certified EMS:

- demonstrating conformance
- management confidence
- improved management of environmental risk
- independent assessment
- compliance with legislative and other requirements
- continual improvement
- reduction in costs
- supply chain pressures.

Bearing in mind the previous environmental topics, employers may wish to become qualified to ISO 14001 for much the same reasons as the ISO 9001 series.

Remember

You must ensure that effluent discharge complies with all conditions given in the trade effluent discharge consent

Water

If you discharge any effluents into a public sewer you must have either prior written authorisation from the Statutory Sewerage Undertaker in the form of a trade effluent consent or have entered into a **Trade Effluent Agreement** with them. In most cases the Statutory Sewerage Undertaker will be your local water company (in England and Wales), or Scottish Water (in Scotland) and the Water Service (in Northern Ireland).

If you discharge any sewage, effluent or contaminated surface water to surface waters or groundwater you must have prior written authorisation from your Environmental Regulator in the form of a **discharge consent**.

Did you know?

In England, Scotland and Wales, if your plant emits excessive (i.e. amounting to a nuisance) levels of grit and dust, local authorities can place limits on these emissions; exceeding the limits may be an offence. This is likely to apply to old plant with a history of complaints or plant run with fuel or procedures it was not designed for

Air

The Clean Air Act 1993 applies to all small businesses that burn fuels in furnaces or boilers or that burn material in the open, including farms, building sites and demolition sites.

You must prevent the emission of dark smoke from any chimney on your premises. This includes chimneys serving furnaces, fixed boilers or industrial plant whether they are attached to buildings or not. There are some instances where dark smoke may be emitted without an offence being committed, such as during start-up conditions, if all practicable steps have been taken to prevent or minimise the emissions.

The burning of tyres is against the law

You must prevent the emission of dark smoke from any industrial or trade premises, e.g. burning tyres and cables. This applies to burning materials on a site that you own or a site where you are working such as a building or demolition site or land used for commercial agriculture or horticulture.

In England, Wales and Scotland it is not necessary for local authorities to have witnessed the emissions of dark smoke for them to take action against you: evidence of the burning of materials that potentially give rise to dark smoke is sufficient (in this way the law aims to stop people creating dark smoke at night and using the lack of visual evidence as a defence!); this does not apply in Northern Ireland.

You must inform your local authority before installing a furnace or a fixed boiler. Any new furnaces or boilers must be able to operate continuously without emitting smoke when burning the type of fuel for which they have been designed. Obtaining a planning permission/building warrant from the local authority for the construction of the chimney or plant is not sufficient: you need the local authority's specific approval under the Clean Air Act or the Clean Air (Northern Ireland) Order 1981.

Remember

In general the burning of waste in the open is an environmentally unsound practice, and you should use less damaging options of waste disposal

Under the Sulphur Content of Liquid Fuels Regulations, you must not use gas oils with a sulphur content higher than 0.2 per cent by weight (this will be reduced further to 0.1 per cent from January 2008). For heavy fuel oils, you must not use heavy fuel oils with a sulphur content higher than 1 per cent sulphur by weight (unless you hold an exemption).

The burning of waste will almost always create smoke (dark or otherwise) and can as a consequence cause nuisance to people in the locality. Combustion under uncontrolled circumstances at low temperature (in comparison to the temperatures in a waste incineration plant) may lead to the release of noxious gases and/or dust and grit in the area.

Residues of harmful chemicals (such as lead paints, tars and oils) that are left in the ashes can be washed into the ground by rain, leading to lasting contamination of the soil or groundwater. Contaminants may also be washed into surface water.

Knowledge check

1. Make a list of the different types of PPE that you could be expected to wear during your normal work, and describe their function.
2. Describe the circumstances in which you could be expected to wear PPE.
3. Describe the four different types of safety signs you may see on a construction site; explain the colour code of these signs.
4. What are the three components in the fire triangle?
5. List the five different categories of fire.
6. What type of fire extinguisher would you use when dealing with an electrical fire?
7. What are the immediate actions you should take on discovering a fire?
8. What are the checks you should make before using a ladder?
9. Why would you choose to use trestles instead of a ladder?
10. What are the risks associated with working in excavations?
11. Describe the process of lifting a heavy object.
12. What are the main hazards in electrical safety?
13. Why do we use reduced voltage equipment on building sites?
14. Describe the safe isolation process.
15. What are the recommended points that you need to look for when choosing test probes?
16. Describe the ABC of resuscitation.
17. Why would you place a casualty in the recovery position?
18. What are Mutagens?
19. To what types of building is an Energy Performance Certificate applicable?
20. Name the two main categories of biomass.
21. Name three categories of Controlled Waste.
22. Does controlled waste require a Waste Transfer Note or a Consignment Note?
23. Accredited to ISO14001, name two benefits of an EMS.

Electrical science

Overview

The work tasks that a competent electrician will undertake are many and varied. However, as well as having practical competence, we also need to know about the operating principles behind the job. The purpose of this chapter is to look at the principles of mathematics, electrical science and electronics necessary to support electrical and electronic installations. This chapter will cover:

- **Basic mathematics**
- **Mechanics**
- **Force, work, energy and power**
- **Electron theory**
- **Resistance**
- **Series and parallel circuits**
- **Electrical energy, work, power and efficiency**
- **Magnetism and electromagnetism**
- **The distribution of electricity**
- **Three-phase supplies**
- **Load balancing**

Basic mathematics

It is impossible to know what people are talking about without understanding the language they are speaking. In science and engineering, mathematics is the language that is used to explain how things work. To come to terms with all things electrical, we need to understand and be able to communicate using this universal language. In this section we will be looking at the following areas:

- SI units: how we describe basic measurement quantities
- powers of 10: mega, pico and the decimal system
- basic rules: getting the right answer
- fractions and percentages: working with parts of the whole
- algebra: formulas for all
- indices: powers of anything
- transposition: re-arranging equations
- triangles and trigonometry: angles on reading the sines
- statistics: ways of showing data.

SI units

Did you know?

SI stands for Système Internationale

Imagine if each country had a different idea of how long a metre is, or how much beer you get in a pint. What if a kilogram in Leeds was different from one in London? A common system for defining properties such as length, temperature, and time is essential if people in different places are to work together.

In the UK and Europe an international system of units for measuring different properties is used, known as SI units. There are seven base units – the main units from which all the other units are created.

Listed below is each quantity (property), and the name given to its base unit. The last column shows the symbol used as a short version of the name.

Base quantity	Base unit	Symbol
Length	metre	m
Mass	kilogram	kg
Time	second	s
Electric current	ampere	A
Temperature	kelvin	K
Amount of substance	mole	mol
Luminous intensity	candela	cd

Table 3.1 SI units

Length	Obviously useful for all sorts things, such as measuring the amount of cable you'll need in an installation
Area	Used to measure a surface such as a sports field, for working out how much floodlighting will be needed
Volume	Used particularly when you are dealing with heating systems, for example, to calculate how much energy is required to heat a hot water cylinder
Mass	Often confused with weight. Mass is 'how much there is' of something. Electricians often want to know how much electrical energy is required to change a mass from one state to another; e.g. to change water into steam, or ice into water
Weight	Related to how gravity has an effect on mass. A mass of 1 kg will weigh about 10 newtons on earth, i.e. it will exert a *force* of 10 N, but rather less on the moon, since gravity is lower there. NB It will still have a mass of 1 kg on the moon since it still contains the same amount of material
Temperature	The SI base unit for temperature is degrees Kelvin (K), but for most purposes we use degrees Celsius (°C). 0 K is −273°C (otherwise known as Absolute Zero or extremely cold!). Sometimes you'll hear people using degrees Fahrenheit (°F), too.

Table 3.2 Base quantities

SI unit prefixes

We often deal in quantities that are much larger or smaller than the base units. If we have to stick with them, the numbers become clumsy and it is easy to make mistakes. For example, the diameter of a human hair is about 0.0009 m, and the average distance of the earth from the sun is around 150000000000 m.

To make life easier, we can alter the symbols (and the quantities they represent) by adding another symbol in front of them (a prefix). These represent base units multiplied or divided by one thousand, one million, etc.

Table 3.3 shows the most common prefixes.

Multiplier	Name	Symbol prefix	As a power of 10
1 000 000 000 000.	Tera	T	1×10^{12}
1 000 000 000.	Giga	G	1×10^{9}
1 000 000.	Mega	M	1×10^{6}
1 000.	kilo	k	1×10^{3}
1.	unit		
0.001	milli	m	1×10^{-3}
0.000 001	micro	µ	1×10^{-6}
0.000 000 001	nano	n	1×10^{-9}
0.000 000 000 001	pico	p	1×10^{-12}

Table 3.3 Common prefixes

Here are some common examples of using the prefixes with the unit symbol:

- km (kilometre = one thousand metres)
- mm (millimetre = one thousandth of a metre)
- MW (megawatt = one million watts)
- *µ*s (microsecond = one millionth of a second)

Powers of 10

So far so good, but what about when you want to do some arithmetic with several quantities? For instance, speed is calculated by dividing the distance covered by the time taken. So an energetic spider might cover 1 m in 1 s, and her speed would therefore be 1 m/s (metre per second or ms^{-1}). But how do we express the speed of a bullet that travels 3 km in 50 ms? This is where an understanding of powers of 10 will help us.

When we write down calculations, the position of the figures shows their size – units, tens, hundreds, thousands, etc.

Example

The number 4123.4 means:

4 times one thousand	4000.0
plus 1 times one hundred	100.0
plus 2 times ten	20.0
plus 3 units	3.0
plus 4 times one-tenth	0.4
Total	4123.4

Going left from the decimal point, each column is 10 times the previous one. Going right from the decimal point, each column is one-tenth of the previous one.

This helps in adding things up, because we can line up the decimal points and know that we are adding the number of 10s in one number to the number of 10s in another number:

Example

	32456.24
+	123.51
Total	32579.75

The fact that each column is one-tenth of the previous column makes it easy to multiply and divide by 10. All you need to do is move the decimal point to the left to divide by 10, and to the right to multiply by 10:

$$123.4 \times 10 = 1234.0$$

$$4321.0 \div 10 = 432.1$$

Multiplying or dividing by 100 just means moving the decimal point to the left or right by two places:

$$6789.345 \times 100 = 678934.5$$

$$6789.345 \div 100 = 67.89345$$

There is a shorthand way of showing a number multiplied by itself. For example 100, which is 10×10, can be written as 10^2 (said as 'ten to the power of two', or 'ten squared').

This can be done as many times as we like:

$$10{,}000 = 10 \times 10 \times 10 \times 10 = 10^4$$

$$1{,}000{,}000 = 10 \times 10 \times 10 \times 10 \times 10 \times 10 = 10^6$$

We can also have negative powers. If there is a minus sign in front of the power, this represents 10 divided by the power.

So $10^{-1} = 1 \div 10^1 = 0.1$

and $10^{-3} = 1 \div 10^3 = 0.001$

When the power is zero, the result is always 1 ($10^{\circ} = 1$).

Powers of 10 are often used to express very large or very small numbers.

Example

$123456.0 = 12345.6 \times 10$ or 12345.6×10^1

$= 1234.56 \times 100$ or 1234.56×10^2

$= 1.23456 \times 10{,}000$ or 1.23456×10^5

Note: Every time the decimal point moves a place to the left, the power goes up by 1.

The same can be done for very small numbers.

Example

$0.000123 = 0.00123 \times \frac{1}{10}$ or 0.00123×10^{-1}

$= 1.23 \times \frac{1}{10000}$ or 1.23×10^{-4}

It is important to understand that in all these examples, the actual numbers do not change at all, only the way we write them down.

In science we can use any power, but in engineering we normally use only powers of 10 in multiples of 3, e.g. 10^{-3}, 10^3, 10^6. We do this because they match the prefixes we use.

If you go back to the table of unit prefixes, you should now be able to see how the powers of ten relate to them:

Mega (M) = 1,000,000 times, which is the same as 10^6

micro (μ) = 0.000 001 times, or 10^{-6}

So just how fast was that speeding bullet? Well:

$3\,\text{km} = 3 \times 10^3\,\text{m} = 3{,}000\,\text{m}$

And:

$50\,\text{ms} = 50 \times 10^{-3}\,\text{s} = 0.05\,\text{s}$

So the speed of the bullet is:

$3\,\text{km}/50\,\text{ms} = 3 \times 10^3/50 \times 10^{-3}\,\text{m/s}$ (metres per second)
$= 3{,}000/0.05\,\text{m/s}$
$= 60{,}000\,\text{m/s}$

This can also be written as:

$60 \times 10^3\,\text{m/s}$ or **60 km/s**

It is important to get used to using these different ways of expressing numbers. Although a calculator is a great help, it's easy to make a mistake, so always check that the result looks right.

Remember

Don't take your calculator for granted! Always check that the answer looks sensible

Basic rules

Unless we all carry out calculations using the same basic rules, we will all get different answers to the same question.

Example

Try this: Work out the answer to the sum:

$(42 \times 4) + (6 \div 3) - 2 = ?$

Annie says it's 56, but Brian says it's 420, and Ali says it's 168. So who's right? What do you think? In fact, Ali has the right answer.

The problem is that they have all followed different rules.

Annie simply worked from left to right:

Incorrect
$42 \times 4 = 168$
$168 + 6 = 174$
$174 \div 3 = 58$
$58 - 2 = 56$

Brian worked from right to left:

Incorrect
$3 - 2 = 1$
$6 \div 1 = 6$
$4 + 6 = 10$
$10 \times 42 = 420$

Ali did the multiplications first, then the divisions, then the addition and subtraction:

Correct
$42 \times 4 = 168$
$6 \div 3 = 2$
$168 + 2 = 170$
$170 - 2 = 168$

(see basic rule 3)

Basic rules

Mathematics needs rules, and we need to know them if we are to get our answers right. Here they are.

Basic rule 1

All numbers are either positive or negative (except 0, of course). Since most things we deal with are positive numbers, we don't usually bother to put '+' in front of them, but we must put '–' in front of negative numbers.

Basic rule 2

'Like signs add, unlike signs subtract'.

For example:

$4 + (+5) = 4 + 5 = 9$ (Like signs add)

$4 + (-5) = 4 - 5 = -1$ (Unlike signs subtract)

$4 - (+5) = 4 - 5 = -1$ (Unlike signs subtract)

$4 - (-5) = 4 + 5 = 9$ (Like signs add)

The same applies when multiplying and dividing:

'Like signs give positive results, unlike signs give negative results'.

For example:

$+4 \times -5 = -20$ (Unlike signs give negative results)

$-4 \times +5 = -20$ (Unlike signs give negative results)

$-4 \times -5 = 20$ (Like signs give positive results)

$20 \div +5 = 4$ (Like signs give positive results)

$-20 \div +5 = -4$ (Unlike signs give negative results)

$20 \div -5 = -4$ (Unlike signs give negative results)

$-20 \div -5 = 4$ (Like signs give positive results)

Basic rule 3

When doing a long calculation, do the sums in the order given in the table.

1	Anything in brackets	( )
2	Percentages and ratios	% :
3	Multiplication	×
4	Division	÷
5	Addition	+
6	Subtraction	–

Table of maths symbols

(Now you can see how Ali got the correct result on page 243.)

When there is a calculation inside brackets, use the same order.

Example:

$4\,(3 + 2 \times 2) = 4\,(3 + 4) = 4 \times 7 = 28$

Two points to note:

(i) When there is a number outside the brackets with no operator (+, –, ×, ÷), a multiplication is assumed, so that $5\,(3 + 9)$ is the same as $5 \times (3 + 9)$. The result is $5 \times 12 = 60$.

(ii) In the above example, you can multiply all the numbers inside the bracket by the number outside, then do the calculation. In other words:

$$5\,(3 + 9) = (5 \times 3) + (5 \times 9)$$
$$= 15 + 45 = 60$$

Fractions

We often need to do calculations with parts of a whole unit. So far we have used the decimal system for this. For example, when you cut a cake into 10 slices, you could say that each piece was 0.1 of the cake. However, we are more likely to talk about having one-tenth of the cake. We call these parts fractions.

A fraction is simply the result of dividing something into smaller, equal parts. For example, if you split a bar of chocolate into four pieces, each piece will be:

$\frac{1}{4}$ of the original bar (or one quarter).

The same chunk of chocolate could also be described as 0.25 of the bar: it doesn't make any difference to the size of the piece. Sometimes we use fractions, sometimes decimals, depending on which is easiest.

Adding and subtracting fractions

You need to know how to handle arithmetic involving different fractions. For example, how much is:

$$\frac{1}{4} + \frac{3}{7}$$

denominator

How do we add quarters to sevenths? We need to express each of the two fractions with the same denominator (the number under the line) so we can add like to like.

We do this all the time with money. Say you get some change in a shop: two 10p coins, one 50p coin and eight 5p coins.

10p is one-tenth of a pound, 50p is half and 5p is one-twentieth, so to find out whether we've got the right change, we could write the coins as fractions of a pound:

$$\frac{2}{10} + \frac{1}{2} + \frac{8}{20}$$

This is quite difficult to work out, and we don't do this, of course. We automatically convert the coins to pence (one-hundredths of a pound) and then add them up:

$$\frac{20}{100} + \frac{50}{100} + \frac{40}{100} = £\frac{110}{100} \text{ or } £1.10$$

What we did (in our heads) was to find a common denominator (pence), then add up the value of each coin in pence. The common denominator is a number into which we can divide each of the original denominators without having any remainder: 10, 2 and 20 all go into 100. Once we've got the same denominator, it's easy to add them.

To add and subtract fractions follow these steps:

1. Choose a common denominator (the number under the line) for both fractions. To do this, we find the lowest possible number that both denominators will go into. In the first example, we choose 28, because both denominators (4 and 7) will go into it.

2. Change each fraction so that the denominators are the same. We do this by multiplying both the top and bottom parts of the fraction by the same number so that the denominator is equal to the one chosen in Step 1. To work out what this number is, divide the chosen common denominator by the original denominator in the fraction. In our example, in the first fraction, 4 goes into 28 seven times, so we need to multiply both top and bottom of the first fraction by 7, and the second fraction by 4 (because 7 goes into 28 four times). Now our sum looks like this:

$$\frac{7}{7} \times \frac{1}{4} + \frac{4}{4} \times \frac{3}{7} = \frac{7}{28} + \frac{12}{28}$$

3. Now we can add the top parts of the fractions (known as the numerator). This is possible because we are adding the same size fractions (twenty-eighths, in this case):

$$\frac{7+12}{28} = \frac{19}{28}$$ numerator

4. One last step might be needed. If the numerator is bigger than the denominator, this means there's more than one whole.
   For example:

   $\frac{3}{2}$

   which is usually written as $1\frac{1}{2}$

   When this happens, we find how many times the denominator will go into the numerator – this is the whole number. What's left becomes the numerator (top number) of the fractional part.

   Another example:

   $\frac{29}{12}$ 12 goes into 29 twice, with 5 left over, so we write it as $2\frac{5}{12}$

Multiplying fractions

Multiplying fractions is much easier than adding and subtracting. You simply multiply all the top numbers (numerators) together to get the new numerator, and do the same with the denominators. So:

$$\frac{1}{4} \times \frac{3}{7} = \frac{1}{4} \times \frac{3}{7} = \frac{3}{28}$$

Dividing fractions

Dividing fractions is nearly as easy: swap the top and bottom numbers in the second fraction, then multiply them together:

$$\frac{1}{4} \div \frac{3}{7} = \frac{1}{4} \times \frac{7}{3} = \frac{1}{4} \times \frac{7}{3} = \frac{7}{12}$$

Did you know?

The word percentage comes from the Latin word *centum*, which means 100; think of 'century' and 'cent'.

Percentages

Percentages are really a special kind of fraction where everything is divided into 100 parts.

The difference from normal fractions is that we don't need to show the denominator when we express things as percentages (it's always 100): we simply use the sign '%' (per cent) to show that the number means 'this number of parts out of one hundred'.

Here's a (big!) bar of chocolate that is divided into five rows of twenty blocks.

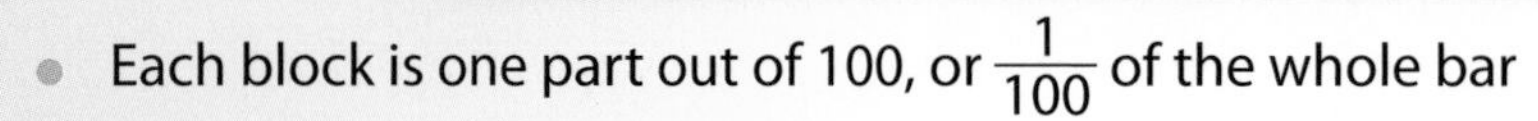

- Each block is one part out of 100, or $\frac{1}{100}$ of the whole bar
- Each row is one row in five, or $\frac{1}{5}$ of the whole bar, and contains 20 blocks
- Expressed as percentages, each block is 1%, and each row is 20%, of the whole bar.

Percentages, fractions and decimals are different ways of expressing parts of a whole. For example, if we broke off two columns of chocolate from one end of the bar, we'd have 10 blocks. So we could say this is: 10% (percentage), $\frac{1}{10}$ (fraction) or 0.1 (decimal).

They are all exactly the same amount of chocolate.

Some common fractions, expressed as percentages:

- $\frac{1}{2}$ as a fraction is the same as 50% (50 parts in 100)
- $\frac{1}{4}$ as a fraction is the same as 25% (25 parts in 100)
- $\frac{1}{10}$ as a fraction is the same as 10% (10 parts in 100)

Example

What is 35% of £18,000?

- Divide £18,000 into 100 parts. Each part (or 1%) is £180
- Now multiply this by 35 = £6,300.

I have a restaurant bill of £72, and I want to leave a 10% tip – how much is it?

- Divide £72 into 100 parts. Each part (or 1%) is 72p
- Therefore 10% = 10 × 72p, or £7.20.

I have a loan of £200, and I have to pay £230 back – what is the interest rate?

- The additional payment is £230 – 200 = £30
- 1% of £200 is £2
- So the percentage is 30 ÷ 2 = 15%

Algebra

People often think that algebra is very difficult, but it is simply a way of writing down a calculation without using specific numbers. Instead, we use letters or symbols to represent different quantities. Using algebra, we can write down relationships between different things, and then later we can replace the symbols with real numbers for a specific example.

As a simple example, let's say we have x girls in a class, and y boys. At the moment, it doesn't matter what numbers x and y represent. Now we can see that the total number of students, which we'll call z, is equal to x + y.

Written as an algebra expression (also known as an equation) this is:

$$x + y = z$$

This obviously works for any class: to calculate z for each class, we simply substitute the correct value of x and y in the equation.

Rules for algebra

In algebra, we don't bother with the multiplication sign. So (D × E) × F is written as (DE)F. In addition and multiplication, it doesn't matter which symbol comes first. So D + E means exactly the same as E + D, and DE is the same as ED.

There are several ways of writing the same thing. So D(E + F) could be written as D × (E + F) or D(F + E). We could also write it as (D × E) + (D × F).

$\frac{D}{E} + \frac{E}{F}$ is the same as $\frac{D+E}{F}$

Here's an example to demonstrate some of these rules.

Example

We want to work out what the expression 8DE – 2EF + DEF is when D = 1, E = 3, and F = 5.

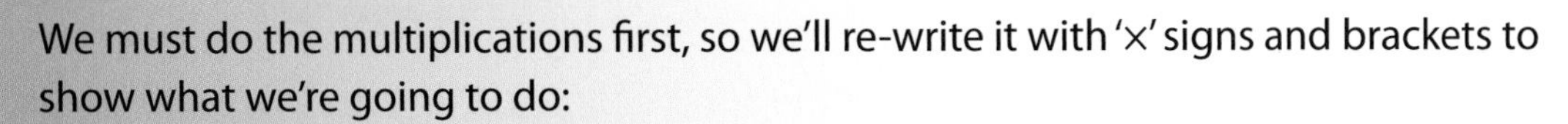

We must do the multiplications first, so we'll re-write it with '×' signs and brackets to show what we're going to do:

$8 \times D \times E - 2 \times E \times F + D \times E \times F$ without brackets

$8 \times (D \times E) - 2 \times (E \times F) + (D \times E \times F)$ with brackets

Now replace the letters with the correct numbers:

$8 \times (1 \times 3) - 2 \times (3 \times 5) + (1 \times 3 \times 5)$

$= 8 \times (3) - 2 \times (15) + (15)$

$= 24 - 30 + 15$

$= \mathbf{9}$

Indices

Remember the table about 'powers of ten'! Well those little numbers were called 'powers' but they are also known as **indices**. If we multiply two identical numbers together, say 5 and 5 the answer is 25 and the process is usually expressed as:

$5 \times 5 = 25$

However, we could express the same calculation as:

$5^2 = 25$

The upper 2 simply means that the lower 5 is multiplied by itself. Sometimes this is referred to as 5 raised to the power of 2. Therefore 5^3 means five multiplied by itself three times, i.e.

$5 \times 5 \times 5 = 125 = 5^3$

Careful – do not make the mistake of thinking $5^3 = 5 \times 3$. It is not!

Example

$3^3 = 3 \times 3 \times 3 = 27$ $\quad 8^3 = 8 \times 8 \times 8 = 512$ $\quad 6^2 = 6 \times 6 = 36$

How about multiplying two numbers together that both have indices? Consider, $4^2 \times 4^2$. This could be shown as $4 \times 4 \times 4 \times 4$ or as 4^4 – which would mean that the indices 2 and 2 have been added together.

So the rule is: when multiplying numbers with indices, simply add the indices.

Example

$4 \times 4^2 = 4^1 \times 4^2$

Add the indices

$= 4^3$

$= 4 \times 4 \times 4$

$= \mathbf{64}$

$5 \times 5^3 \times 5^2 = 5^1 \times 5^3 \times 5^2$

Add the indices

$= 5^6$

$= 5 \times 5 \times 5 \times 5 \times 5 \times 5$

$= \mathbf{15{,}625}$

How about dividing numbers that have indices.

Example

$$\frac{2^5}{2^3} = \frac{2 \times 2 \times 2 \times 2 \times 2}{2 \times 2 \times 2}$$

If we cancel the 2s we get:

$$\frac{2^5}{2^3} = \frac{2 \times 2 \times \cancel{2} \times \cancel{2} \times \cancel{2}}{\cancel{2} \times \cancel{2} \times \cancel{2}} = 2 \times 2$$

This gives:

$2 \times 2 = \mathbf{2^2}$

So the rule is: when dividing numbers with indices simply subtract the indices.

Example

$\frac{5^3 \times 1}{5^2}$ is the same as $\frac{5^3}{5^2}$

Subtracting the indices (i.e. 3 – 2) gives us:

$\frac{5^3}{5^2} = 5^1 = \mathbf{5}$

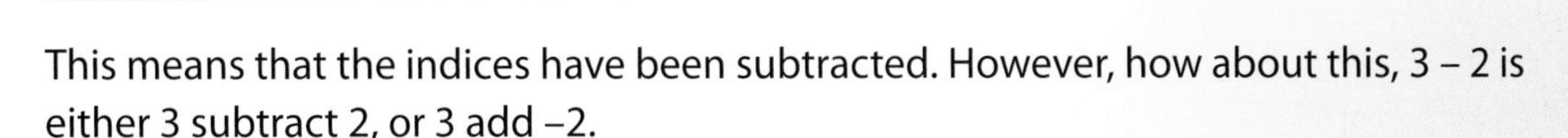

This means that the indices have been subtracted. However, how about this, 3 – 2 is either 3 subtract 2, or 3 add –2.

Remember the addition of indices goes with multiplication. So from this we can see that 5^3 divided by 5^2 is the same as 5^3 multiplied by 5^{-2}.

Therefore:

$\frac{1}{5^2}$ is the same as 5^{-2}

Further examples of this could be:

$\frac{1}{4^3} = 4^{-3}$ $\qquad \frac{1}{2^6} = 2^{-6}$

We can now see that indices can be moved above or below the line providing the sign is changed.

Example

$$\frac{5^6 \times 5^7 \times 5^{-3}}{5^4 \times 5^2} = \frac{5^{13} \times 5^{-3}}{5^6} = \frac{5^{10}}{5^6} = 5^{10} \times 5^{-6} = 5^4 = 5 \times 5 \times 5 \times 5 = \mathbf{625}$$

Example

$(5^4 \times 5^{-6}) \times (5^{-4} \times 5^3) = 5^4 \times 5^{-6} \times 5^{-4} \times 5^3$

Adding all the indices (all the parts are multiplied) we get:

$5^{(4-6-4+3)} = 5^{-3} = \frac{1}{5^3} = \frac{1}{125} = \mathbf{0.008}$

Transposition

Transposition is a method using principles of mathematics, which will enable you to rearrange a formula or equation so that you can find an unknown quantity.

There is one important rule that must always be followed … WITHOUT FAIL! What you do to one side of the equation, you must do to the other side.

Example

Transpose this formula (re-arrange it) to make Y the subject (the one we want):

$$X = Y + Z$$

First, think of transposition as being a pair of scales and remember that each side of the scales (each side of the equals sign) must be balanced. Secondly, when we want to remove something, we perform an opposite operation.

Let us show you. To get Y by itself, we must remove Z. But, as Z has been added to Y, we need to perform 'an opposite operation'. In other words we need to subtract Z from both sides of the equation to keep it balanced.

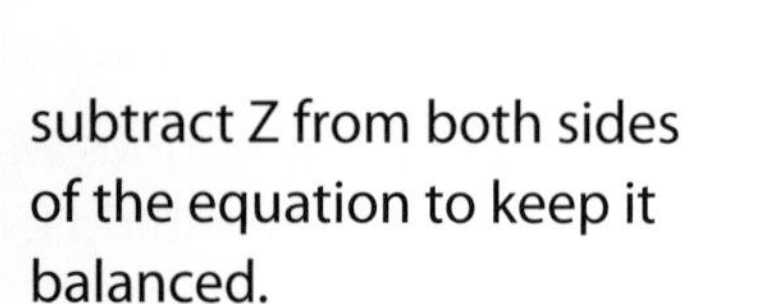

So:

$$X = Y + Z$$

now becomes:

$$X - Z = Y + Z - Z$$

As + Z – Z cancels itself out, you are left with:

$$X - Z = Y$$

Example

Transpose this: A = R – LS so that R is the subject of the equation.

Well, to get R by itself, we need to get rid of LS. This is currently being subtracted from R, so we need to add it to both sides.

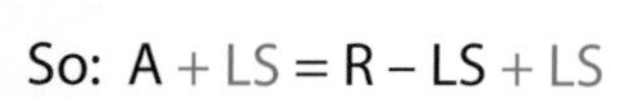

So: $A + LS = R - LS + LS$

Remember: – LS + LS cancels out.

Therefore, complete the calculation and we have an answer of:
$A + LS = R$

The previous examples showed how a formula made up of addition and subtraction could be changed. However, we can view multiplication and division in much the same way.

Example

Transpose the formula $V = I \times R$ to make I the subject.

If we follow through with our idea of opposite operations, we can see that at the moment I has been multiplied by R.

Therefore, to leave I by itself, we must divide by R on both sides.

So: $V = I \times R$

Now becomes: $\frac{V}{R} = \frac{I \times R}{R}$

Carry out this calculation and you are left with an answer of: $\frac{V}{R} = I$

How about one more example?

Example

This equation describes how resistance is related to length, area and type of material in a conductor (you will come across this equation later in this section).

$R = \frac{\rho L}{A}$ Transpose it to find L.

First we need to get rid of the A.

As it is currently dividing into ρL, we need to multiply both sides by A.

Therefore, $R = \frac{\rho L}{A}$

Becomes: $R \times A = \frac{\rho L \times A}{A}$

Giving us: $R \times A = \rho L$

We now need to get rid of the symbol ρ

This is currently multiplying L, so we must divide by ρ on both sides.

Therefore: $R \times A = \rho L$

Becomes: $\frac{R \times A}{\rho} = \frac{\rho L}{\rho}$

The two ρ on the right hand side cancel,

Giving us: $\frac{R \times A}{\rho} = L$

Triangles and trigonometry

You might think this subject isn't relevant to being an electrician, but there are areas of electrical science where knowledge of angles, sines and trigonometry is very useful, particularly in alternating current (a.c.) theory. This section covers some basic principles.

Angles

An angle is the size of the opening between two lines.

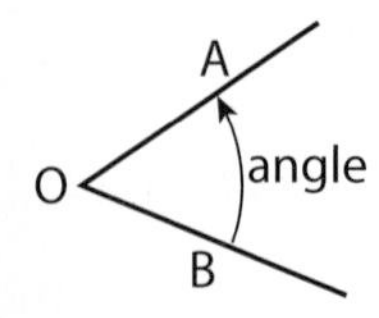

Both lines start at point O, and the length of the lines makes no difference to the angle. Even if line B was twice as long, the angle would be the same.

If we rotate line A (with one end still fixed to point O) anti-clockwise, the angle will get bigger. Eventually, it will be on top of line B, and the angle will be zero. We divide this complete turn into 360 parts, called degrees, with the symbol ° as shorthand.

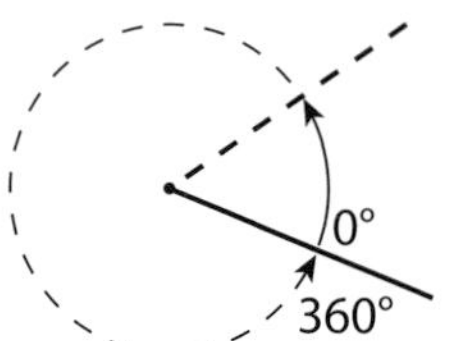

Angles on a straight line

In the diagrams so far, you can see that we think of a straight line as an angle of 180° (which is not surprising as there are 360° in a circle). You can see this clearly on a protractor. Starting at one side and moving round, we end up at the 180° point.

It stands to reason, then, that if we have angles drawn on a straight line, they must add up to 180°.

Angle A is 50°, and angle B is 130°, which adds up to 180°.

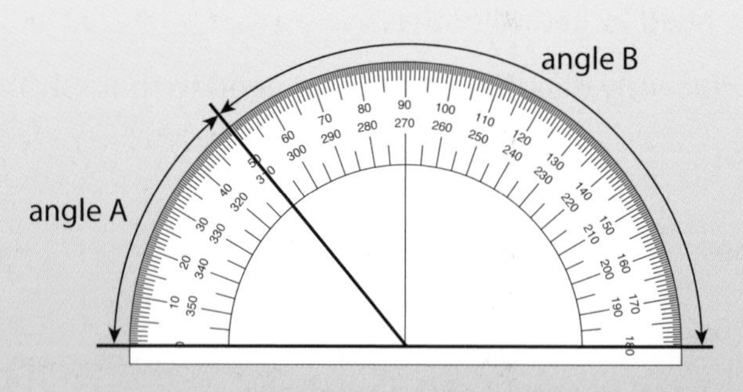

Angle sum of a triangle

In any triangle there are three internal angles at the corners and these must add up to 180°. That is good news, because it means that if we are given two of the angles, we can easily find the third.

For example, if one angle is 40° and another 60°, then the third must be 80°. Try another one.

In this example one angle is 83° the other is 45° and so the third must be 52°.

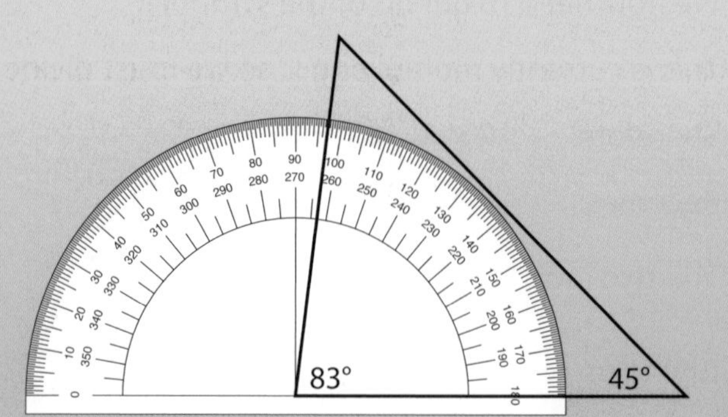

Acute angle

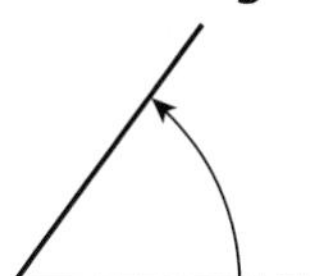

Right angle

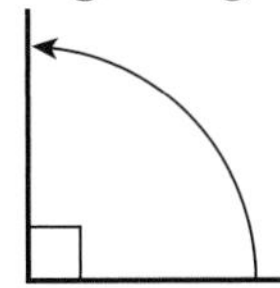

The small square is often used in drawings to show a right angle

Obtuse angle

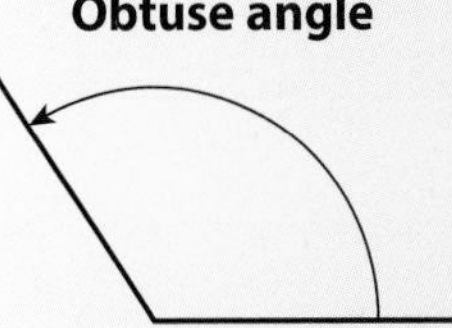

Straight line

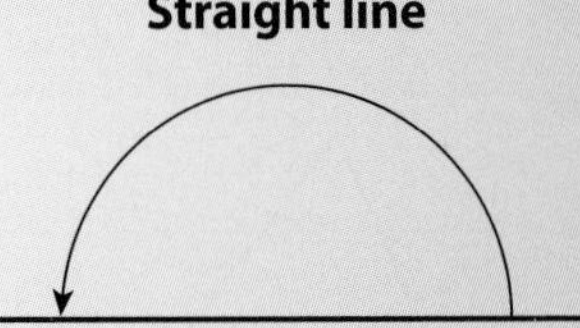

Pythagoras' theorem

This is probably one of the most useful formulas that you will ever be given and it can actually be seen in everyday use. In technical language, Pythagoras' theorem states that:

> *For a right-angled triangle, the square of the hypotenuse is equal to the sum of the squares on the other two sides.*

Which means that if you know the length of two of the sides, you can find out the length of the third. Let's have a look at a right-angled triangle.

We can see that the right angle (often represented as a square in the corner) is like an arrowhead and it always points at the longest side, which is called the **hypotenuse**.

For ease, let's call the hypotenuse side (H) and the other sides (A) and (B). Remember the formula:

$H^2 = A^2 + B^2$

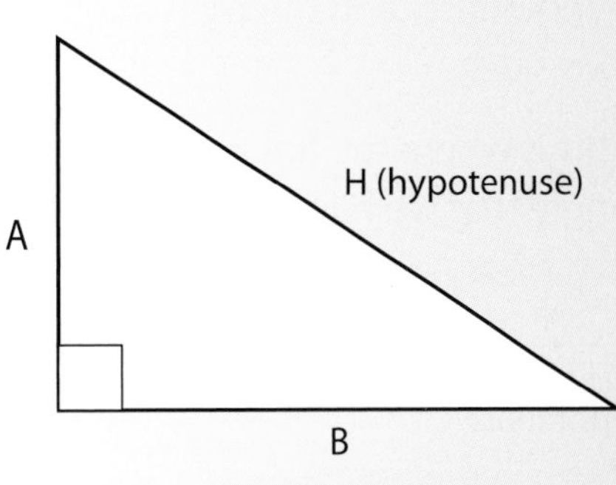

Right-angled triangle

Did you know?

The quickest and easiest way to find the value of H is to tap the value of H^2 into a calculator and hit the square root button ($\sqrt{\ }$).

We said that it could be used every day

Well how do they get the sides of a building straight? Answer: If you know that one side wall is 30 metres long and the other is 40 metres, then if you use Pythagoras to work out the diagonal (hypotenuse) then you can get both walls exactly straight by getting their ends to touch the diagonal and form a triangle, as in the diagram.

Using the formula:

$H^2 = A^2 + B^2$

$H^2 = 30^2 + 40^2$

$H^2 = 900 + 1600$

Therefore $H^2 = 2500$

Therefore $H = \sqrt{2500}$

Therefore $H = 50\text{m}$

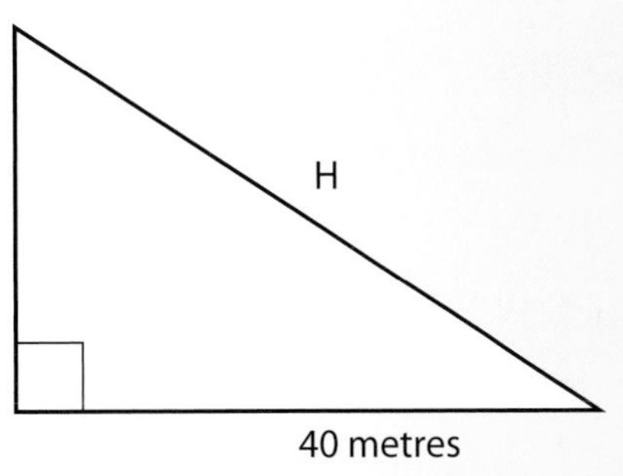

Remember

For a right-angled triangle, the square of the hypotenuse is equal to the sum of the squares on the other two sides – or my name's not Pythagoras!

Five types of triangle

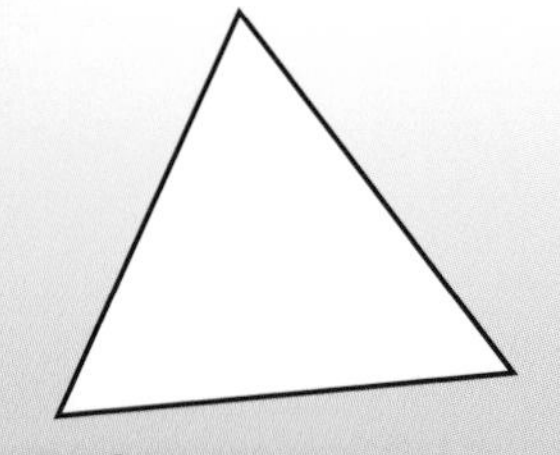

Scalene
Every angle is less than 90°

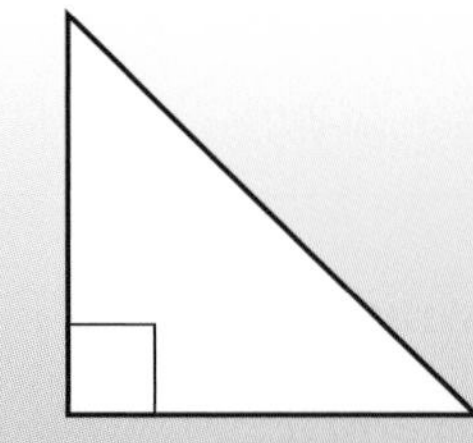

Right-angled
One angle is 90°

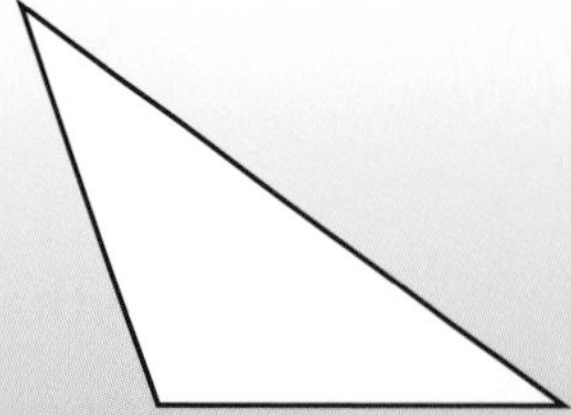

Obtuse
One angle is greater than 90°

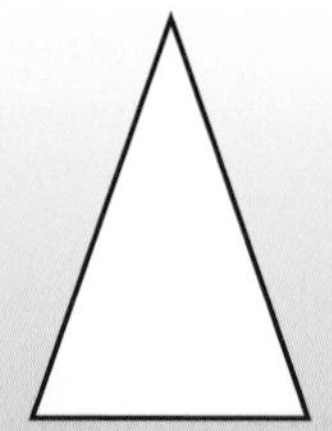

Isosceles
Two equal sides and two equal angles

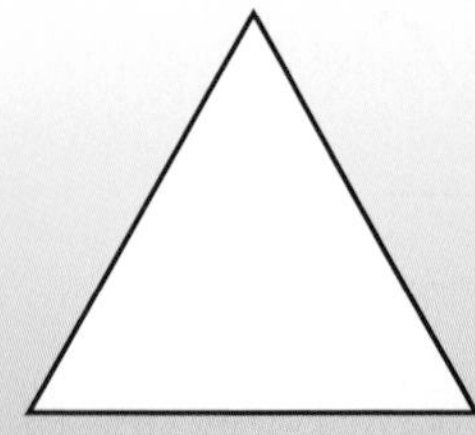

Equilateral
All sides are equal length and all angles are 60°

Trigonometry

Trigonometry is all about the relationship between the angles and sides of triangles. We have just discovered that in a right-angled triangle, we call the long side the hypotenuse, and the right angle 'points' at it. We now need to find a name for the other two sides.

This is where the fun starts, because the names of the other two sides depend on the angle that we have to find, or intend to use.

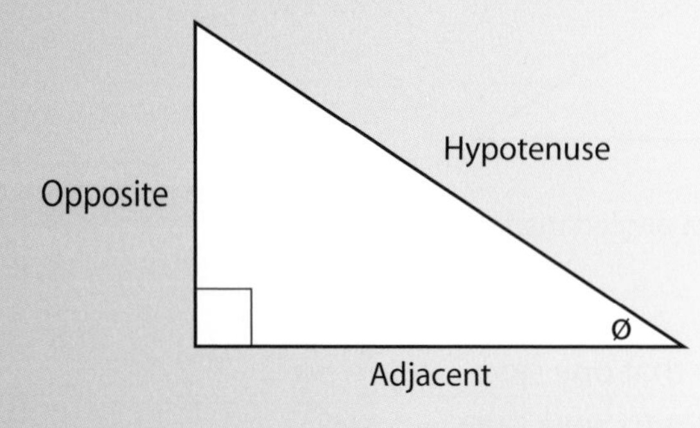

It goes a bit like this: the side which is opposite to the angle being considered is called the **opposite** and the side which is next to the angle under consideration and the right angle, is called the **adjacent**.

In the drawing, Ø is the angle to be considered. So, think of it as being like a torch and the beam is shining on to the opposite side.

We already know that the longest side is the hypotenuse, so the one that is left must be the adjacent.

The next step relates to the following strange names **tangent, sine** and **cosine**. These are used to show the ratio between angles and sides. You choose which one you need depending upon the information that you have been given. So:

$$\mathbf{S}\text{ine } Ø = \frac{\mathbf{O}\text{pposite}}{\mathbf{H}\text{ypotenuse}} \qquad \mathbf{C}\text{osine } Ø = \frac{\mathbf{A}\text{djacent}}{\mathbf{H}\text{ypotenuse}} \qquad \mathbf{T}\text{angent } Ø = \frac{\mathbf{O}\text{pposite}}{\mathbf{A}\text{djacent}}$$

To help you remember them, try to remember the following name: SOH CAH TOA. It represents the letters that we have highlighted in each formula, i.e.:

SOH sine = opposite over hypotenuse

CAH cosine = adjacent over hypotenuse

TOA tangent = opposite over adjacent

Let us have a look at some examples. **Remember**: the formula that you will use depends upon the information that you have been given.

Example 1

From the diagram, what is the value of angle Ø?

Because in this example we are given details about the opposite and adjacent sides, we will therefore use the tangent formula. The shaded area indicates a typical calculator function for use with these calculations. Be careful to ensure that you have a calculator with these functions.

$$\text{Tangent } \text{Ø} = \frac{\text{Opposite}}{\text{Adjacent}}$$

$$\text{Tangent } \text{Ø} = \frac{7}{5} = 1.4$$

We have now found that the angle has a tangent of 1.4. Use your calculator by entering 1.4, then press the INV key and then the TAN key. The answer should be 54.5°. On some calculators, you can use the TAN^{-1} function.

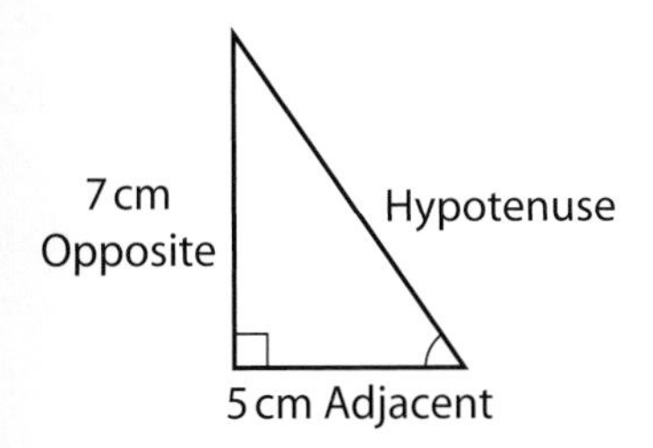

Example 2

From the diagram, what is the value of angle Ø?

This time we have information about the opposite and hypotenuse and so we use the sine formula.

$$\text{Sine } \text{Ø} = \frac{\text{Opposite}}{\text{Hypotenuse}}$$

$$\text{Sine } \text{Ø} = \frac{3}{8} = 0.375$$

Now find the angle that has a sine of 0.375. Use your calculator and enter 0.375, then press the INV key and then the SIN key or SIN^{-1} key.

The answer should be 22°.

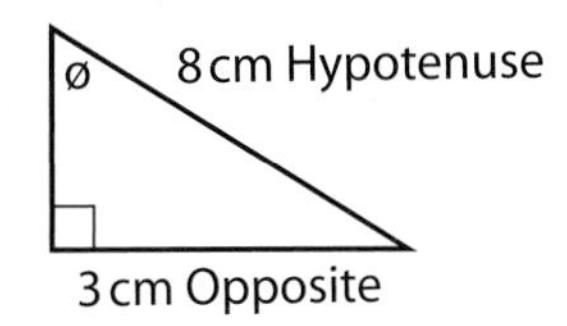

Example 3

From the diagram, what is the value of angle Ø?

This time we have information about the adjacent and hypotenuse and so we use the cosine formula.

$$\text{Cosine } \text{Ø} = \frac{\text{Adjacent}}{\text{Hypotenuse}}$$

$$\text{Cosine } \text{Ø} = \frac{9}{11} = 0.81818$$

Now find the angle that has a cosine of 0.81818. Use your calculator and enter 0.81818, then press the INV key and then the COS or COS^{-1} key.

The answer should be 35.1°. Example 1

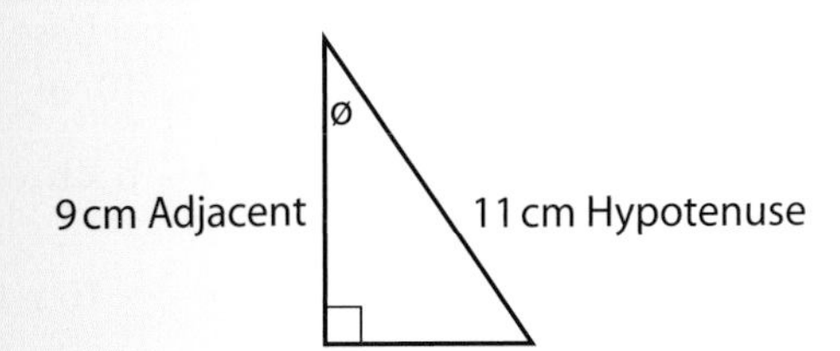

In the previous examples we were using trigonometry to find out an angle. We can also use it to work out the length of a side of the triangle.

Example 4

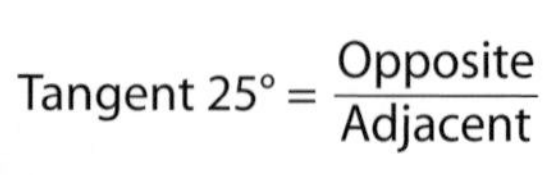

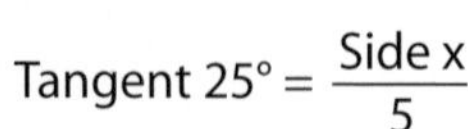

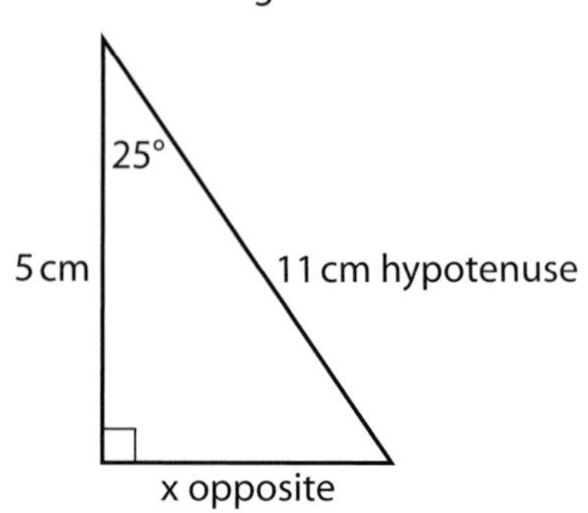

From the diagram, find the length of side x.

In a similar way to the previous examples, the formula that we use depends upon the information that we have been given. In this question, we have been told to find the length of side × and this is opposite the angle that we have been given. We are also told that one side is 5cm long. This is the adjacent side.

Remember, if we are given details about the opposite and adjacent sides, then we use the tangent formula.

We now need to transpose (there is that word again) the formula to make side × the subject. So:

Side x = Tangent 25° × 5cm

Using your calculator, enter 25 and then press the TAN key. This will give you 0.4663. Now multiply this by 5 and the answer should be 2.331.

In other words, side x (opposite) of the triangle is 2.33 cm long.

Area of a triangle

Calculating the area of a triangle is quite simple. You use the formula:

$$\text{Area (A)} = \frac{1}{2}\,\text{base} \times \text{height} \quad \text{or} \quad \frac{\text{base} \times \text{height}}{2}$$

Here's an example:

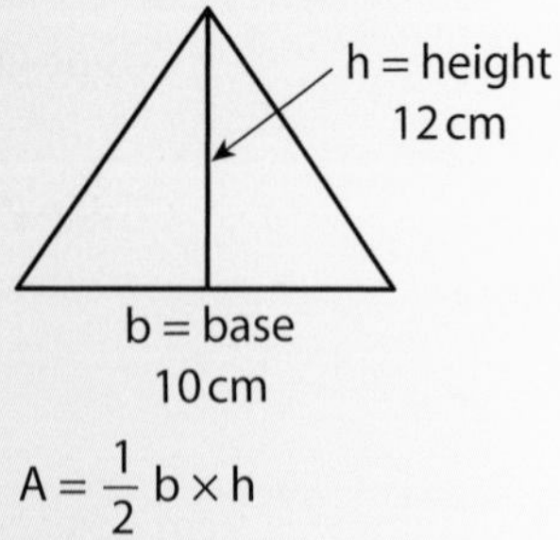

$$A = \frac{1}{2}\,b \times h$$

$$= \frac{1}{2} \times 10 \times 12$$

$$= 60\,\text{cm}^2$$

or

$$A = \frac{b \times h}{2}$$

$$= \frac{10 \times 12}{2}$$

$$= \frac{120}{2}$$

$$= 60\,\text{cm}^2$$

Statistics

Charts

There are many ways to record and display data. This section describes some of the basic techniques. Starting with some data, suppose that in an electrical contracting company the number of people employed on various jobs is as given in the table below.

> **Definition**
>
> **Data** – facts and statistics used for information and analysis

Type of staff	No. Employed	Percentage
A Electricians	14	35%
B Apprentices	12	30%
C Clerical staff	8	20%
D Labourers	4	10%
E Managers	2	5%
Total	40	100%

Table 3.4 Electrical contracting company staffing statistics

The information in the table can be shown in several ways:

Pie chart

The diagram below shows the proportions as angles, the completed circle representing the total number employed.

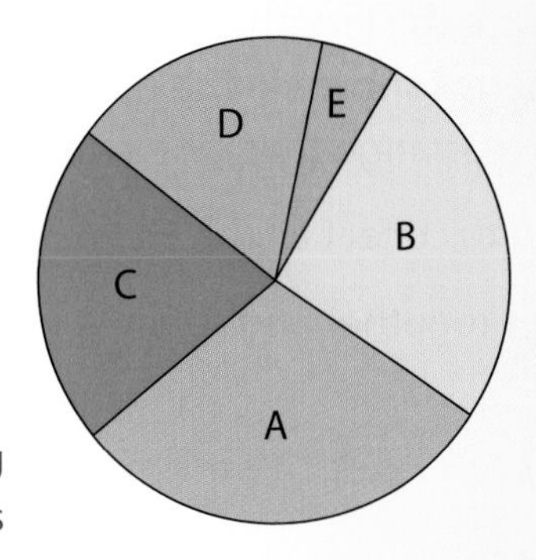

Figure 3.1 Pie chart of contracting company staffing statistics

Thus for electricians the angle is

$$\frac{14}{40} \times 360 = 126°$$

And for apprentices,

$$\frac{12}{40} \times 360 = 108°$$

The column bar chart

This method relies on heights to convey the proportions. The total height of the diagram representing 100%.

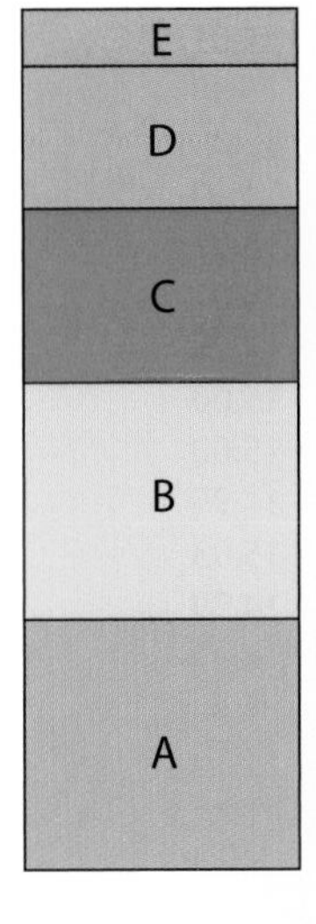

Figure 3.2 Column bar chart

The bar chart

This chart gives a better comparison of the various types of personnel employed but it does not readily display the total number employed in the factory. The bars can be shown vertically, or horizontally.

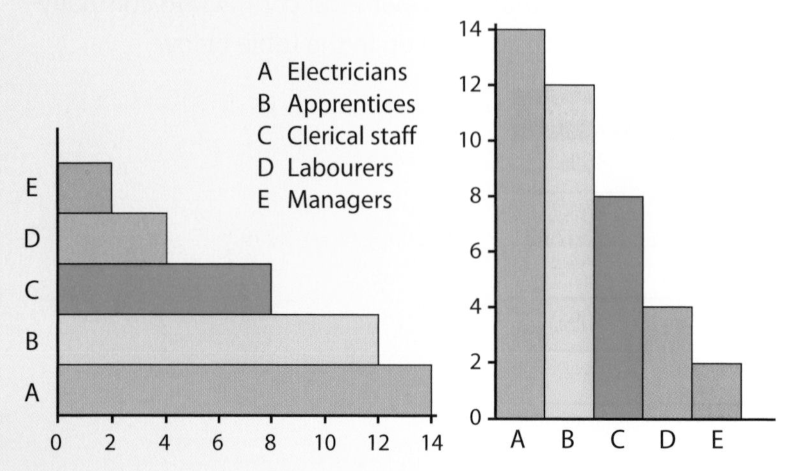

Figure 3.3 Horizontal and vertical bar charts

Frequency distributions and tally charts

Suppose we measure the length of one hundred, 15 mm screws. We may get the readings in millimetres shown in the table. These figures do not mean very much as they stand, but we can rearrange them into what is called frequency distribution.

To do this we need to collect all the 14.96 mm screw readings together and all the 14.97 mm readings together and so on. A tally chart, (Figure 3.5), is the best way of doing this.

15.02	15.00	15.00	15.01	15.01
15.01	14.99	14.99	15.00	14.99
15.01	15.01	14.99	14.98	15.03
15.01	15.00	15.00	15.02	15.03
15.01	15.00	14.98	15.02	15.04
14.98	14.97	14.99	15.00	14.98
15.00	15.00	14.98	14.99	15.00
14.98	15.00	14.99	14.97	15.01
15.01	15.00	15.03	14.98	14.98
15.01	14.99	15.00	15.02	15.00
14.98	14.98	15.00	14.96	14.99
15.03	15.02	15.01	15.03	15.01
14.99	15.04	15.02	15.01	15.01
15.01	15.01	14.98	15.02	14.99
15.01	15.01	14.99	15.02	15.00
15.03	14.97	14.97	15.00	15.00
14.99	15.00	14.99	14.99	14.99
15.02	15.00	15.00	15.00	15.03
15.01	14.99	15.00	14.96	14.99
15.01	14.99	15.03	15.01	14.99

Figure 3.4 Lengths of screws

Length (mm)	Number of screws with this length	Frequency
14.96	II	2
14.97	IIII	4
14.98	𝍸 I	11
14.99	𝍸 𝍸 𝍸 𝍸	20
15.00	𝍸 𝍸 𝍸 𝍸 III	23
15.01	𝍸 𝍸 𝍸 𝍸 I	21
15.02	𝍸 IIII	9
15.03	𝍸 III	8
15.04	II	2

Figure 3.5 Tally chart

Each time a measurement arises a tally mark is placed opposite the appropriate measurement. The fifth mark is usually made diagonally, which orders the tally marks into groups of five. This is purely to make counting easier.

When the tally marks are complete, the marks are counted and the numerical value recorded in the column headed Frequency. The frequency is the number of times each measurement occurs.

From the tally chart we can see that the measurement 14.96 occurs twice (that is, it has a frequency of 2), the measurement 14.97 occurs four times (a frequency of 4) and so on.

The histogram

The frequency distribution becomes even more understandable if we draw a diagram to represent it. The best type of diagram consists of a set of columns, each of the same width whose height represents the frequency.

When you look at the histogram opposite the pattern of the variation is easy to understand, most of the values being grouped near the centre of the diagram with a few values more widely scattered.

Length (mm)	Number of screws with this length	Frequency
14.96	II	2
14.97	IIII	4
14.98	卌 I	11
14.99	卌 卌 卌 卌	20
15.00	卌 卌 卌 卌 III	23
15.01	卌 卌 卌 卌 I	21
15.02	卌 IIII	9
15.03	卌 III	8
15.04	II	2

Figure 3.6 Histogram

Grouped Data

When dealing with a large number of observations it is often useful to group the data into classes or categories. We can then work out the number of items belonging to each class so we can obtain a class frequency. The diagram shows the results of a test given to 200 apprentices (the maximum mark being 20).

From the table, we can see that the first class consists of marks from 1 to 5. 22 apprentices obtained marks in this range and the class frequency is therefore 22. A histogram of this data can be drawn, as shown below right. By using the midpoints of the class intervals as the centres of the rectangles.

Mark	Frequency
1–5	22
6–10	55
11–15	93
16–20	30

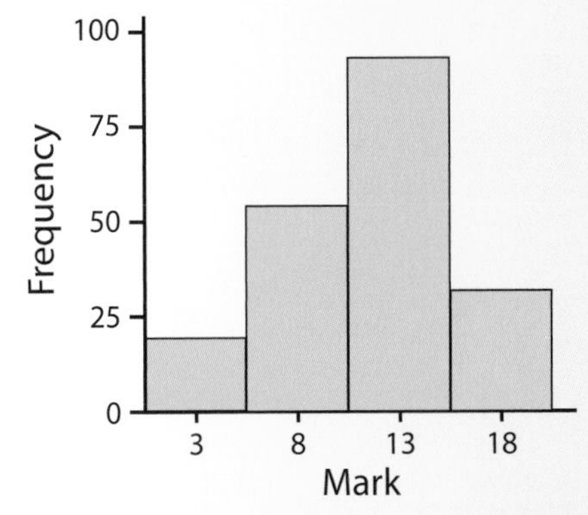

Figure 3.7 Grouped data

Statistical averages

Mean

This is the commonest type of average and it is determined by adding up all the items in the set and dividing the result by the number of items:

$$\text{Mean} = \frac{\text{The total of items}}{\text{The number of items}}$$

Example

The marks of an apprentice in four examinations were 86, 72, 68 and 78. Find the mean of his marks.

$$\text{Mean} = \frac{86 + 72 + 68 + 78}{4}$$

$$\text{Mean} = \frac{304}{4} = 76$$

Median

If a distribution is arranged so that all the items are in ascending (or descending) order of size, the median is the value that is half way along the series. Generally there will be an equal number of items below and above the median. If there is an even number of items the median is found by taking the average of the two middle items.

Example

The median of 3, 4, 4, 5, 6, 8, 8, 9, 10 is 6

The median of 3, 3, 5, 7, 9, 10, 13, 15 is $\left(\frac{7 + 9}{2}\right) = 8$

Mode

The observation, which occurs most frequently in a distribution, is called the mode. Looking at example 2 above, number 3 occurs most often, so 3 is the mode.

Mechanics

The difference between mass and weight

Before we start talking about mechanics, we need to understand a very important concept. That is, the difference between weight (a force) and mass.

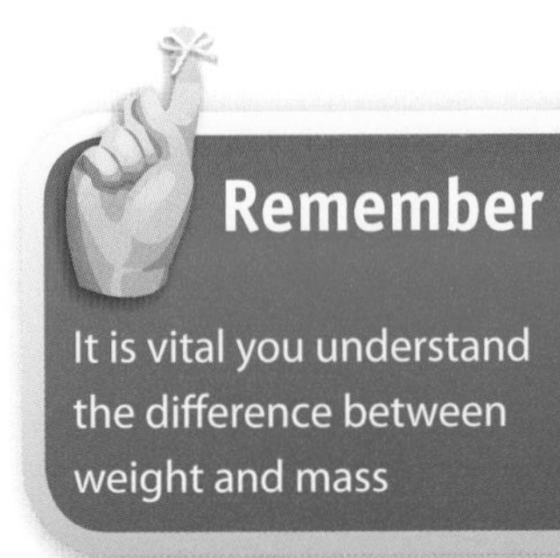

- **Mass**
  This is simply the amount of stuff or matter contained in an object. Assuming we do not cut or change the object, the mass of an object will stay the same wherever we are.

The unit of mass is the kilogram (kg)

- **Weight**
  This is a force and depends on how gravity pulls on a mass. This can vary according to where we are (the higher above sea level you go, the less you weigh). The change in weight is tiny but can be measured with very sensitive and expensive scientific equipment.

The unit of weight, and force in general, is the newton (N)

On earth, if we disregard the effect of height above sea level, the weight acting on 1 kg of mass is equal to 9.81 newtons (N). So, **1 kg weighs 9.81 N**. In many situations this can be rounded up to 10 N.

The human race is very inventive. We have devised many means of overcoming simple problems, such as lifting a heavy object and moving it from one place to another. In this section we will be looking at the following areas:

- simple machines for lifting and handling
- force, work, energy and power
- efficiency.

Simple machines for lifting and handling

A simple machine is a device that helps us to perform our work more easily when a force is applied to it. A screw, wheel and axle and lever are all simple machines.

A machine also allows us to use a smaller force to overcome a larger force and can also help us change the direction of the force and work with a faster speed. The most common simple machines are shown as follows.

In medieval times, siege engines were used to hurl rocks at enemy castles. A siege engine is simply a class 1 lever

Levers

Levers let us use a small force to apply a larger force to an object. They are grouped into three classes, depending on the position of the fulcrum (the pivot).

Class 1

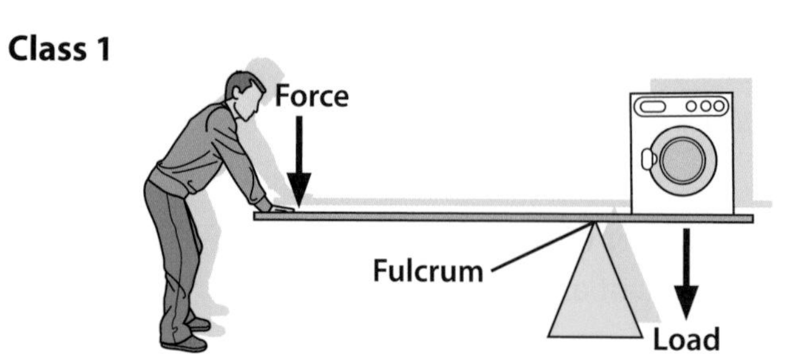

The fulcrum is between the force and the load, like a seesaw.

Class 2

The fulcrum is at one end, the force at the other end, and the load is in the middle. A wheelbarrow is a good example.

Class 3

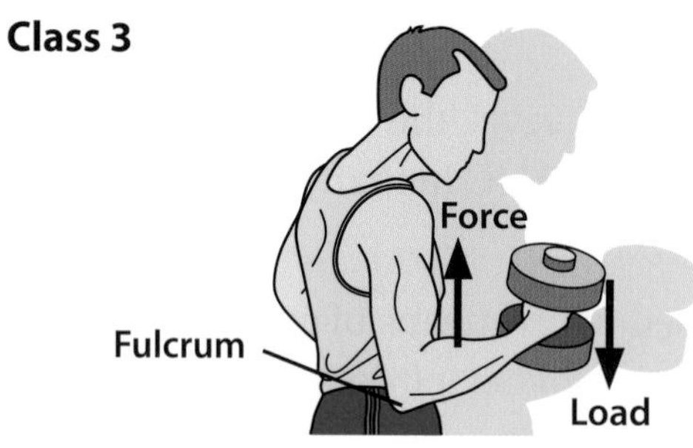

The fulcrum is at one end, the load at the other end and the force in the middle, like a human forearm.

Figure 3.8 Classes of levers

A small force at a long distance will produce a larger force close to the pivot

$10 \times 2 = F \times 0.5$

$F = \frac{10 \times 2}{0.5} \quad \frac{20}{0.5} = 40\,N$

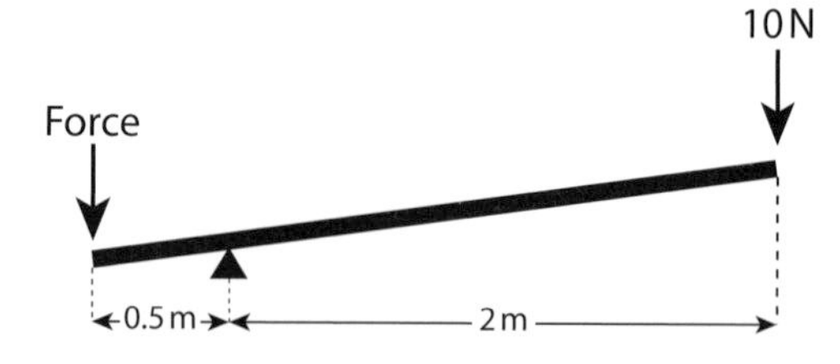

Figure 3.9 Large force close to pivot

Gears

Gears are wheels with teeth; the teeth of one gear fit snugly into those around it. You can use gears to slow things down or speed them up, to change direction, or to control several things at once. The gears, when placed together, have to be fitted into the teeth. Each gear in a series changes the direction of rotation of the previous gear. The smaller gear will always turn faster than the larger gear and in doing so, turns more times.

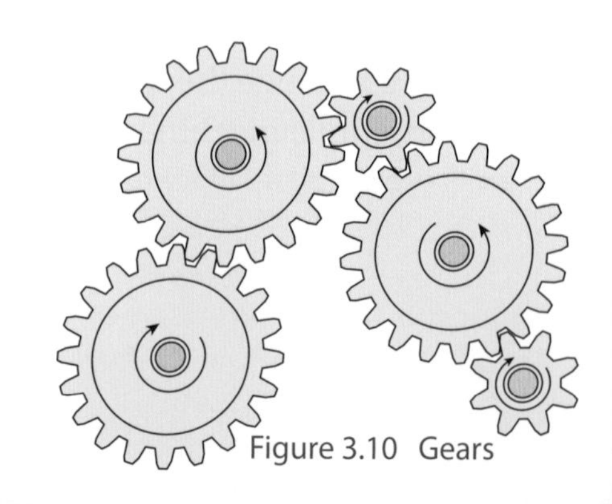

Figure 3.10 Gears

The inclined plane

The inclined plane is the simplest machine of all, as it is basically a ramp or sloping surface. The shortest distance between two points is generally taken as a straight line, but it is easier to move a heavy object to a higher point by using stairs or a ramp. If you think of the height of a mountain, the shortest distance is straight up from the bottom to the top. However, we always build a road on a mountain as a slowly winding inclined plane from bottom to top.

As an electrician, you will use the inclined plane most days in the form of a screw, which is simply an inclined plane wound around a central cylinder.

So the inclined plane works by saving effort, but you must move things a greater distance.

Pulleys

A pulley is made with a rope, belt or chain wrapped around a wheel and can be used to lift a heavy object (load). A pulley changes the direction of the force, making it easier to lift things. There are two main types of pulleys: the single fixed pulley and the moveable pulley.

A **single fixed pulley** is the only pulley that uses more effort than the load to lift the load from the ground. The fixed pulley, when attached to an unmoveable object e.g. a ceiling or wall, acts as a first class lever with the fulcrum being located at the axis but with a minor change – the bar becomes a rope. The advantage of the fixed pulley is that you do not have to pull or push the pulley up and down. The disadvantage is that you have to apply more effort than the load.

A **moveable pulley** is one that moves with the load. The moveable pulley allows the effort to be less than the weight of the load. The moveable pulley also acts as a second class lever. The load is between the fulcrum and the effort.

There are many combinations of pulleys, the most common being the block and tackle, that use the two main types as their principle of operation.

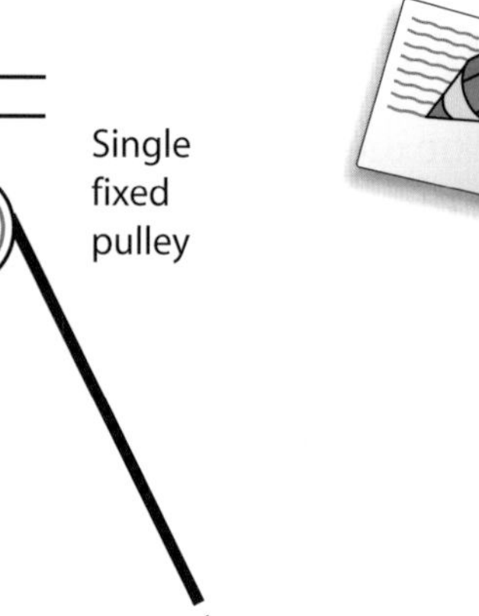

Example

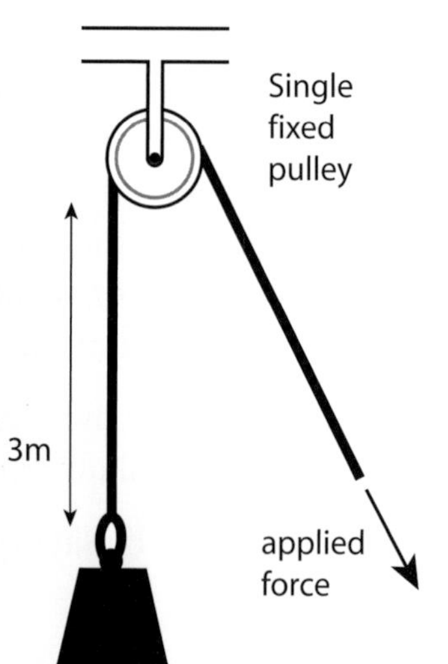

Let us look at examples of the two main types to understand their operating principles. Imagine that you have the arrangement of a 20 newtons (N) weight suspended from a rope, but actually resting on the ground as shown in the diagram.

In this example, if we want to have the load suspended in the air above the ground, then we have to apply an upward force of 20 N to the rope in the direction of the arrow. If the rope was 3 m long and we wanted to lift the weight up 3 m above the ground, we would have to pull in 3m of rope to do it.

Now imagine that we add a single fixed pulley to the scenario, as shown in the diagram. We have not really changed anything in our favour. The only thing that has changed is the direction of the force we have to apply to lift the load. We would still have to apply 20 N of force to suspend the load above the ground, and would still have to reel in 3 m of rope in order to lift the weight 3 m above the ground. This type of system gives us the convenience of pulling downwards instead of lifting.

This diagram shows the arrangement if we add a second, moveable pulley. This new arrangement now changes things in our favour because effectively the load is now suspended by two ropes rather than one. That means the weight is split equally between the two ropes, so each one holds only half the weight, or 10 N. That means that if you want to hold the weight suspended in the air, you only have to apply 10 N of force (the ceiling exerting the other 10 N of force on the other end of the rope). However, if you want to lift the weight 3 m above the ground, then you have to reel in twice as much rope – i.e. 6 m of rope must be pulled in.

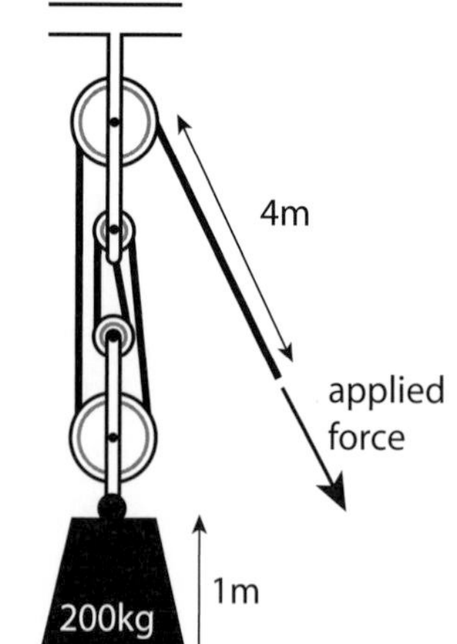

The more pulleys we have the easier it is to lift heavy objects. As rope is pulled from the top pulley wheel, the load and the bottom pulley wheel are lifted. If 2 m of rope are pulled through, the load will only rise 1 m (there are two ropes holding the load and both have to shorten by the same amount).

With pulley systems, to calculate the effort required to lift the load, we divide the load by the number of ropes (excluding the rope connected to the effort). The diagram shows a four pulley system, where the person lifting the 200 kg mass or 2000N load (remember 1 kg weighs 10 N) has to exert a pull equal to only 500 N (i.e. 2000 N divided by 4 ropes).

Mechanical advantage

The common theme behind all of these machines is that because of the machine, we can increase our ability and gain an advantage over nature. This is a relationship between the effort needed to lift something (input) and the load itself (output) and we call this ratio the **mechanical advantage**. Consequently, when a machine can put out more force than is put in, the machine is said to give a good mechanical advantage. Mechanical advantage can be calculated by dividing the load by the effort. There are no units for mechanical advantage, it is just a number.

$$\text{Mechanical Advantage (MA)} = \frac{\text{Load}}{\text{Effort}}$$

Remember

Mechanical advantage is just a number. It has no units

Example

Using the previous diagram, what is the mechanical advantage of the pulley system?

$$\textbf{MA} = \frac{\text{Load}}{\text{Effort}} = \frac{2000\ \text{N}}{500\ \text{N}} = 4$$

In a lever, an effort of 10 N is used to move a load of 50 N. What is the mechanical advantage of the lever?

$$\textbf{MA} = \frac{\text{Load}}{\text{Effort}} = \frac{50}{10} = 5$$

This effectively means that for this lever, any effort will move a load that is five times larger. To summarise:

- **Where MA is greater than 1:** The machine is used to magnify the effort force (e.g. a class 1 lever)
- **Where MA is equal to 1:** The machine is normally used to change the direction of the effort force (e.g. a fixed pulley)
- **Where MA is less than 1:** The machine is used to increase the distance an object moves or the speed at which it moves (e.g. the siege machine).

Velocity ratio

Sometimes machines translate a small amount of movement into a larger amount (or vice versa). For example, in Figure 3.11, a small movement of the piston causes the load to move a much greater distance. This property is known as the velocity ratio, and is found by dividing the distance moved by the effort by the distance moved by the load in the same period of time. There are no units for velocity ratio, it is just a number.

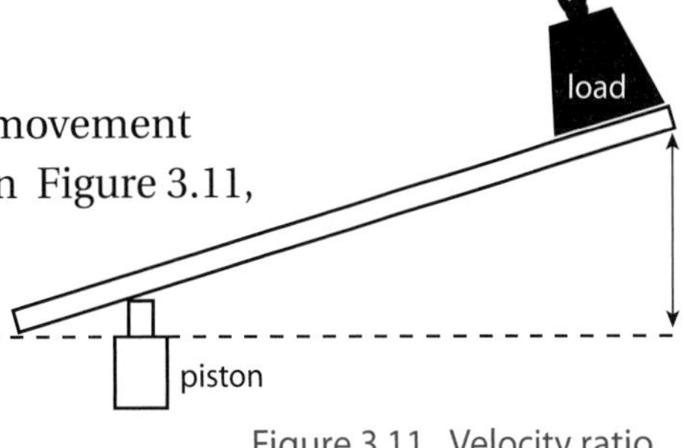

Figure 3.11 Velocity ratio

$$\text{Velocity Ratio (VR)} = \frac{\text{Distance effort moves}}{\text{Distance load moves}}$$

Remember

The VR of any pulley system is equal to the number of pulley wheels

Example

In the diagram, the piston moves 1 m to move the load 5 m. The velocity ratio is:

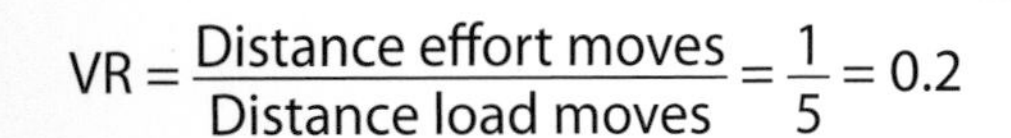

$$\text{VR} = \frac{\text{Distance effort moves}}{\text{Distance load moves}} = \frac{1}{5} = 0.2$$

Force, work, energy and power

Remember

Mass and weight are **not** the same. Mass is the amount of material in an object. Weight is a force – e.g. a person who weighs a certain amount on Earth would weigh less on the moon due to the decreased gravitational force, but they'll still have the same mass

Force

Force is a push or pull that acts on an object. If the force is greater than the opposing force, the object will change motion or shape. Obvious examples of forces are gravity and the wind. Force is measured in **newtons**.

The presence of a force is measured by its effect on a body, e.g. a heavy wind can cause a stationary football to start rolling; or a car colliding with a wall causes the front of the car to deform and the occupants of the car to be forced forwards towards the windscreen (hence the use of seat belts).

Equally, gravitational force will cause objects to fall towards the earth. Therefore, a spring will extend if we attach a weight to it, because gravity is acting on the weight.

As the force of gravity acts on any mass, such a mass tends to accelerate and exert a force that depends upon the mass and the acceleration due to gravity. This acceleration due to gravity is agreed worldwide as being 9.81 m/s^2 at sea level and therefore a mass of 1 kg will exert a force of 9.81 N.

Expressed as a formula:

Force (N) = Mass × Acceleration

(Note: In calculations, it is often assumed for ease that the value of acceleration is taken as 10m/s^2.)

Remember

Do not assume that the acceleration due to gravity is = 10m/s^2. It is 9.81m/s^2. Only take it as 10m/s^2 if you are told to do so in a question, or if you are doing a rough calculation for your own purposes

Work

If an object is moved, then work is said to have been done. The unit of work done is the joule. Work done is the relationship between the effort (force) used to move an object and the distance that the object is moved. Expressed as a formula:

Work done (J) = Force (N) × Distance (m)

Example

A distribution board has a mass of 50 kg. How much work is done when it is moved 10 m?

Work = Force × Distance
= (50 × 9.81) × 10
= 490.5 × 10
= 4905 J

Energy

Definition

Friction – force that opposes motion

Energy, measured in joules, is the ability to do work, or to cause something to move or the ability to cause change. Machines cannot work without energy and we are unable to get more work out of a machine than the energy we put into it. This is due mainly to **friction**. Friction occurs when two substances rub together. Try rubbing your hands together. Did you feel them get warmer?

Work produced (output) is usually less than the energy used (input). Energy can be transferred from one form to another, but energy cannot be created or destroyed. The loss of energy by friction usually ends up as heat.

There may be many forms of energy, but there are only two types:

- **Potential Energy** (Energy of Position or Stored Energy)
- **Kinetic Energy** (Energy due to the motion of an object).

Some forms of energy are: solar, electrical, heat, light, chemical, mechanical, wind, water, muscles and nuclear.

Potential energy

Anything may have stored energy, giving it the potential to cause change if certain conditions are met. The amount of potential energy something has depends on its position or condition. A brick on the top of scaffolding has potential energy because it could fall – due to gravity. The bow used to propel an arrow has no energy in its normal position, but draw the bow back and it now possesses a stored potential energy. A change in its condition (releasing it) can cause change (propelling the arrow).

Potential energy due to height above the Earth's surface is called gravitational potential energy, and the greater the height, the greater the potential energy.

There is a direct relation between gravitational potential energy and the mass of an object; more massive objects have greater gravitational potential energy. There is also a direct relation between gravitational potential energy and the height of an object. The higher an object is above the earth, the greater its gravitational potential energy. These relationships are expressed by the following equation:

$$PE_{grav} = \text{mass of an object} \times \text{gravitational acceleration} \times \text{height}$$

$$\mathbf{PE_{grav} = m \times g \times h}$$

Another example of potential energy is the spring inside a watch. The wound spring transforms potential energy to kinetic energy of the wheels and cogs etc. as it unwinds.

Kinetic energy

Kinetic energy is energy in the form of motion and the greater the mass of a moving object, the more kinetic energy it has.

Power

When we do work in a mechanical system, the energy we put into the system does not appear instantaneously. It takes a certain time to move an object, lift a weight etc. The power that we put into a system must depend not only on the amount of work we do but also how fast we carry out the work.

To try to understand this, think of a 100-metre runner and a marathon runner. The sprinter has a burst of energy for maybe 10 seconds or so whereas the marathon runner may only use up slightly more energy but at a much slower pace. Let's face it, both events would leave you feeling shattered! But it is usual to say that the sprinter had greater power because he used his energy very quickly.

We usually say that:

Power = the rate of doing work.

In terms of equations we can say that:

$$\text{Power (P)} = \frac{\text{Work done (W)}}{\text{Time taken to do that work (t)}}$$

or

$$\text{Power (P)} = \frac{\text{Energy used (E)}}{\text{Time taken to do that work (t)}}$$

Remember

Be really careful, the shorthand for work is W and the units for power is W. Do not get them confused with each other

In terms of units, energy or work is measured in joules (J) and time is measured in seconds (s). Power is measured in joules per second or J/s known as watts (W). Also 1000 watts (W) = 1 kilowatt (kW).

To try to fix this in your mind, let us go back and have another look at an example we did earlier:

Example

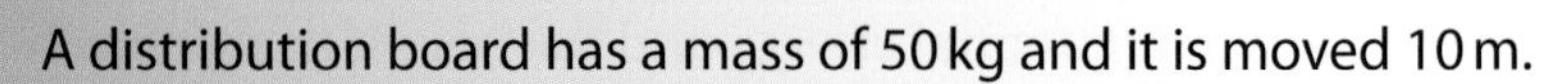

A distribution board has a mass of 50 kg and it is moved 10 m.

$$\text{Work} = \text{Force} \times \text{Distance} = (50 \times 9.81) \times 10 = 490.5 \times 10 = 4905\ \text{J}$$

That is where we got to last time. Now what if it took 20 s to move the distribution board by 10 m. How much power did we put into moving it?

In this case:

$$\text{P} = \frac{\text{Work done in moving the distribution board by 10m (W)}}{\text{The time it took to move the distribution board (t)}}$$

Now, W = 4905 J, and t = 20 s. So:

$$\text{P} = \frac{4905}{20}$$

Therefore: **P = 245.25 W**

Efficiency

We often think of machines as having an input and an output. Try instead of thinking of a machine as having two outputs, one that is wanted and one that is not and is therefore wasted. The greater the unwanted component, the less efficient the machine is.

In all machines, the power at the input is greater then the power output, because of losses that occur in the machine such as friction, heat or vibration. This difference, expressed as the ratio of output power over input power, is called the **efficiency** of the machine. The symbol sometimes used for efficiency is the Greek letter η (eta).

That is:

$$\text{Efficiency} = \frac{\text{Output Power}}{\text{Input Power}}$$

To give the efficiency as a percentage, which is usually more convenient and understandable, we can say that

$$\%\ \text{efficiency} = \frac{\text{Output Power}}{\text{Input Power}} \times 100$$

We will now run through a series of steps to end up with a final equation for efficiency. It is quite complicated, but try to follow the steps. The main thing, though, is to remember the final formula.

Now, for any machine:

$$\text{Work done at the input} = \text{Effort} \times \text{Distance the effort moves (force} \times \text{distance)}$$

And:

$$\text{Work done at the output} = \text{Load} \times \text{Distance the effort moves (force} \times \text{distance)}$$

Dividing these two equations gives us:

$$\frac{\text{Work at Output}}{\text{Work at Input}} = \frac{\text{Load} \times \text{Distance moved by load}}{\text{Effort} \times \text{Distance moved by effort}} = \text{Efficiency}$$

Which can be rewritten as:

$$\frac{\text{Work at Output}}{\text{Work at Input}} = \frac{\text{Load}}{\text{Effort}} \times \frac{\text{Distance moved by load}}{\text{Distance moved by effort}} = \text{Efficiency}$$

Now, you already know that:

$$\text{Mechanical Advantage (MA)} = \frac{\text{Load}}{\text{Effort}}$$

And that:

$$\text{Velocity Ratio (VR)} = \frac{\text{Distance moved by effort}}{\text{Distance moved by load}}$$

So:

$$\frac{1}{\text{VR}} = \frac{\text{Distance moved by effort}}{\text{Distance moved by load}}$$

So:

$$\text{Efficiency} = \frac{\text{Work at Output}}{\text{Work at Input}} = \frac{\text{Load}}{\text{Effort}} \times \frac{\text{Distance moved by load}}{\text{Distance moved by effort}} = \frac{\text{MA} \times 1}{\text{VR}}$$

Therefore:

$$\mathbf{Efficiency} = \frac{\mathbf{Mechanical\ Advantage}}{\mathbf{Velocity\ Ratio}} = \frac{\mathbf{MA}}{\mathbf{VR}} \quad \text{or} \quad \mathbf{\%\ Efficiency} = \frac{\mathbf{MA}}{\mathbf{VR}} \times \mathbf{100}$$

That was quite tricky, but you must remember the final equation above.

If a machine has low efficiency, this does not mean it is of limited use. A car jack, for example, has to overcome a great deal of friction and therefore has a low efficiency, but it is still a very useful tool as a small effort allows us to lift the weight of a car when changing a tyre.

Let us look at a couple of examples that will help us get to grips with some of these concepts.

Example 1

In the following diagram, a trolley containing lighting fittings is pulled at constant speed along an inclined plane to the height shown. Assume that the value of the acceleration due to gravity is 10 m/s². If the mass of the loaded cart is 3.0 kg and the height shown is 0.45 m, then what is the potential energy of the loaded cart at the height shown?

$PE = m \times g \times h$

$PE = 3 \times 10 \times 0.45$

$\mathbf{PE = 13.5\,J}$

If a force of 15.0 N was used to drag the trolley along the incline for a distance of 0.90m, then how much work was done on the loaded trolley?

$W = F \times d$

$W = 15 \times 0.9$

$\mathbf{W = 13.5\,J}$

Example 2

A motor control panel arrives on site. It is removed from the transporter's lorry using a block and tackle that has five pulley wheels. Establish the percentage efficiency of this system given that the effort required to lift the load was 200 N, the panel has a mass of 80 kg and acceleration due to gravity is 10 m/s².

$$\text{The load} = \text{Mass} \times \text{Acceleration due to gravity}$$

$$= 80 \times 10$$

$$= 800\,\text{N}$$

$$\text{Mechanical Advantage} = \frac{\text{Load}}{\text{Effort}} = \frac{800\,\text{N}}{200\,\text{N}} = 4$$

Remembering that velocity ratio is equal to the number of pulley wheels

$$\text{Velocity Ratio} = \text{the number of pulley wheels} = 5$$

$$\text{Efficiency} = \frac{\text{Mechanical Advantage}}{\text{Velocity Ratio}} = \frac{4}{5} = 0.8$$

$$\text{So, \% efficiency} = 0.8 \times 100 = 80\%$$

Therefore the system is 80 per cent efficient.

Electron theory

This section is where we really start to look at electricity and electrical circuits in detail. You cannot even think of becoming an electrician unless you have a sound knowledge of the principles involved starting with the atomic theory of matter and how this gives rise to an electric current.

In this section we will be looking at the following areas:

- molecules and atoms
- the electric circuit
- the causes of an electric current
- the effects of an electric current
- resistance.

Molecules and atoms

Every substance is composed of molecules, which in turn are made up of atoms.

Atoms are not solid but consist of even smaller particles. At the centre of each atom is the **nucleus**, which is made up from particles known as **protons** and **neutrons**. Protons are said to possess a positive charge (+), and neutrons are electrically neutral. The neutrons act as a type of 'glue' which holds the nucleus together.

You probably know that:

- **like charges repel each other** (+ **and** + **or** – **and** –)
- **unlike charges attract each other** (+ **and** –)**.**

So in a world without neutrons, the positively charged **protons** would repel each other and the nucleus would fly apart. It is the job of the neutrons to hold the nucleus together. Since neutrons are electrically neutral, they play no part in the electrical properties of atoms.

The remaining particles in an atom are known as **electrons**. These circle in orbits around the nucleus and are said to possess a negative charge (–).

All atoms possess equal numbers of protons and electrons. Thus the positive and negative charges are cancelled out, leaving the atom electrically neutral. In some cases it is possible to remove or add an electron to a neutral atom and leave it with a net positive or negative charge. Such atoms are then known as **ions**.

So what is the relationship between protons and electrons and how do they form an atom?

Perhaps the simplest explanation is to look to our solar system – where we have a central star (the sun) around which are the orbiting planets. In the atom the protons and neutrons form a central nucleus (sun) and the electrons are the orbiting particles (planets).

The three states of matter

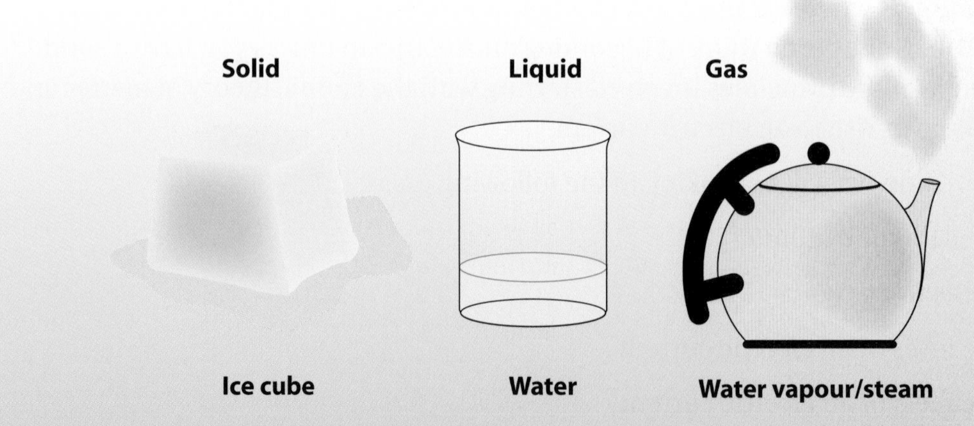

Molecules are always in a state of rapid motion, but when they are densely packed together, this movement is restricted and the substance formed by these molecules is solid. When the molecules of a substance are less tightly bound, there is a great deal of free movement and the substance is a liquid. Finally when the molecule movement is almost unrestricted, the substance can expand and contract in any direction and is a gas.

The simplest atom is that of hydrogen which has one proton and one electron. Figure 3.12 shows the hydrogen atom.

Electrons are arranged in layers at varying distances from the nucleus; those nearest to the nucleus are more strongly held in place than those farthest away. These distant electrons are easily moved from their orbits and so are free to join those of another atom, whose own distant electrons may in turn leave to join another atom and so on.

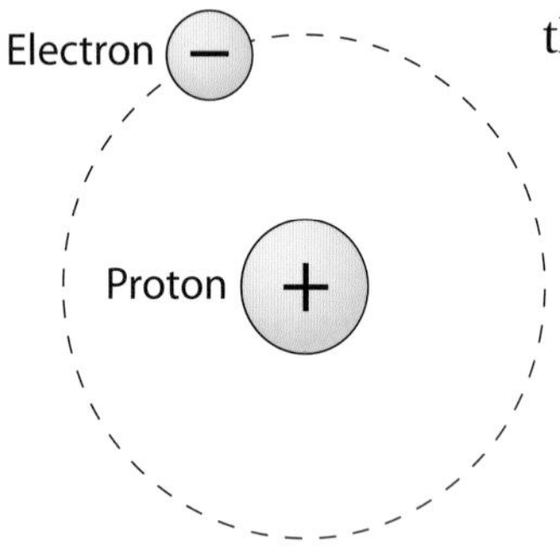

Fig 3.12 Hydrogen atom

It is these wandering or 'free' electrons moving about the molecular structure of a material that give rise to electricity.

We call a material that allows the movement of free electrons a conductor and one which does not, an insulator.

Remember

Electricity is the movement of free electrons along a suitable material. These materials are known as conductors and consequently have a low resistance to the flow of the electrons

Conductors

Try to think of **conductors** as being materials that have their atoms packed together loosely and this therefore allows the free electrons to move through them.

Gold and silver are among the best conductors, but cost inhibits their use. The following table is a guide to the most common conductors and what they are used for.

Aluminium (Al)	• Low cost and weight • Not very flexible • Used for large power cables
Brass (alloy of Copper and Zinc)	• Easily machined • Corrosion resistant • Used for terminals and plug pins
Carbon (C)	• Hard • Low friction in contact with other materials • Used for machine brushes
Copper (Cu)	• Good conductor • Soft and ductile • Used in most cables and busbar systems
Iron/Steel (Fe)	• Good conductor • Corrodes • Used for conduit, trunking and equipment enclosure
Lead (Pb)	• Flexible • Corrosion resistant • Used as an earth and as sheath of a cable
Mercury (Hg)	• Liquid at room temperature • Quickly vaporises • Used for contacts • Vapour used for lighting lamps
Sodium (Na)	• Quickly vaporises • Vapour used in lighting lamps
Tungsten (W)	• Extremely ductile • Used for filaments in light bulbs

Table 3.5 Common conductors

Insulators

Something has to stop electricity from leaking everywhere; otherwise we would get an electric shock every time we used a piece of electrical equipment. The materials that we use to do this are called insulators.

Think of an insulator as being a material whose atoms are so tightly packed together that there is no room for the free electrons to move through them.

Surprisingly, one insulator that is used in cable manufacture is paper! Others are shown in Table 3.6.

Rubber/plastic	• Very flexible • Easily affected by temperature • Used in cable insulation
Impregnated paper	• Stiff and **hygroscopic** • Unaffected by moderate temperature • Used in large cables
Magnesium oxide	• Powder, therefore requires a containing sheath • Very hygroscopic • Resistant to high temperature • Used in cables for alarms and emergency lighting
Mica	• Unaffected by high temperature • Used for kettle and toaster elements
Porcelain	• Hard and brittle • Easily cleaned • Used for carriers and overhead line insulators
Rigid plastic	• Less brittle and less costly than porcelain • Used in manufacture of switches and sockets

Table 3.6 Common insulators

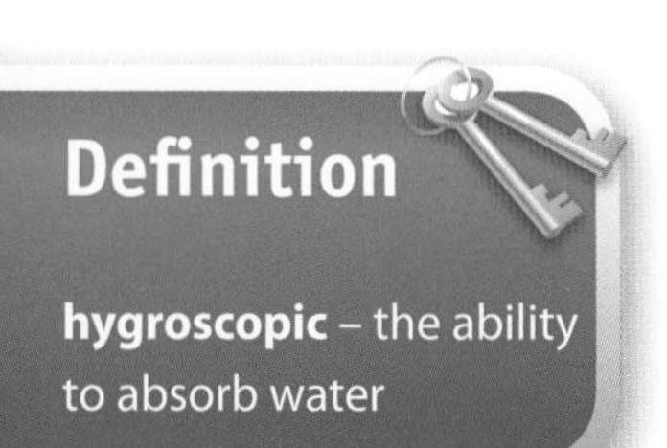

Definition

hygroscopic – the ability to absorb water

Measuring electricity

This would seem a simple task, apart from the fact that electricity is invisible. But what exactly shall we measure?

Remember

Think of a coulomb as an imaginary bucket of electrons

Electricity is simply the flow of free electrons along a conductor, so it would seem obvious to measure the number of electrons moving along the conductor. However, the electron is far too small to be of any practical use. So we group a number of electrons together and then measure the number of groups of electrons moving along. This grouping is known as a **coulomb**, and contains an unimaginable 6,240,000,000,000,000,000 electrons (give or take a couple).

A plumber will measure the amount of water flowing in gallons not drops, as drops are too small a unit of measurement. If the plumber wishes to know how much water is being used at any one time, in other words 'the rate of flow' of the water, this would be measured in gallons per second.

The movement of the water is thought of as its current.

Remember

One ampere equals one coulomb of electrons passing by every second

Similarly, the electrician may wish to know the amount of electrons flowing at any one time (rate of flow of electrons). In electricity, just as with water, this rate of flow of electrons is called the current and is defined as being one coulomb (an imaginary bucket full) of electrons passing by every second.

If one coulomb of electrons passes along the conductor every second, we say that the current flowing along is a current of one ampere. We use the symbol I to represent current.

In other words, one ampere equals one 'imaginary bucket full' of electrons passing by every second.

The electric circuit

We now know, generally speaking, that electricity is the movement of electron charges along a conductor and that the rate of flow is known as the current. But what makes the electron charges move?

Battery

In this circuit, the battery has an internal chemical reaction that provides what is known as an **electromotive force** (e.m.f. for short), which will push the electrons along the conducting wire and into the lamp.

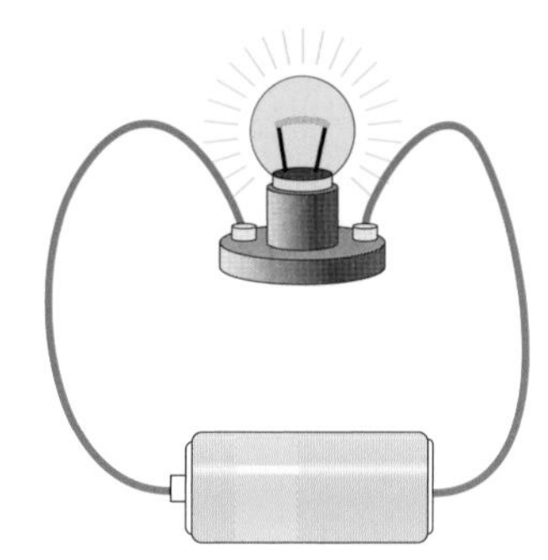

Figure 3.13 A simple circuit

In other words, a battery is a chemically-fuelled charge pump and, just like every other pump in the world, the battery does not supply the 'thing' that it pumps. When a battery runs down it's because its chemical 'fuel' is exhausted, not because any charges have been lost.

The electrons will then pass through the lamp's filament, causing it to heat up and glow and then leave via the second conductor, returning to the battery and thus completing the circuit. If either of the two wires becomes broken or disconnected the flow of electricity will be interrupted and the lamp will go out.

It is this principle that we use to control electricity in a circuit. By inserting a switch into one of the wires connected to the lamp, we can physically 'break the circuit' with the switch and thus switch the lamp off and on.

To summarise, for practical purposes a working circuit should:

- have a source of supply (such as the battery)
- have a device (fuse/MCB) to protect the circuit
- contain conductors through which current can flow
- be a complete circuit
- have a load (such as a lamp) that needs current to make it work
- have a switch to control the supply to the equipment (load).

Electron flow and conventional current flow

An electromotive force is needed to cause this flow of electrons. This has the quantity symbol **E** and the unit symbol **V** (volt). Any apparatus which produces an e.m.f. (such as a battery) is called a power source and it will require wires or cables to be attached to its terminals to form a basic circuit.

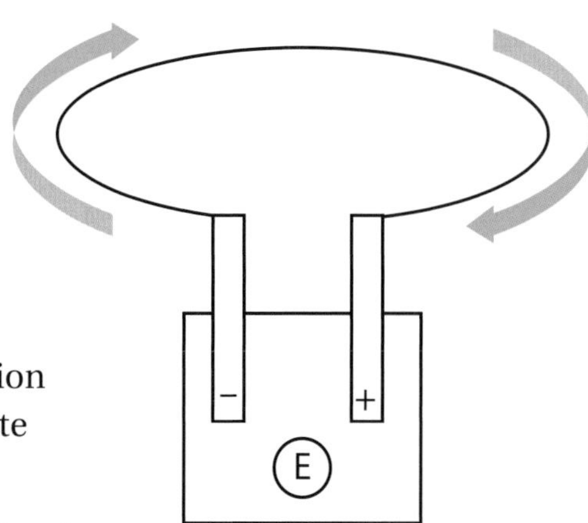

Figure 3.14 Electron flow through an electrolyte

If we take two dissimilar metal plates and place them in a chemical solution (an electrolyte) a reaction will take place in which electrons from one plate travel through the electrolyte and collect on the other plate.

One plate now has an excess of electrons, which will make it more negative than positive. The other plate will now have an excess of protons, which makes it more positive than negative. This process is the basis of how a simple battery or cell works.

Remember

Be careful, in circuit drawings we always show the current as flowing from the positive to the negative. You now know this is not the case

Now select a piece of wire as a conductor (which we already know will have free outer electrons) and connect this wire to the ends of the plate as shown in Figure 3.14.

Since unlike charges are attracted towards each other while like charges repel each other, you can see that the negative electrons will move from the negative plate, through the conductor towards the positive plate.

This drift of free electrons is what we know as electricity and this process will continue until the chemical action of the battery is exhausted and there is no longer a difference between the plates.

Note: The electron flow is actually negative to positive through the conductor. The conventional current flow is from positive to negative.

Potential difference

Remember

A current cannot exist without a potential difference

In the previous section, the chemical energy within the battery is used to do work on a charge to move it from the negative terminal to the positive terminal (the internal circuit). This chemical reaction provides what we refer to as the e.m.f. We could say that this is a measure of the amount of joules of work required to push one coulomb along the circuit. It is measured in joules/coulomb, more commonly referred to as the volt, thus:

One volt = One joule/coulomb

The battery is acting as an energy conversion system, converting chemical energy into electric potential energy. This work increases the potential energy of the charge and thus its electric potential. The charge is moving from a 'low potential' terminal

Basic electric principles

Consider the diagram:

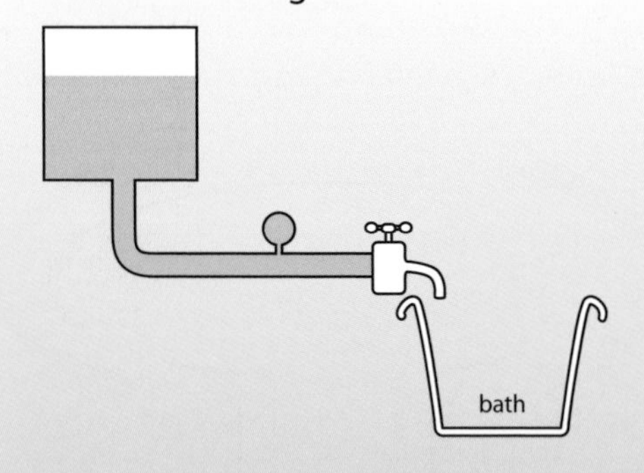

In most water systems, the force that will drive water through the pipes is gravity. However, for water to come out of a tap there has to be a difference in level between the **supply** (the water tank) and the tap.

This gives us a **potential difference**.

The tap acts like a **switch**. When on, the gauge in the water line will measure the amount of water flowing past it every second, in other words, the **current**. Electrically, we would call this gauge an **ammeter**. In our example, the current is dictated by the **potential difference** in levels and total **resistance** to water flow provided by the pipe and tap, which is our **load**. A more useful load could be a paddle wheel in the path of the water to turn a shaft, producing a water driven motor.

In electrical terms, a lamp only requires a small current and, thus, we can afford to use a small diameter 'pipe' (i.e. cable). An item such as a heater would require a bigger cable to take the larger current, otherwise our 'pipe' would provide too much resistance, the current would be reduced and our heater would not work properly.

to a 'high potential' terminal inside the internal circuit of the battery. Once there it will then move through the external circuit (the conductor and equipment), before returning to the low potential terminal. The difference between our terminals is referred to as the **potential difference** and without it there can be no flow of charge.

As our charge moves through the external circuit, it can meet different types of component each of which acts as an energy conversion system, for example, the lamp in Figure 3.13. Here the moving charge is doing work upon the lamp to produce different forms of energy: heat and light. However, in doing so it is losing its electric potential energy and therefore on leaving the lamp, it is less energised. Think of a marathon runner starting fresh and full of energy. As the race progresses, the runner uses up energy until at the end the runner has no energy left.

Looking again at Figure 3.13, we therefore have a point just prior to entering the lamp (or any circuit element) that has a higher electric potential when compared to a point just after leaving the lamp. This loss of potential across a circuit component is also called the volt drop and we will discuss this later.

The causes of an electric current

We need an electromotive force (e.m.f.) to drive electrons through a conductor. The principal sources of an e.m.f. can be classed as being:

- chemical
- thermal
- magnetic.

Chemical

When we take two electrodes of dissimilar metal and immerse them in an electrolyte, we have effectively created a battery. So as we have seen earlier, the chemical reactions in the battery cause an electric current to flow.

Thermal

When a closed circuit consists of two junctions, each made between two different metals, a potential difference will occur if the two junctions are at different temperatures. This is known as the Seebeck effect, based upon Seebeck's discovery of this phenomenon in 1821.

If we now connect a voltmeter to one end (the cold end) and apply heat to the other, then our reading will depend upon the difference in temperature between the two ends. When we have two metals arranged in this pattern, we have a thermocouple.

We can apply this to the measurement of temperatures, with the 'hot end' being placed inside the equipment (such as an oven or hot water system) and the 'cold end' connected to a meter that has been located in a suitable remote position.

Magnetic

Under certain circumstances, a magnetic field can be responsible for the flow of an electric current. We call this situation **electromagnetic induction**. If a conductor is moved through a magnetic field, then an e.m.f. will be induced in it. Provided that a closed circuit exists, this e.m.f. will then cause an electric current.

The effects of an electric current

The effects are categorised in the exact same way as the causes, namely:

- chemical
- thermal
- magnetic.

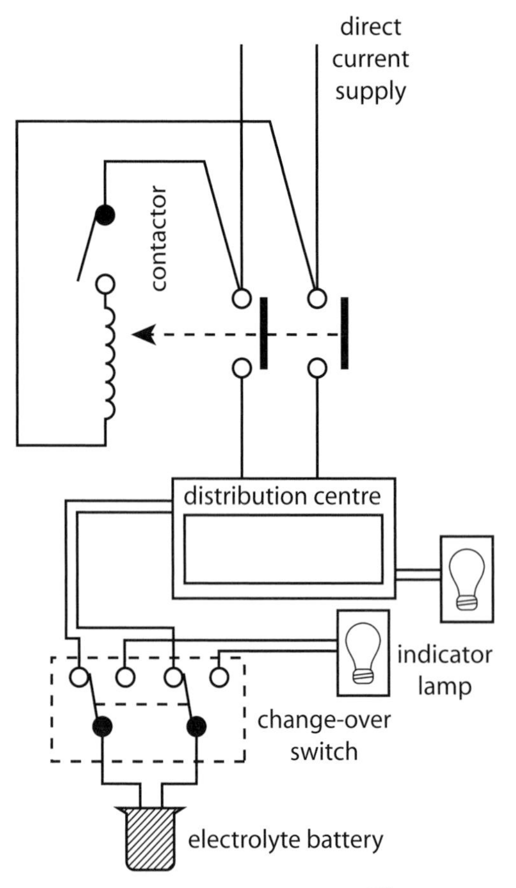

Figure 3.15 d.c. supply

In Figure 3.15, a d.c. supply enters a contactor. When we close the switch on the contactor the coil is energised and becomes an electromagnet, thus pulling anything containing iron in the magnetic field towards it. This allows the main supply to flow through to the distribution centre.

From the distribution centre a supply is taken to a change-over switch. In its current position, this switch allows current to flow into the electrolyte (dilute sulphuric acid and water) via one of the two lead plates. The current returns to the distribution centre via the other lead plate.

Also fed from the distribution centre is a filament lamp. We could have equally used an electric fire. This is because when current flows through a conductor heat is generated. The amount of heat varies according to circumstances, such as the conductor size. If sized correctly, we can make the conductor glow white-hot (a lamp) or red-hot (a fire).

If we run the system like this for a few minutes and then switch off the contactor, obviously we would see our filament lamp go out. However, if we now move our change-over switch into its other position, we would see that our indicator lamp would glow for a short while.

If we were to look at the lead plates, we would see that one of the plates has become discoloured. This is because the current has caused a chemical reaction, changing the lead into an oxide of lead. In this respect the plates acted as a form of re-chargeable battery, also known as a secondary cell.

Figure 3.16 indicates how various equipment relate to these effects of current as their principle of operation.

Chemical effect	**Heating effect**	**Magnetic effect**
Cells	Filament lamp	Bell
Batteries	Heater	Relay
Electro-plating	Cooker	Contactor
	Iron	Motors
	Fuse	Transformers
	Circuit breaker	Circuit breaker
	Kettle	

Figure 3.16 Principles of operation

Resistance

So far, we have considered the amount of electrons flowing in a conductor every second and the force that is pushing them along the conductor but does anything interfere with this flow? Think of the Marathon runner, would he rather be running on a brand new athletics track or through a field of sticky mud which is four feet deep?

Obviously the new track would be the easiest to run on, because it will offer the least amount of opposition to his progress. In other words, the new track would offer less resistance to his progress than the muddy field. Like the muddy field, in electrical circuits – electrical conductors, connections and known resistors will offer a level of resistance to the electrons trying to flow through them.

You could also try thinking of resistance as hurdles that electrons have to jump over on their way around the circuit. The more hurdles there are, then the longer it will take to get around the circuit.

There is a scientific law that we can apply to resistance.

Ohm's Law

So far we have established that **current** is the amount of electrons flowing by every second in a conductor and that a force known as the e.m.f. (or **voltage**) is pushing them. We now also know that the conductor will try to oppose the current, by offering a **resistance** to the flow of electrons.

Ohm's Law, the means by which these three topics are linked together, is probably the most important electrical concept that you will need to understand and is stated as follows:

The current flowing in a circuit is directly proportional to the voltage applied to the circuit, and indirectly proportional to the resistance of the circuit, provided that the temperature affecting the circuit remains constant.

Ohm's Law was named after the nineteenth century German physicist G.S. Ohm who researched how current, potential difference and resistance are related to each other.

In simple language we could re-write Ohm's Law as follows: The amount of electrons passing by every second will depend upon how hard we push them, and what obstacles are put in their way.

Remember

ARemember Ohm's Law,

$I = \frac{V}{R}$

We can prove this is true, because if we increase the voltage (push harder), then we must increase the number of electrons that we can get out at the other end.

Try flicking a coin along the desk. The harder you flick it, the further it travels along the desk. This is what we mean by **directly proportional**. If one thing goes up (voltage), then so will the other thing (current).

Equally we could prove that if we increase the resistance (put more obstacles in the way), then this will reduce the amount of electrons that we can get along the wire.

This time put an obstacle in front of the coin before you flick it. If flicked at the same strength, it will not go as far as it did before. This is what we mean by **indirectly proportional**. If one thing goes up (resistance), then the other thing will go down (current).

Ohm's Law can be expressed by the following formula:

$$\text{Current (I)} = \frac{\text{Voltage (V)}}{\text{Resistance (R)}}$$

Resistivity

Did you know?

Even if we removed all resistors from the circuit, we would still have some resistance caused by the actual conductors such as the connecting wires

Conductor resistance

Electrons find it easier to move along some materials than others and each material has its own resistance to the electron flow. This individual material resistance is **resistivity**, represented by the Greek symbol rho (ρ)and measured in micro ohm millimetres ($\mu\Omega$mm).

How long is it? Would you rather run for 100 metres or 25 miles? Yes? Well so would the electron!

What is its cross-sectional area (csa)? Which is easier, to walk along a 3 metre high corridor, or to crawl along a 1 metre high pipe on your stomach?

To summarise: The amount of electrons that can flow along a conductor will be affected by how far they have to travel, what material they have to travel through and how big the object is that they are travelling along.

As an electrical formula, this is expressed as follows:

$$\text{Resistance} = \frac{\text{Resistivity} \times \text{Length}}{\text{Cross-Sectional Area}}$$

or

$$R = \frac{\rho \times L}{A}$$

We find the value of resistivity for each material, by first measuring the resistance of a 1 metre cube of the material. Then, as cable dimensions are measured in square millimetres (e.g. 2.5 mm^2), this figure is divided down to give the value of a 1 millimetre cube.

This resistivity, as we found out earlier, is given in $\mu\Omega$mm, or in other words we will encounter a resistance of so many millionths of an ohm for every millimetre forward that we travel through the conductor.

The accepted value for copper is 17.8 $\mu\Omega$mm. The accepted value for aluminium is 28.5 $\mu\Omega$mm.

Let us now look at a typical question involving resistivity.

Example 1

Find the resistance of the field coil of a motor where the conductor cross sectional area (csa) is 2 mm², the length of wire is 4000 m and the material resistivity is 18 $\mu\Omega$mm.

$$R = \frac{\rho \times L}{A}$$

Problem 1 *Problem 2*

$$R = \frac{18}{1{,}000{,}000} \times \frac{4{,}000{,}000}{2}$$

What has happened here?

Problem 1: Well, if you remember, the value of ρ is given in millionths of an ohm millimetre. If we have 18 $\mu\Omega$mm, then we have 18 millionths of an ohm and we therefore write it as: 18 divided by one million, or:

$$\frac{18}{1{,}000{,}000}$$

Problem 2: Remember, when we are doing calculations, all units should be the same. Well, the length is in metres, but everything else is in millimetres. Therefore, as all units must be the same and as there are 1000 mm in a metre, note that 4000 m has now become 4,000,000 mm.

So back to the calculation:

$$R = \frac{18}{1{,}000{,}000} \times \frac{4{,}000{,}000}{2}$$

Therefore:

$$R = \frac{72{,}000{,}000}{2{,}000{,}000} = 36\,\Omega$$

Another way of doing this calculation, **without the calculator**, would have been to cancel the zeros down (division):

$$R = \frac{18}{1{,}\cancel{000{,}000}} \times \frac{4{,}\cancel{000{,}000}}{2}$$

Leaving us with:

$$R = \frac{18 \times 4}{2} = \mathbf{36\,\Omega}$$

Remember

All measurements should be of the same type, i.e. metres or millimetres. In these examples, millimetres have been used, hence the cable length is multiplied by 10^3. The resistivity is multiplied by 10^{-6} to provide R in Ø rather then µØ.

Try some further examples.

Example 2

A copper conductor has a resistivity of 17.8 $\mu\Omega$mm and a csa of 2.5 mm^2. What will be the resistance of a 30 m length of this conductor?

If:

$$R = \frac{\rho \times L}{A}$$

Then:

$$R = \frac{17.8 \times 30 \times 10^{-6}}{2.5 \times 10^{3}}$$

So:

$$R = \mathbf{0.2136\,\Omega}$$

Example 3

A copper conductor has a resistivity of 17.8 $\mu\Omega$mm and is 1.785 mm in diameter. What will be the resistance of a 75 m length of this conductor? To enable this question to be answered, we must first convert the diameter into the csa. This is carried out by using one of the following formulas, which you may remember from your school days.

(a) $CSA = \frac{\pi d^2}{4}$ Where d = diameter

Or:

(b) $CSA = \pi r^2$ Where r = radius and $\pi = \mathbf{3.142}$

Using the first formula:

(a) $$CSA = \frac{\pi d^2}{4}$$

Step 1: Put in the correct values:

$$CSA = \frac{3.142 \times 1.785 \times 1.785}{4}$$

Step 2: Multiply the top line:

$$CSA = \frac{10.01}{4}$$

Step 3: Divide by 4

$$CSA = \mathbf{2.5\,mm^2}$$

Therefore, using this method, the csa is 2.5 mm^2.

Using the second formula:

$$CSA = \pi r^2$$

Step 1: Put in the correct values:

$$CSA = 3.142 \times 0.8925 \times 0.8925$$

Step 2: Multiply out

$$CSA = \mathbf{2.5\,mm^2}$$

Using the second method, the csa is still 2.5 mm^2. We can now proceed with the example:

$$R = \frac{\rho \times L}{A}$$

Step 1: Put in the correct values:

$$R = \frac{17.8 \times 10^{-6} \times 75 \times 10^{3}}{2.5}$$

Step 2: Calculate out the top line:

$$R = \frac{17.8 \times 10^{-3} \times 75}{2.5}$$

Which is the same as:

$$R = \frac{17.8 \times 75}{2.5 \times 10^{3}}$$

So:

$$= \frac{1335}{2500}$$

Therefore: $\mathbf{R = 0.534\,\Omega}$

Series and parallel circuits

We have now started looking seriously at circuits in terms of what is in them and how current, resistance and potential difference are all related. However, a circuit can contain many resistors and they can be connected in many ways. In this section, we will be applying Ohm's Law and looking at:

- series circuits
- parallel circuits
- parallel-series circuits
- voltage drop.

Series circuits

If a number of resistors are connected together end to end and then connected to a battery, the current can only take one route through the circuit. This type of connection is called a series circuit.

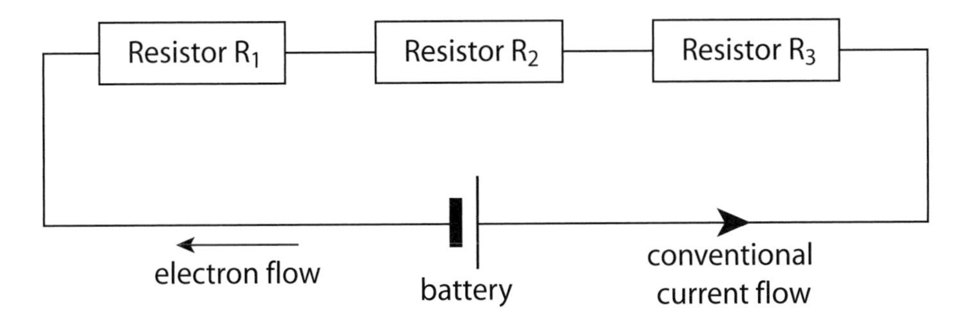

Figure 3.17 Series circuit

Features of a series circuit

- The total circuit resistance (R_t) is the sum of all the individual resistors. In our diagram, this means:

  $R_t = R_1 + R_2 + R_3$

- The total circuit current (I) is the supply voltage divided by the total resistance. You'll recognise this as Ohm's Law:

  $I = \frac{V}{R}$

- The current will have the same value at every point in the circuit.
- The potential difference across each resistor is proportional to its resistance. If we think back to Ohm's Law, we use voltage to push the electrons through a resistor. How much we use depends upon the size of the resistor. The bigger the resistor, the more we use. Therefore:

  $V_1 = I \times R_1 \qquad V_2 = I \times R_2 \qquad V_3 = I \times R_3$

- The supply voltage (V) will be equal to the sum of the potential differences across each resistor. In other words, if we add up the p.d. across each resistor (the amount of volts 'dropped' across each resistor), it should come to the value of the supply voltage. We show this as:
  $V = V_1 + V_2 + V_3$
- The total power in a series circuit is equal to the sum of the individual powers used by each resistor.

Calculation with a series circuit

Example

Two resistors of 6.2 Ω and 3.8 Ω are connected in series with a 12 V battery as shown.

We want to calculate:

(a) total resistance

(b) total current flowing

(c) the potential difference (p.d.) across each resistor.

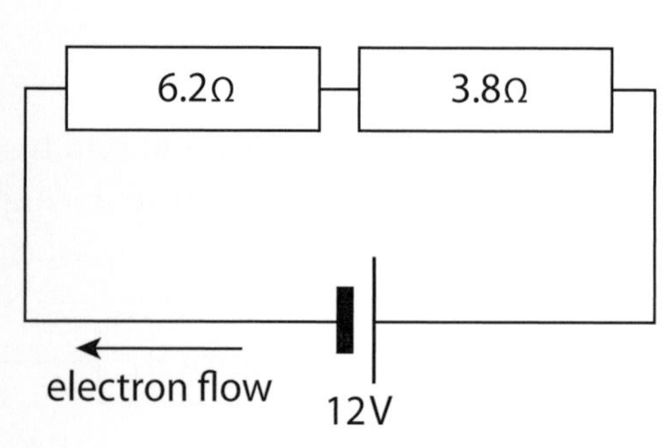

Figure 3.18 Series circuit

(a) Total resistance

For series circuits, the total resistance is the sum of the individual resistors:

$R_t = R_1 + R_2 = 6.2 + 3.8 = 10$ ohms

(b) Total current

Using Ohm's Law:

$I = \frac{\text{Voltage}}{\text{Resistance}} = \frac{12}{10} = 1.2$ ampere

(c) The p.d. across each resistor

$V = I \times R$, therefore:

Across R_1: $V_1 = I \times R_1 = 1.2 \times 6.2 = 7.44$ volts

Across R_2: $V_2 = I \times R_2 = 1.2 \times 3.8 = 4.56$ volts

Parallel circuits

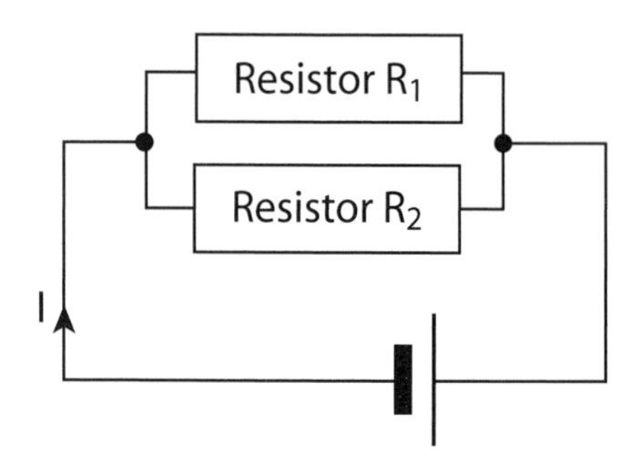

Figure 3.19 Parallel circuit

If a number of resistors are connected together so that there are two or more routes for the current to flow, as shown in Figure 3.18, then they are said to be connected in parallel.

In this type of connection, the total current splits up and divides itself amongst the different branches of the circuit. However, note that the pressure pushing the electrons along (voltage), will be the same through each of the branches. Therefore any branch of a parallel circuit can be disconnected without affecting the other remaining branches.

Explanation

If we think about our definition of Ohm's Law, we now know that the amount of electrons passing by (current) depends upon how hard we are pushing.

We have said that, in a parallel circuit, voltage is the same through each branch. Try to push two identical pencils in the same direction as the current flow towards a point on the circuit where the two branches split (shown as black circles on the drawing). When they reach that point, one pencil will travel towards R_1 and the other towards R_2. But look how the force pushing the pencils has stayed the same.

However, how easily a pencil can then pass through a branch will depend upon the size of the obstacle in its way (the resistance of a resistor).

In summary, the following rules will apply to a parallel circuit:

- the total circuit current (I) is found by adding together the current through each of the branches:

  $I = I_1 + I_2 + I_3$

- the same potential difference will occur across each branch of the circuit:

  $V = V_1 = V_2 = V_3$

- where resistors are connected in parallel and, for the purpose of calculation, it is easier if the group of resistors is replaced by one total resistor (R_t). So:

  $$\frac{1}{R_t} = \frac{1}{R_1} + \frac{1}{R_2} + \frac{1}{R_3}$$

Calculations with parallel circuit

Examples

Three resistors of 16 Ω, 24 Ω and 48 Ω are connected across a 240 V supply. There are two ways to find out the total circuit current.

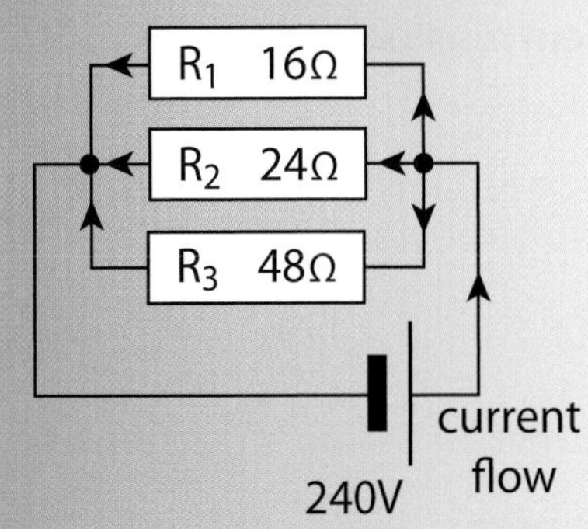

Method 1
Find the equivalent resistance, then use Ohm's Law:

$$\frac{1}{R_t} = \frac{1}{R_1} + \frac{1}{R_2} + \frac{1}{R_3}$$

Therefore:

$$\frac{1}{R_t} = \frac{1}{16} + \frac{1}{24} + \frac{1}{48}$$

And therefore:

$$\frac{1}{R_t} = \frac{3+2+1}{48}$$

Giving us:

$$\frac{1}{R_t} = \frac{6}{48}$$

Rearranging the equation:

$$R_t = \frac{48}{6}$$

and thus, $R_t = 8\,\Omega$

Now using the formula:

$$I = \frac{V}{R}$$

We can say that:

$$I = \frac{240}{8}$$

And therefore:

$\mathbf{I = 30\,A}$

Method 2

Find the current through each resistor and then add them together.

Now, for R_1:

$$I_1 = \frac{V}{R_1}$$

Gives:

$$I_1 = \frac{240}{16}$$

So:

$$I_1 = 15\,A$$

For R_2:

$$I_2 = \frac{V}{R_2}$$

Gives:

$$I_2 = \frac{240}{24}$$

So:

$$I_2 = 10\,A$$

For R_3:

$$I_3 = \frac{V}{R_3}$$

Gives:

$$I_3 = \frac{240}{48}$$

So:

$$I_3 = 5\,A$$

As:

$$I_t = I_1 + I_2 + I_3$$

Then:

$$I_t = 15 + 10 + 5$$
$$\mathbf{= 30\,A}$$

Series/parallel circuits

This type of circuit combines the series and parallel circuits as shown in the diagram. To calculate the total resistance in a combined circuit, we must first calculate the resistance of the parallel group. Then, having found the equivalent value for the parallel group, we simply treat the circuit as being made up of series connected resistors and now add this value to any series resistors in the circuit, thus giving us the total resistance for the whole of the network.

Here is a worked example.

Example

Calculate the total resistance of this circuit and the current flowing through the circuit, when the applied voltage is 110V.

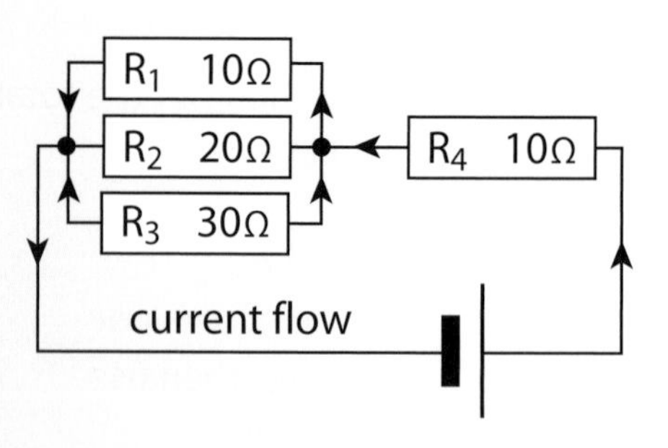

Step1: Find the equivalent resistance of the parallel group (R_p)

$$\frac{1}{R_p} = \frac{1}{R_1} + \frac{1}{R_2} + \frac{1}{R_3}$$

$$\frac{1}{R_p} = \frac{1}{10} + \frac{1}{20} + \frac{1}{30}$$

$$\frac{1}{R_p} = \frac{6+4+2}{60}$$

$$\frac{1}{R_p} = \frac{11}{60}$$

Therefore: $R_p = \mathbf{5.45\,\Omega}$

Step 2: Add the equivalent resistor to the series resistor R_4

$$R_t = R_p + R_4$$

$$R_t = 5.45 + 10$$

$$R_t = 15.45\,\Omega$$

Step 3: Calculate the current

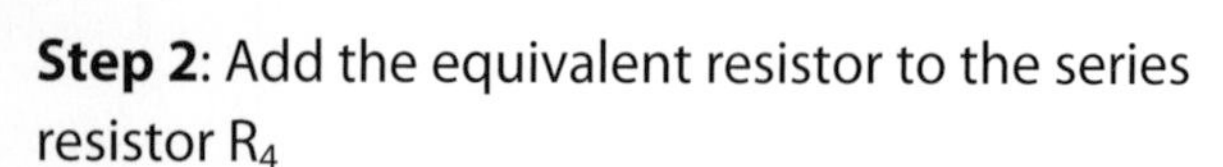

$$I = \frac{V}{R_t} = \frac{110}{15.45} = \mathbf{7.12\ amperes}$$

Voltage drop

Cables in a circuit are similar to resistors, in that the longer a conductor is, the higher its resistance becomes and thus the greater the voltage drop.

Applying Ohm's Law (using the circuit current and the conductor resistance), it is possible to determine the actual voltage drop. To determine voltage drop quickly in circuit cables, BS 7671 and cable manufacturer data include tables of voltage drop in cable conductors. The tables list the voltage drop in terms of (mV/A/m) and are listed as conductor feed and return, e.g. for two single core cables or one two-core cable.

Regulation 525.01.02 states that the voltage drop between the origin of the installation (usually the supply terminals) and a socket-outlet (or the terminals of the fixed current using equipment) shall not exceed four per cent (4%) of the nominal voltage of supply.

Example

A low-voltage radial circuit is arranged as shown in the diagram. It is wired throughout with 50 mm copper cable, for which the voltage drop is given as 0.95 mV/A/m.

We want to calculate:

(a) The current in each section
(b) The volt drop in each section
(c) The supply voltage.

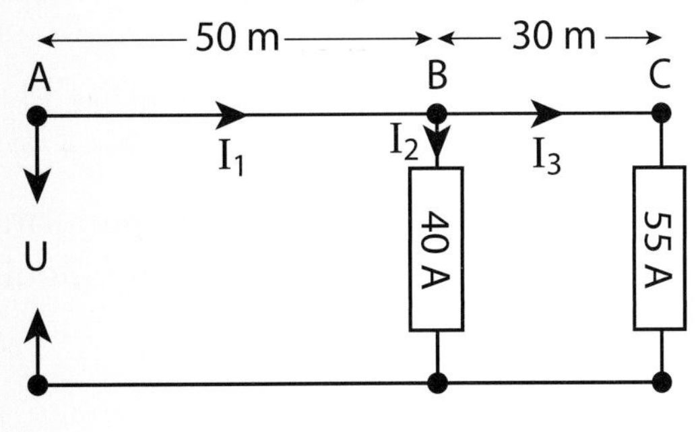

Low-voltage radial circuit

Step 1: Determine the current flowing in section A–B and the circuit length in section A–B:

Current = 40 A + 55 A = 95 A in section A–B

(The length of cable in section A–B is 50 m.)

Step 2: Calculate the volt drop in section A–B by applying the following formula:

$$\text{volt drop} = \frac{\text{mV/A/m} \times I \times m}{1000}$$

Step 3:

$$\text{volt drop in section A–B} = \frac{0.95 \times 95 \times 50}{1000}$$

Giving us 4.5125 volts.

Step 4: Determine the current flowing in section B–C and the circuit length B–C:

The current = 55 A and the length of section B is 30 m.

Step 5: Calculate the volt drop in section B–C by applying the formula used in step 2 above:

$$\text{volt drop in section B–C} = \frac{0.95 \times 55 \times 30}{1000}$$

Giving us 1.5675 volts.

The total volt drop is now:

4.5125 + 1.5675 = **6.08 volts**

Therefore the supply voltage is:

200V + 6.08 = **206.08 volts**

The question to be answered was 'will this voltage drop be smaller than the four per cent allowed by BS 7671?'.

Four per cent of 200V is 8 volts, so this cable will satisfy the requirements of the Regulations. If the volt drop was larger than four per cent it may be necessary to change the csa of the cable, thus reducing the resistance and so lowering the volt drop.

Other options are to reduce the length of cable where possible, or you may reduce the load. If necessary, any one of these alternatives might have to be considered so that you satisfy the requirements of BS 7671.

Electrical energy, work, power and efficiency

Electrical circuits require energy to perform work, they heat things up, they have loads for work to be done on – so why should they not be treated in a similar way to mechanical machines?

Power

We know that electrons are pushed along a conductor by a force called the e.m.f.. Now consider the electrical units of work and power. This is because energy and work are interchangeable, in that we use up energy to complete work. Both are measured in terms of force and distance.

If a force is required to move an object some distance, then work has been done and some energy has been used to do it. The greater the distance and the heavier the object, then the greater the amount of work done.

We already know that:

Energy (or Work done) = Distance moved × Force required

Power may then be stated as being, 'the rate at which we do work' and it is measured in watts. So:

$$\textbf{Power} = \frac{\textbf{Energy (or Work done)}}{\textbf{Time taken}} = \frac{\textbf{Distance moved} \times \textbf{Force required}}{\textbf{Time taken}}$$

For example, we could drill two holes in a wall – one using a hand drill, and the other with an electric drill. When we have finished, the work done will be the same in both cases, in other words there will be two identical holes in the wall, but the electric drill will do it more quickly because its power is greater.

If power is therefore considered to be the ratio of work done against the time taken to do the work, we may express this as follows:

$$\text{Power (P)} = \frac{\text{Work done (W)}}{\text{Time taken (t)}} = \frac{\text{Energy used}}{\text{Time}}$$

The units are:

$$\text{watts} = \frac{\text{joules}}{\text{seconds}}$$

In an earlier chapter, we considered the e.m.f. and defined it as being the amount of joules of work necessary to move one coulomb of electricity around the circuit. We then said that it was measured in joules per coulomb, also known as the volt. **Noting that 1 volt = 1 joule/coulomb** and rearranging the formula, this could be expressed as:

$$\text{joules} = \text{volt} \times \text{coulombs}$$

and since:

$$\text{coulombs} = \text{amperes} \times \text{seconds}$$

we can substitute this to get:

$$\text{joules} = \text{volts} \times \text{amperes} \times \text{seconds}$$

and, since joules are the units of work:

$$\text{Work} = V \times I \times t \text{ (joules)}$$

Taking this one step further, we can show how we arrive at some of our electrical formulae. It goes as follows.

If:

$$\text{Power} = \frac{\text{work}}{\text{second}} \text{ this means: } P = \frac{V \times I \times t}{t}$$

So cancelling the tees:

$$P = V \times I = I \times V$$

And in Ohm's Law:

$$V = I \times R$$

Thus:

$$P = I \times (I \times R) = \mathbf{I^2 \times R}$$

And in Ohm's Law:

$$I = \frac{V}{R}$$

Thus:

$$P = \frac{V \times V}{R} = \frac{V^2}{R}$$

Simple when you know how! And when you have practised changing equations dozens of times!

Some examples of power calculations

Example 1

A 100 Ω resistor is connected to a 10 V d.c. supply. What will be the power dissipated in it?

$$P = \frac{V^2}{R} \text{ therefore } \frac{10 \times 10}{100} = \mathbf{1\,W}$$

Example 2

How much energy is supplied to a 100 W resistor that is connected to a 150 V supply for one hour?

$$P = \frac{V^2}{R} \text{ therefore } \frac{150 \times 150}{100} = \mathbf{225\,W}$$

Now:

$$E = P \times t$$

Therefore:

$$E = \text{energy supplied}$$
$$= 225 \times 3600 \text{ joules}$$

So:

$$E = \mathbf{810{,}000\ joules}$$

Note: That as all time measurements are given in seconds, we have to change hours or minutes into seconds. So, in the example:

$$1 \text{ hour} = 60 \text{ minutes} = 60 \times 60 \text{ seconds} = \mathbf{3{,}600\,s}$$

Kilowatt hour

It should be noted that the joule is far too small a unit for sensible energy measurement. For most applications, we use something called the kilowatt hour.

The kilowatt hour could be defined as the amount of energy used when one kilowatt (1,000 watts) of power has been used for a time of one hour (3,600 seconds).

From this we can see that:

1 joule (J) = 1 watt (W) for one second (s)

1000 joules (J) = 1 kilowatt (kW) for one second

In one hour there are 3,600 seconds. Therefore:

3,600 s × 1,000 J = 1 kW for one hour (kWh)

So:

1 kWh = 3.6 × 10⁶ J

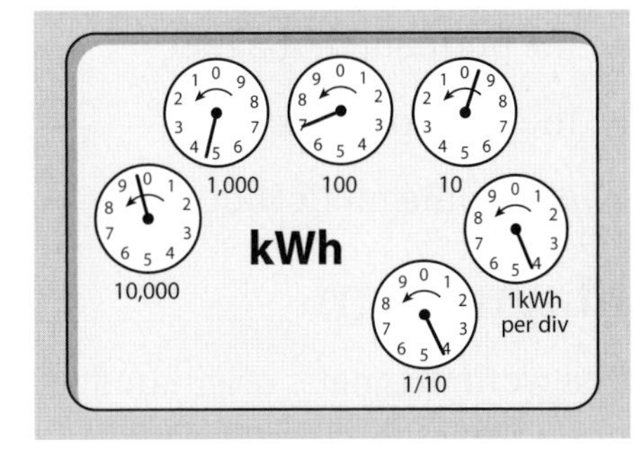

Figure 3.20 Typical electric meter dials

The kilowatt hour is the unit used by the electrical supply companies to charge their customers for the supply of electrical energy. Have a look in your house. You will see that the electric meter is measuring in kWh. However, these are more often referred to as Units by the time they appear on your bill!

Efficiency

We have looked at efficiency in the Mechanics section. The calculations and theory for efficiency applied to electrical circuits are very similar.

You already know that:

$$\textbf{Percentage efficiency} = \frac{\textbf{Output}}{\textbf{Input}} \times \textbf{100}$$

Let us have a look at two examples.

Example 1

Calculate the efficiency of a water heater if the output in kilowatt-hours is 25 kWh and the input energy is 30 kWh.

$$\text{Efficiency (\%)} = \frac{\text{Output}}{\text{Input}} \times 100 = \frac{25}{30} \times 100 = \mathbf{83.33\%}$$

Example 2

The power output from a generator is 2700 W and the power required to drive it is 3500 W. Calculate the percentage efficiency of the generator.

$$\text{Efficiency (\%)} = \frac{\text{Output}}{\text{Input}} \times 100 = \frac{2700}{3500} \times 100 = \mathbf{77.1\%}$$

Magnetism and electromagnetism

We can really go no further with circuit theory until we have looked more closely at the magnetic behaviour of materials and the way this affects the interaction between electrical currents and magnetic fields.

In this section we will be looking at the following areas:

- the permanent magnet
- the electromagnet
- electromagnetic induction.

Magnetism

The word magnetic originated with the ancient Greeks, who found natural rocks possessing this characteristic.

Magnetic rocks such as magnetite, an iron ore, occur naturally. The Chinese observed the effects of magnetism as early as 2600 BC when they saw that stones like magnetite, when freely suspended, had a tendency to assume a north and south direction. Because magnetic stones aligned themselves north–south, they were referred to as lodestones or leading stones.

Magnetism is hard to define – we all know what its effects are: the attraction or repulsion of a material by another material, but why does this happen? And why do we only see it in some materials, notably metals and particularly iron? The physics behind this is too complex to go into here, but it's useful to remember that magnetism is a fundamental force (like gravity) and it arises due to the movement of electrical charge. Magnetism is seen whenever electrically charged particles are in motion.

Materials that are attracted by a magnet, such as iron, steel, nickel and cobalt, have the ability to become magnetised. These are called magnetic materials.

For the electrician, we say that a magnet is any device that produces an external magnetic field and, for the purpose of this book, we are only interested in two types of magnet:

- the permanent magnet
- the electromagnet (temporary magnet)

The permanent magnet

A permanent magnet is a material that when inserted into a strong magnetic field will not only begin to exhibit a magnetic field of its own, but also continue to exhibit a magnetic field once it has been removed from the original field.

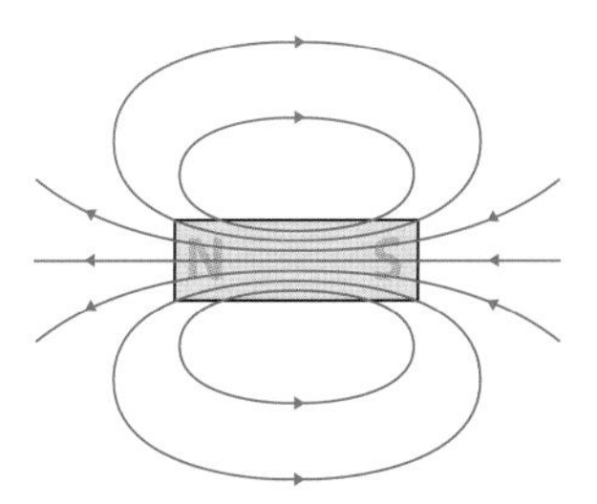

Figure 3.21 Bar magnet

This remaining field would allow the magnet to exert force (the ability to attract or repel) on other magnetic materials. This magnetic field would then be continuous without weakening, as long as the material is not subjected to a change in environment (temperature, de-magnetising field, etc.).

The ability to continue exhibiting a field while withstanding different environments helps to define the capabilities and types of applications in which a magnet can be successfully used.

Magnetic fields from permanent magnets arise from two atomic sources: the spin and orbital motions of electrons. Therefore, the magnetic characteristics of a material can change when alloyed with other elements.

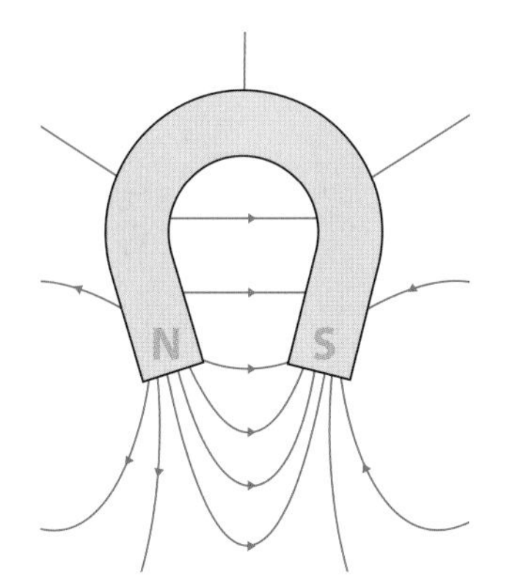

Figure 3.22 Horseshoe magnet

For example, a non-magnetic material such as aluminium can become magnetic in materials such as alnico or manganese-aluminium-carbon.

When a ferromagnetic material (a material containing iron) is magnetised in one direction, it will not relax back to zero magnetisation when the imposed magnetising field is removed. The amount of magnetisation it retains is called its remanence.

It must be driven back to zero by a field in the opposite direction; the amount of reverse driving field required to de-magnetise it is called its coercivity.

We have probably all experienced at some time the effect of a permanent magnet (although we cannot see the magnetic field with the naked eye), even if it was just to leave a message on the fridge door.

The magnetic field looks like a series of closed loops that start at one end (pole) of the magnet, arrive at the other and then pass through the magnet to the original start point.

Did you ever do the famous experiment at school where you took a magnet, placed it on a piece of paper and then sprinkled iron filings over it? If you did, you would see that it looks a bit like the diagram based on a bar magnet (Figure 3.21).

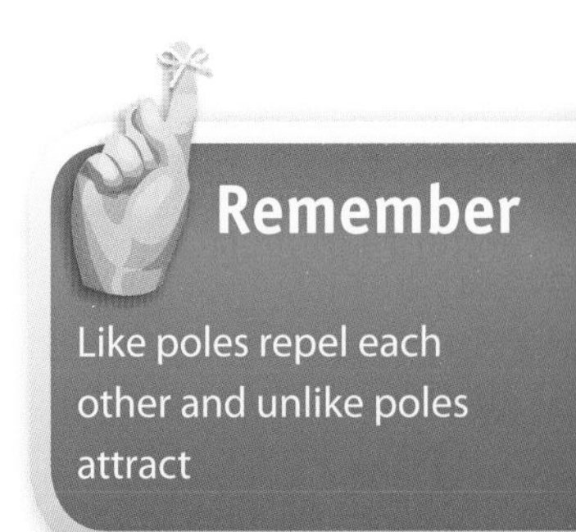

Each one of these lines is called a line of magnetic flux and has the following properties:

- they will never cross, but may become distorted
- they will always try to return to their original shape
- they will always form a closed loop
- outside the magnet they run north to south
- the higher the number of lines of magnetic flux, the stronger the magnet.

If we could count the lines, we could establish the magnetic flux (which we measure in webers), and we would find that the more lines that there were, the stronger the magnet would be. In other words the bigger the magnet the bigger the flux produced.

The strength of the magnetic field at any point is calculated by counting the number of lines that we have at that point and this is then called the **flux density** (measuring webers/square metre, which are given the unit title of a tesla).

We define this by saying that: If one weber of magnetic flux was spread evenly over a cross-sectional area of one square metre, then we have a flux density of one tesla. In other words the flux density depends upon the amount of magnetic flux lines and the area to which they are applied.

We use the following formula to express this:

$$\text{Flux density B (tesla)} = \frac{\text{(magnetic flux)}}{\text{(csa)}} = \frac{\Phi}{A}\ \text{(webers/m}^2\text{)}$$

Here is an example to help you understand what is going on.

The field pole of a motor has an area of 5 cm² and carries a flux of 80 μWb. What will be the flux density?

Remembering the formula:

$$\text{Flux density B (tesla)} = \frac{\text{(magnetic flux)}}{\text{(csa)}} = \frac{\Phi}{A}\ \text{(webers/m}^2\text{)}$$

We need to make allowance for the area being given in cm² and the flux in mWb.

Therefore:

$$\text{Flux density B (tesla)} = \frac{\text{(magnetic flux)}}{\text{(csa)}} = \frac{80 \times 10^{-6}\,\text{Wb}}{5 \times 10^{-4}\,\text{m}^2} = 0.16\text{T}$$

The electromagnet

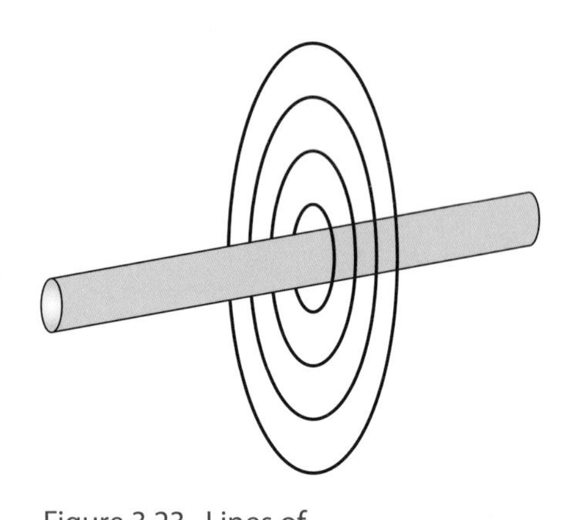

Figure 3.23 Lines of magnetic force set up around a conductor

An electromagnet is produced where there is an electric current flowing through a conductor, as a magnetic field is produced around the conductor. This magnetic field is proportional to the current being carried, since the larger the current, the greater the magnetic field.

An electromagnet is defined as being a temporary magnet because the magnetic field can only exist while there is a current flowing. If we have a typically shaped conductor such as a wire, the magnetic field looks like concentric circles and these are along the whole length of the conductor. However, the direction of the field depends on the direction of the current.

The Screw Rule

The direction of the magnetic field is traditionally determined using the 'Screw Rule'.

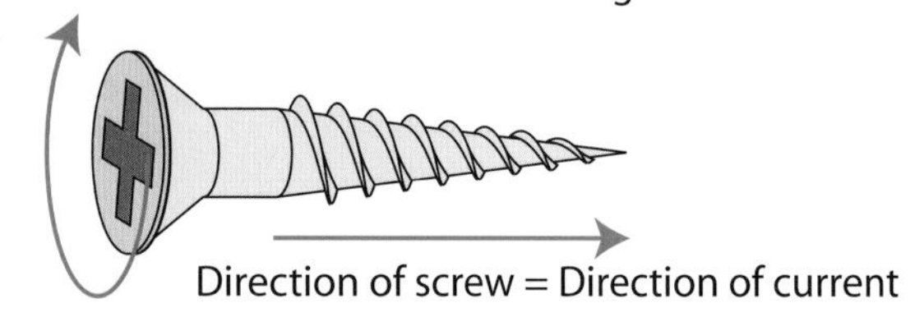

Figure 3.24 The screw rule

In the Screw Rule we think of a normal right-hand threaded screw. The movement of the tip represents the direction of the current through a straight conductor and the direction of rotation of the screw represents the corresponding direction of rotation of the magnetic field:

Rotation of screw = rotation of magnetic field

What is a relay?

To get the basic concept across, let's look at the world-famous one-way switch again. In any typical lighting circuit, if we want to put the light on in a room, the switch is operated by your finger. When we do this, we are closing the internal switch contact and the contact is mechanically held in place across the terminals. Consequently when we take our finger off the switch, it remains in position and the light stays on.

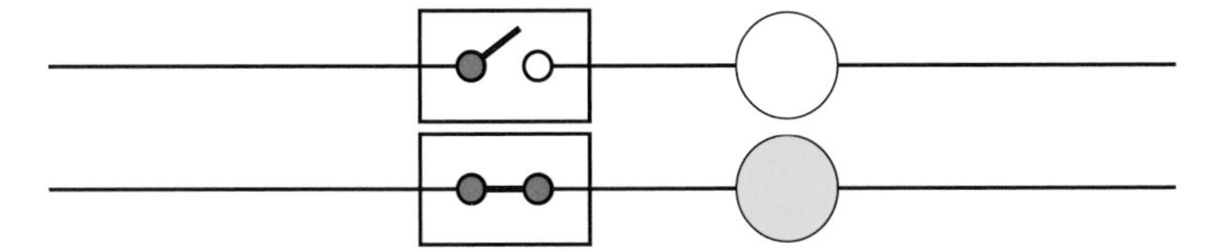

Figure 3.25 One-way switch – on position

But let's say that we don't want that sort of switch. Instead we want to control the light by using a relay. The concept is similar if you think of a relay as being an assembly that contains a one-way switch and a coil.

We'll draw the switch contact in a slightly different way this time, but the idea is exactly the same, i.e. electricity will pass from one terminal to the other when the contact is closed.

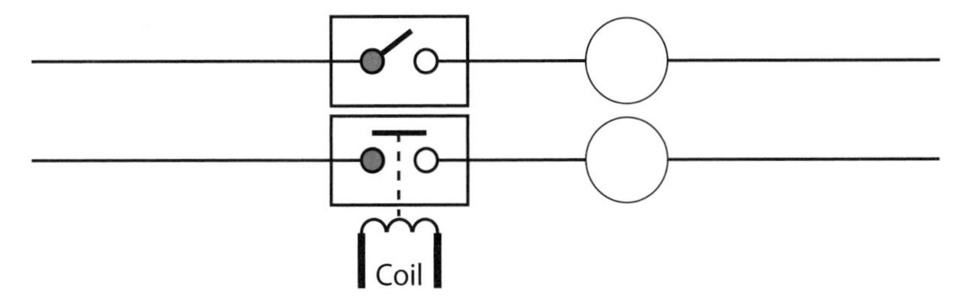

Figure 3.26 One-way switch and coil – off position

If we were now to energise the coil, the resulting magnetic field would pull the contact across the two terminals, thus closing the circuit and the light would come on. Except this time instead of the switch contact being held in place mechanically, it is being held in place by the magnetic field produced by the coil in the relay. It will only remain this way while the coil is energised.

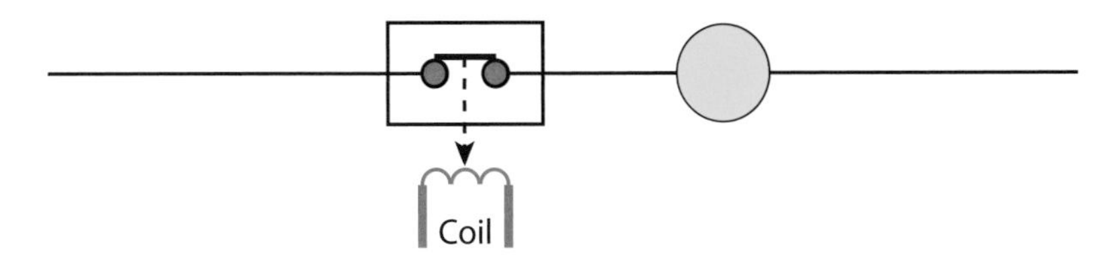

Figure 3.27 When the coil is energised, the switch is on

We describe this type of relay as having 'normally open' contacts, in that when the coil is de-energised, the contact opens and no electricity can pass through the relay.

It is possible to have a relay where the exact opposite function takes place, i.e. when the coil is energised the contact is pulled away from the terminals. In such a relay the supply would normally be passing through the closed contact, and operating the coil will break the circuit. We say that such a relay has 'normally closed' contacts.

Now that we understand the concept, we can accurately say that a relay is an electro-mechanical switch that uses an electromagnet to create a magnetic field to open or close one *or many sets* of contacts.

Applications

Relays can be used to:

- control a high–voltage circuit with a low–voltage signal, as in some types of modem
- control a high–current circuit with a low–current signal, as in all the lights in the hall of a leisure centre being controlled from a 5 A switch in reception
- control a mains–powered device from a low–voltage switch.

When choosing a relay there are several things to consider:

Coil voltage – This indicates how much voltage (230 V, 24 V) and what kind (a.c. or d.c.) must be applied to energise the coil. Make sure that the coil voltage matches the supply fed into it.

Contact ratings – This indicates how heavy a load the relay can control.

Contact arrangement – There are many kinds of switches, so there are many kinds of relays. The contact geometry indicates how many poles there are, and how they open and close.

For example, a changeover relay has one moving contact and two fixed contacts. One of these is normally closed when the relay is switched off, and the other is normally open. Energising the coil causes the normally open contact to close and the normally closed contact to open.

So far we have looked at the force on an object in a magnetic field produced by a permanent magnet or electromagnet, but what happens when we place a current-carrying conductor (a wire that has a current flowing through it) inside a magnetic field?

Let us have a look at this next.

Remember

Relays are powerful as they can be used to switch current between circuits or turn a circuit on and off

Did you know?

In a relay, because the coil-energising circuit is completely separate to the contact circuit(s), we can operate a high-voltage circuit with a very low, safe, switching circuit

Force on a current-carrying conductor in a magnetic field

Nearly all motors work on the basic principle that when a current-carrying conductor is placed in a magnetic field it experiences a force.

If we place the current carrying conductor into the magnetic field as shown in the next diagram, you can see that the current is going away from you, therefore the field is clockwise. This can be best remembered by imagining a corkscrew being twisted into a cork; you need to turn the corkscrew to the right (this symbolises the magnetic field) and the corkscrew is moving away from you, which symbolises the current flowing away from you.

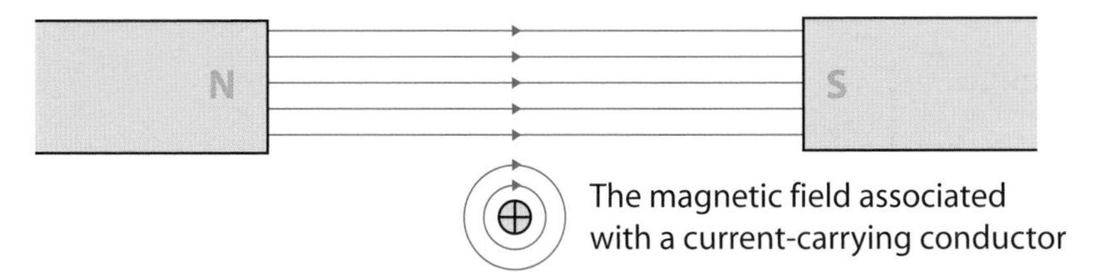

Figure 3.28 The magnetic field associated with two fixed poles

In Figure 3.29, note the following:

- the main field now becomes distorted
- the field is weaker below the conductor because the two fields are in opposition
- the field is stronger above the conductor because the two fields are in the same direction and aid each other. Consequently the force moves the conductor downwards.

If either the current through the conductor or the direction of the magnetic field between the poles is reversed, the force acting on the conductor tends to move it in the reverse direction (Figure 3.30).

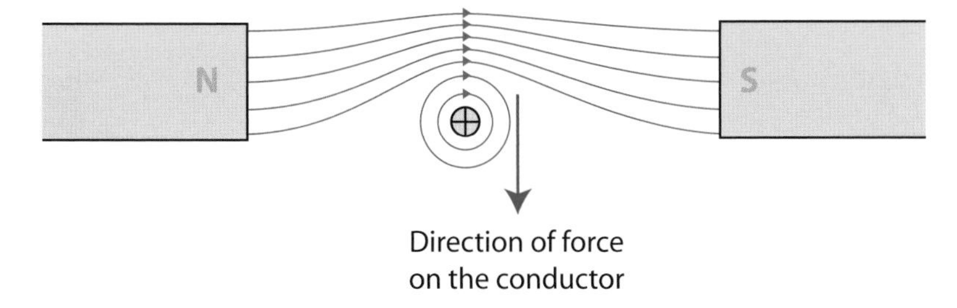

Figure 3.29 The magnetic field when the conductor is placed between the poles

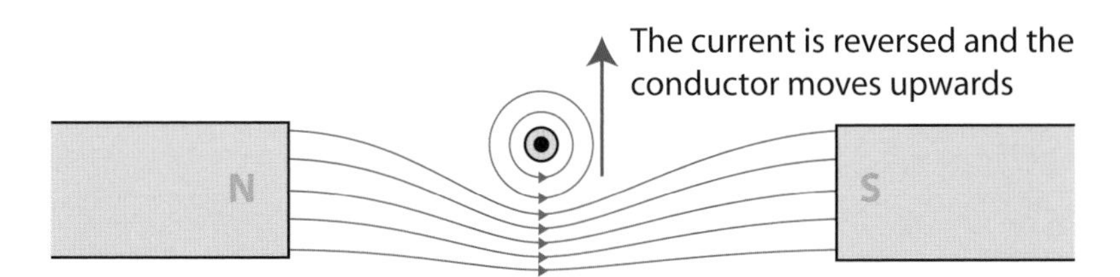

Figure 3.30 Reversing the current
⊕ indicates the current is flowing away from you;
⊙ indicates the current is flowing towards you

The direction in which a current carrying-conductor tends to move when it is placed in a magnetic field can be determined by **Fleming's left-hand (motor) rule**. This rule states that if the first finger, the second finger and the thumb of the left hand are held at right angles to each other as shown in Figure 3.31, then with the first finger pointing in the direction of the Field (N to S), and the second finger pointing in the direction of the current in the conductor, then the thumb will indicate the direction in which the conductor tends to move.

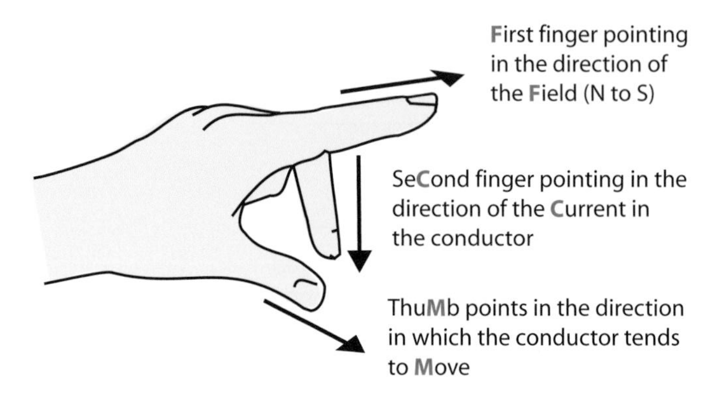

Figure 3.31 Fleming's left-hand rule

Calculating the force on a conductor

The force that moves the current-carrying conductor that is placed in a magnetic field depends on the strength of the magnetic flux density (B), the magnitude of the current flowing in the conductor (I), and the length of the conductor in the magnetic field (l).

The following equation expresses this relationship:

Force (F) = $B \times I \times l$

where B is in tesla, l is in metres, I is in amperes, F is in newtons.

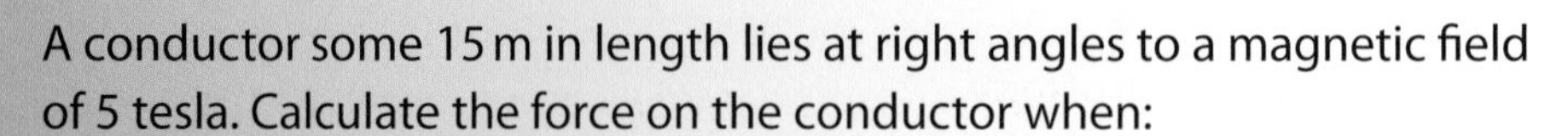

A conductor some 15 m in length lies at right angles to a magnetic field of 5 tesla. Calculate the force on the conductor when:

(a) 15 A flows in the coil

(b) 25 A flows in the coil

(c) 50 A flows in the coil.

Answer: Using the formula $F = B \times I \times l$:

$F = 5 \times 15 \times 15 = 1125\,N$

$F = 5 \times 25 \times 15 = 1875\,N$

$F = 5 \times 50 \times 15 = 3750\,N$.

Example 2

A conductor 0.25 m long situated in, and at right angles to, a magnetic field experiences a force of 5 N when a current through it is 50 A. Calculate the flux density.

Answer: Transpose the formula $F = B \times I \times l$ for (B):

$$B = \frac{F}{I \times l}$$

Substitute the known values into the equation:

$$B = \frac{5}{0.25 \times 50} = 0.4T$$

The solenoid

A **solenoid** is a long hollow cylinder around which we wind a uniform coil of wire. When a current is sent through the wire, a magnetic field is created inside the cylinder.

The solenoid usually has a length that is several times its diameter. The wire is closely wound around the outside of a long cylinder in the form of a helix with a small pitch. The magnetic field created inside the cylinder is quite uniform, especially far from the ends of the solenoid. The larger the ratio of the length to the diameter, the more uniform the field near the middle.

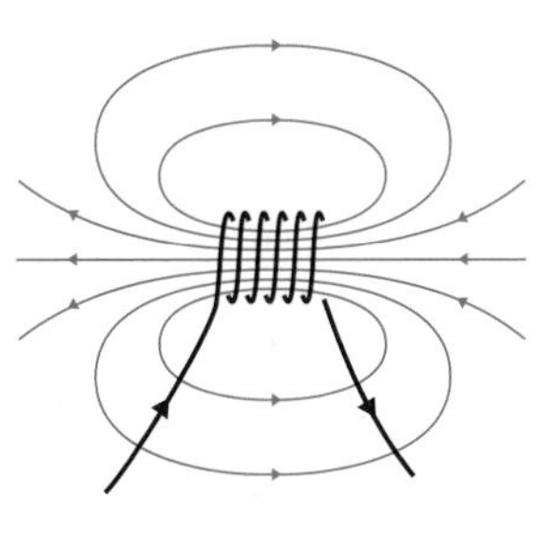

Figure 3.32 Magnetic field of solenoid

Essentially, the magnetic field produced by a solenoid is similar to that of a bar magnet.

If an iron rod were placed partly inside a solenoid and the current turned on, the rod will be drawn into the solenoid by the resulting magnetic field. This motion can be used to move a lever or operate a latch to open a door and is most commonly seen in use inside a door bell.

It is important to note that with a solenoid, we can use an electric switch to energise the solenoid and therefore produce a mechanical action at a remote location, e.g. the doorbell.

Now, rather than having a current-carrying conductor placed in a magnetic field causing it to move, what if we took a conductor with no current flowing in it and instead moved the conductor through the magnetic field?

Instead of having the current causing the motion, we will now have the motion causing the current. So now we have a magnetic field responsible for the flow of an electric current. We call this situation electromagnetic induction.

Electromagnetic induction

Stated simply: if a conductor is moved through a magnetic field, provided there is a closed circuit, then a current will flow through it.

We know that we need a 'force' to drive electrons along a conductor, and we can say that an e.m.f. must be producing the current. In this situation we are causing an e.m.f. and this is known as the induced e.m.f. and it will have the same direction as the flowing current.

If we were to pass an electric current through a conductor this would generate a uniform magnetic field around the conductor and at right angles to the conductor. The strength of this magnetic field is directly proportional to the current flowing in the conductor.

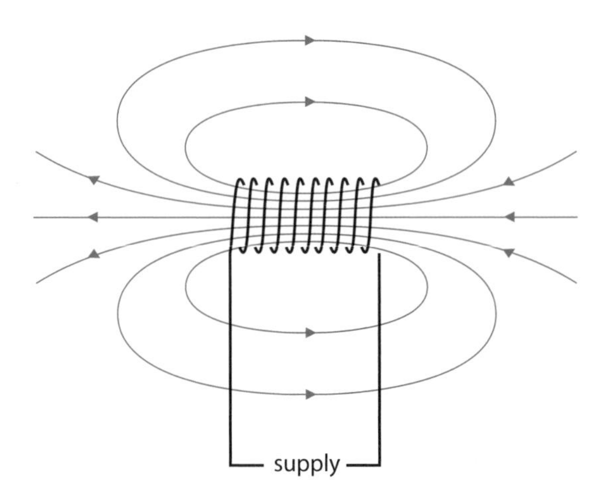

Figure 3.33 A solenoid

The strength of this magnetic field can be further increased by coiling the conductor to form a solenoid.

If the coil were connected to a d.c. supply the only resistance to the current flow would be the resistance of the conductor itself; however, if the coil is connected to an a.c. supply the situation must be looked at differently.

Any change in the magnetic environment of a coil of wire will cause a voltage (e.m.f.) to be 'induced' in the coil. No matter how the change is produced, the voltage will be generated. The change could be produced by changing the magnetic field strength, moving a magnet toward or away from the coil, moving the coil into or out of the magnetic field, rotating the coil relative to the magnet, etc.

The alternating current creates the effect of a continuously changing magnetic field inside the coil; this effect reacts with the flow of current and opposes it.

Inductance is typified by the behaviour of a coil of wire in resisting any change of electric current through the coil. The SI unit of inductance is known as the Henry, symbol H. The symbol for inductance is L.

The unit of inductance is given to be the rate of change of current in a circuit of 1 amp per second, which produces an induced electromotive force of 1 volt.

Values of inductors range from about 0.1 microhenry, written as 0.1 μH, to 10 henries (H).

The inductance of a coil can be altered by:

- changing the number of turns of wire on the coil
- changing the material composition of the core (air, iron or steel)
- changing the diameter of the coil
- changing the material composition of the coil.

Thinking of the solenoid we have just been talking about, you realise that an e.m.f. is induced only when we have a changing situation. What could change in the solenoid setup? Well, the following could change:

- the number of turns in the coil (N)
- the rate of change of current flowing in the coil ($\frac{\Delta I}{\Delta t}$) – how quickly the current alternates in the coil
- the rate of change of magnetic flux ($\frac{\Delta \Phi}{\Delta t}$) – how quickly the magnetic flux changes.

In the 19th century a rather clever scientist named Michael Faraday spent a lot of time looking at magnetic induction. He came up with a law that tells how much e.m.f. is induced when a conductor is moving in a magnetic field. We will have a look at this and try to make it as simple as possible.

Faraday found that, for a conductor, the induced e.m.f. is given by:

$$\text{e.m.f.} = -(\frac{\Delta \Phi}{\Delta t})$$

So we can find the induced e.m.f. by knowing the rate of change of flux. This is simply the same as how quickly the conductor cuts the lines of flux.

So, for a coil of N turns:

$$\text{e.m.f.} = -N(\frac{\Delta \Phi}{\Delta t})$$

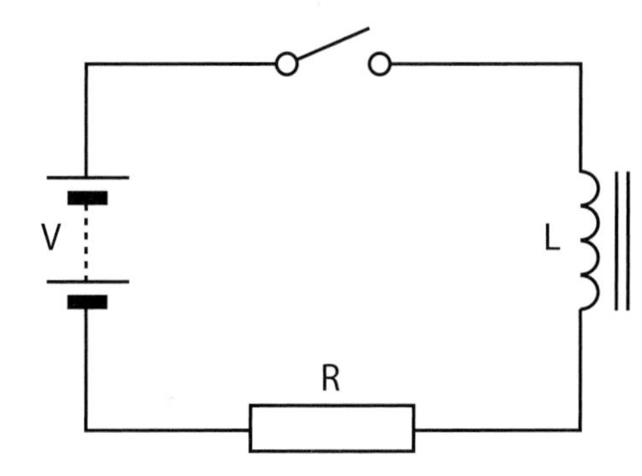

So we can find the induced e.m.f. by knowing the number of turns and rate of change of flux and

$$\text{e.m.f.} = -L(\frac{\Delta I}{\Delta t})$$

So we can find the induced e.m.f. by knowing the inductance and rate of change of current.

These equations are true for both self and mutual inductance.

Note: In all these equations there is a negative (–) sign. This is because any induced e.m.f. will always be in opposition to the changes that created it.

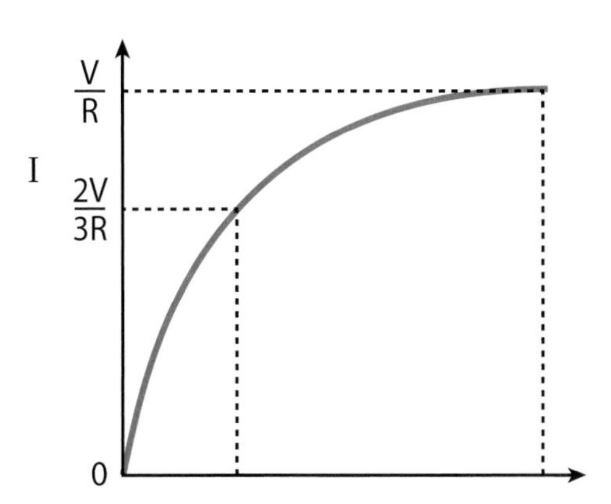

Figure 3.34 Effects of inductance

When a number of inductors need to be connected together to form an equivalent inductance they follow the same rules as for resistors:

- to increase inductance, connect inductors in series
- to decrease inductance and increase the current rating, connect inductors in parallel.

To show the effects of inductance we can plot a graph of current against time.

It takes time to build up to maximum current; however, this is important when connected to an a.c. supply because the rate of change of current with time can be calculated and adjusted so that a smoothing effect can be produced in the a.c.. If the coil is suddenly switched off, the magnetic field collapses and a high voltage is induced across the circuit. This effect is used for starting fluorescent tube circuits.

Using the effects of magnetism

There are many applications based on the effects of magnetism. Following our previous pages, let us start by exploring the movement of a conductor through a magnetic field as it applies to the single-phase a.c. generator (or alternator).

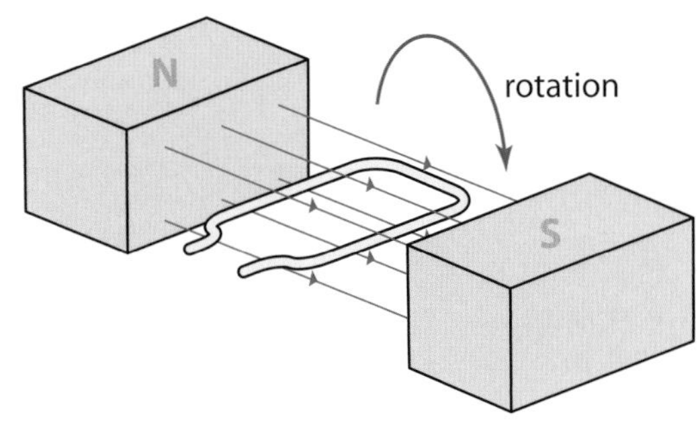

Figure 3.35

If we were to form our conductor into a loop and then mount it so that it could rotate between two permanent magnets, it would look like Figure 3.35.

In this position, it could be said that the loop is lying in between the lines of magnetism (magnetic flux). If we were to look at a side view of this arrangement towards the ends of our loop, Figure 3.36 shows what it would look like.

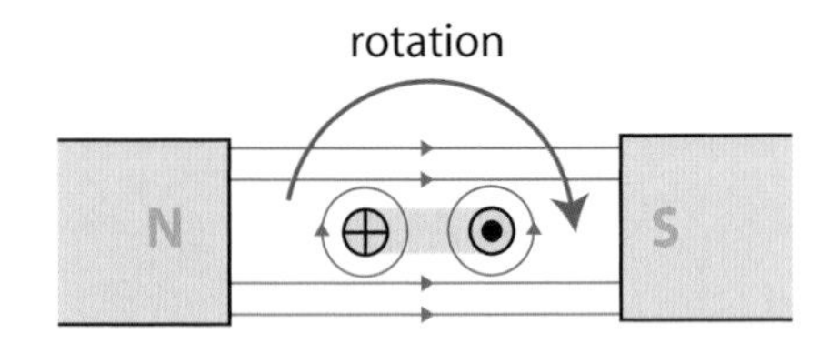

Figure 3.36

Therefore, as the loop is not interfering with any lines of flux, we say that it is not 'cutting' any lines of flux. However, as we slowly start to rotate the loop in the direction indicated, it will start to pass through the lines of flux. When this happens, we say that we are cutting through the lines of flux and as we do so, we start to induce an e.m.f.

The maximum number of lines that are cut through will occur when the loop has moved through 90° and the maximum induced e.m.f. in this direction will therefore occur at this point. Keep rotating the loop and the number will once again reduce to zero as we are again lying between the lines of flux. The loop has now completed what is known as the positive half cycle.

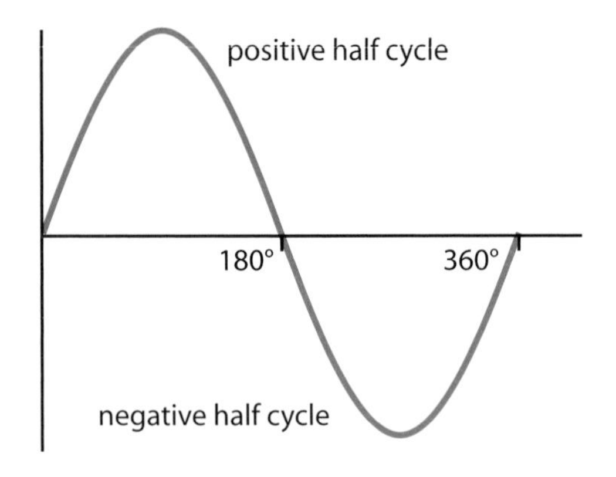

Figure 3.37 Sine wave

Repeat the process and an e.m.f. will be induced in the opposite direction (the negative half cycle) until the loop returns to its original starting position.

If we were to plot this full 360° revolution (cycle) of the loop as a graph, we would see the e.m.f. induced in the loop as in Figure 3.37. This shape is known as a **sine wave**.

As we can see, the sine wave shows the e.m.f. rising from zero as we start to cut through more lines of flux, to its maximum after 90° of rotation. This is known as the **peak value**.

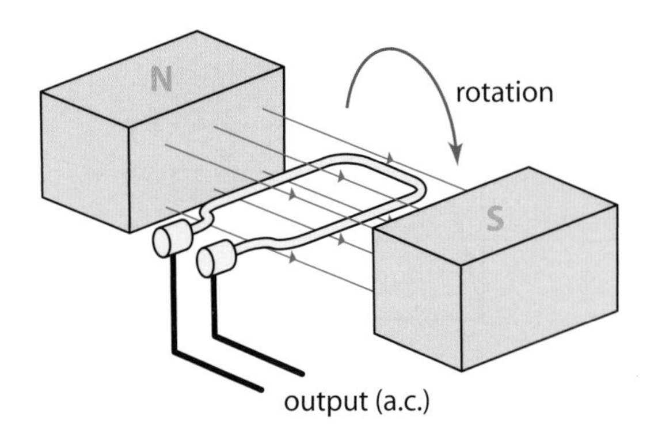

Figure 3.38 Ends of the loop are connected via slip rings

However, after completing 180° (half a rotation or cycle), the e.m.f. passes through zero and then changes direction. As we said earlier, we sometimes refer to these as being the positive and negative half cycles.

The opposite directions of the induced e.m.f. will still drive a current through a conductor, but that current will alternate as the loop rotates through the magnetic field. It will be flowing in the same direction as the induced e.m.f. Consequently, the current will rise and fall in the same way as the induced e.m.f. and when this happens, we say that they are in phase with each other. To access this a.c. output, the ends of the loop are connected via slip rings as shown in Figure 3.38.

We say that such a device is called an a.c. generator or alternator and is producing an alternating current (**a.c.**). The number of complete revolutions (cycles) that occur each second is known as the frequency, measured in hertz (Hz) and given the symbol f. The frequency of the supply in this country is 50 Hz.

Originally **d.c.** was used as the main method of transmitting electricity. But as the technology developed, a.c. became the preferred choice for two main reasons:

- **Reason 1**:
  Transformers make it easy to adjust a.c. to a higher or lower voltage very efficiently. This is useful, because to transmit at high voltage reduces current and power loss and therefore allows smaller cable sizes and a reduction in costs. Transformers do not work for d.c.. Adjusting d.c. voltages requires converting the d.c. to a.c., adjusting the resulting a.c. voltage with a transformer, and then converting the adjusted a.c. voltage to a corresponding d.c. voltage. Clearly, adjusting d.c. voltage is more complicated, and not surprisingly more expensive, than adjusting a.c. voltages.
- **Reason 2**:
  Good a.c. motors (and generators) are easier and cheaper to build than good d.c. motors. Although motors are available for either a.c. or d.c., the structure and characteristics of a.c. and d.c. motors are quite different. a.c. makes it easy to produce a magnetic field whose direction rotates rapidly in space. Any electric conductor placed within the rotating magnetic field rotates with the field. Consequently, a metal armature rotates with the rotating magnetic field with little slippage and, through a shaft attached to the armature, can deliver mechanical power to a mechanical load such as a fan or a water pump.

  Called an a.c. induction motor, it is a reasonably simple means of converting electric power to mechanical power. However, d.c. motors rely on a complex mechanical system of brushes and commutator switches. The mechanical complexity of d.c. motors, consequently, not only makes them more expensive to manufacture than a.c. motors, but also more expensive to maintain.

The distribution of electricity

In this section, we are going to take a basic look at how we get electricity from the power station to the customer. This involves several processes, from generation to transmission and distribution. The following areas will be looked at:

- generation
- transmission
- distribution
- three-phase systems
- delta connections
- star connections
- neutral currents
- load balancing.

Generation

Electricity is generated in power stations. The shaft of a three-phase alternator (a.c. generator) is turned, in the majority of cases, by using steam. Most electricity in the UK is produced by this method. Water is heated until it becomes high-pressure steam, which is forced onto the vanes of a steam turbine, which in turn rotates the alternator. A variety of energy sources can be used to heat the water in the first place. The more popular ones are coal, gas, oil and nuclear power.

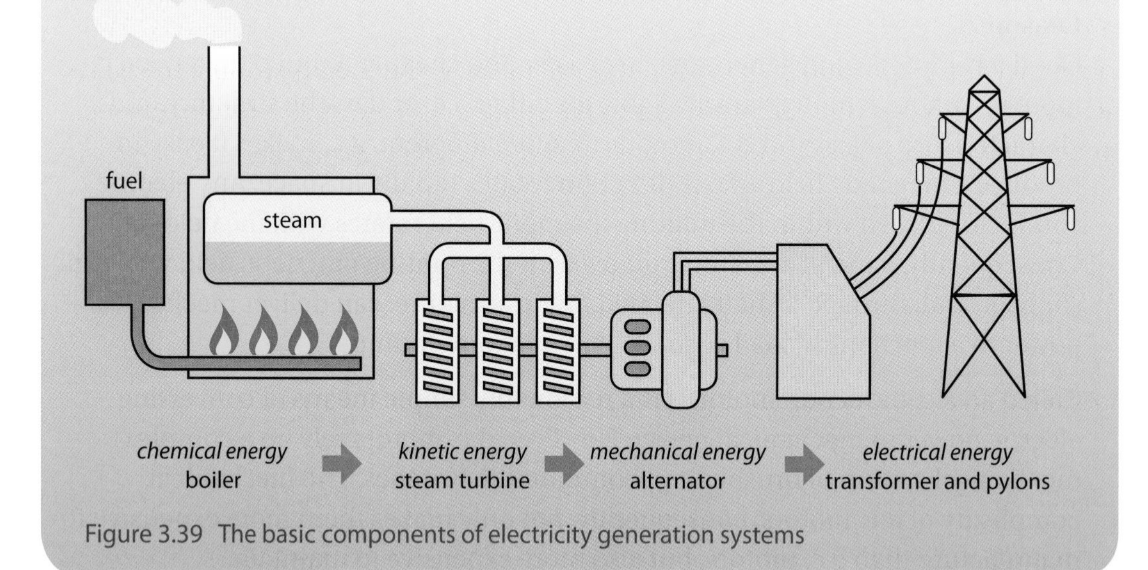

Figure 3.39 The basic components of electricity generation systems

Transmission

In Figure 3.39 electricity goes from the alternator to a transformer. This is because the output of most alternators is about 25,000 V (25 kV) and it must be transformed to:

- 400 kV, 275 kV for the super grid
- 132 kV for the original grid
- 66 kV and 33 kV for secondary transmission
- 11 kV for high-voltage distribution
- 415/400 V for commercial consumer supplies
- 240/230 V for domestic consumer supplies.

Electricity is transmitted at very high voltage values in order to compensate for the power losses that occur in the power lines. Transmission at low voltage values would necessitate the installation of very large cables and switchgear indeed. At this point, the electricity is fed into the National Grid system.

The National Grid is a network of nearly 5,000 miles of overhead and underground power lines that link power stations together and are interconnected throughout the country. The concept is that should a fault develop in any one of the contributing power stations or transmission lines, then electricity can be requested from another station on the system.

Electricity is transmitted around the grid, mainly via steel-cored aluminium conductors, which are suspended from steel pylons.

We do this for three main reasons:

1. The cost of installing cables underground is excessive.
2. Air is a very cheap and readily available insulator.
3. Air also acts as a coolant for the heat being generated in the conductors.

Electricity is then 'taken' from the National Grid via a series of appropriately located sub-stations. These will eventually transform the grid supply back down to 11 kV and then distribute electricity at this level to a series of local sub-stations. It is their job to take the 11 kV supply, transform it down to 400V and then distribute this via a network of underground radial circuits to the customer. However, in rural areas we sometimes see this distribution take place using overhead lines.

It is also at this point that we see the introduction of the neutral conductor. This is normally done by connecting the secondary winding of the transformer in star and then connecting the star point to earth via an earth electrode beneath the sub-station.

Did you know?

Sub-stations are dotted throughout our cities. These small brick buildings are normally connected together on a ring circuit basis

Distribution to the customer

Once the electricity has left the local sub-station, it will eventually arrive at the customer. This is called the main intake position.

There are many different sizes of installation, but generally speaking we will find certain items at every main intake position. These items, which belong to the supply company, are:

- a sealed overcurrent device that protects the supply company's cable
- an energy metering system to determine the customer's electricity usage.

It is after this point that we say we have reached the consumer's installation.

The consumer's installation must be controlled by a main switch, which must be located as close as possible to the supply company equipment and be capable of isolating all phase conductors. In the average domestic installation, this device is merged with the means of distributing and protecting the final circuits in what we know as the consumer unit.

Remember

Electricity is the movement of electrons along a conductor. The rate of electron flow, the current, is then maintained by a force known as the e.m.f.

Three-phase supplies

So far, everything that we have looked at has revolved around a single-phase circuit. In such a circuit, we normally use two conductors, where one delivers current and one returns it. You could logically assume that we would therefore need six conductors for a three-phase system, with two being used per phase. However, in reality we only use three or four conductors, depending upon the type of connection that is being used.

We call these connections either **star** or **delta**.

Remember that current flows along one conductor and returns along another called the neutral.

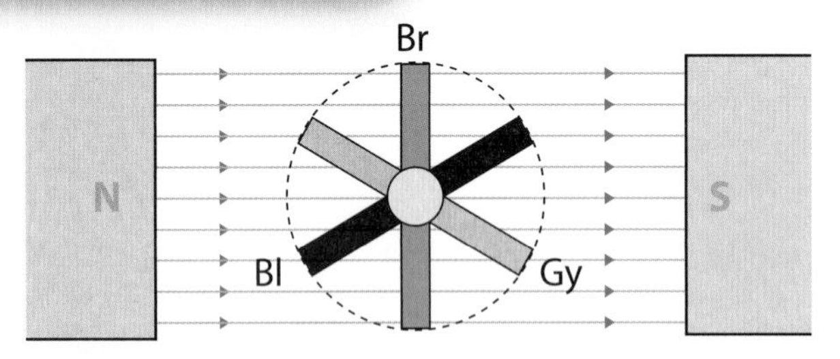

Figure 3.40

But what if there were no neutral? And what exactly is the neutral conductor for? Keeping things simple, when we generate an e.m.f., we do so by spinning a loop of wire inside a magnetic field. To get three phases, we just spin three loops inside the magnetic field. Each loop will be mounted on the same rotating shaft, but they'll be 120° apart.

In the diagram, each loop will create an identical sinusoidal waveform, or in other words, three identical voltages, each 120° apart.

Whether or not we need a neutral will now depend upon the load. If we accept that the e.m.f. being generated by each loop pushes current down the conductors (lines), then we would find that where we have a balanced three-phase system, i.e. one where the current in each of the phases (lines) is the same, then by phasor addition, we would find that the resultant current is zero.

If the current is zero then we do not need a neutral, because the neutral is used to carry the current in an out-of-balance system. Let us now look at the different types of connection.

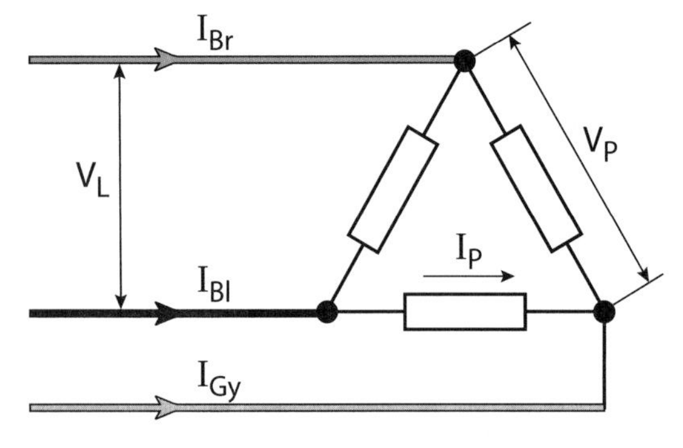

Figure 3.41 Delta connection

Delta connection

We tend to use the delta connection when we have a balanced load. This is because there is no need for a neutral connection and therefore only three wires are needed. We tend to find that this configuration is used for power transmission from power stations or to connect the windings of a three-phase motor.

In Figure 3.41, we have shown a three-phase load, which has been delta connected. You can see that each leg of the load is connected across two of the lines, e.g. Br–Gy, Gy–Bl and Bl–Br. We refer to the connection between phases as being the line voltage and have shown this on the drawing as V_L.

Equally, if each line voltage is pushing current along, we refer to these currents as being line currents, which are represented on the drawing as I_{Br}, I_{Gy} or I_{Bl}. These line currents are calculated as being the phasor sum of two phase currents, which are shown on the drawing as I_P and represent the current in each leg of the load. Similarly, the voltage across each leg of the load is referred to as the phase voltage (V_P).

In a delta connected balanced three-phase load, we are then able to state the following formulae:

$V_L = V_P \qquad I_L = \sqrt{3} \times I_P$

Note that a load connected in delta would draw three times the line current and consequently three times as much power as the same load connected in star. For this reason, induction motors are sometimes connected in **star–delta**.

This means they start off in a star connection (with a reduced starting current) and are then switched to delta. In doing so we reduce the heat that would otherwise be generated in the windings.

Star connection

Although we can have a balanced load connected in star (three-wire), we tend to use the star connection when we have an unbalanced load, i.e. one where the current in each of the phases is different. In this circumstance, one end of each of the three star connected loops is connected to a central point and it is then from this point that we take our neutral connection, which in turn is normally connected to Earth. This is the three-phase four-wire system.

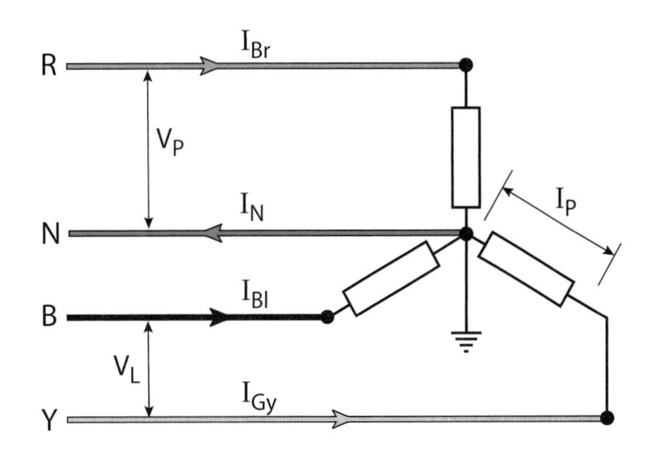

Figure 3.42 Star connection

Another advantage of the star connected system is that it allows us to have two voltages – one when we connect between any two phases (400V) and another when we connect between any phase and neutral (230V). You should note that we will also have 230V between any phase and earth.

In Figure 3.42, we have shown a three-phase load that has been star connected.

As with delta, we refer to the connection made between phases as the line voltage and have shown this on the drawing as V_L. However, unlike delta, the **phase voltage** exists between any phase conductor and the neutral conductor and we have shown this as V_P. Our line currents have been represented by I_{Br}, I_{Gy} and I_{Bl} with the phase currents being represented by I_P.

In a star connected load, the line currents and phase currents are the same, but the line voltage (400V) is greater than the phase voltage (230V).

In a star connected load, we are therefore able to state the following formulae:

$I_L = I_P \qquad V_L = \sqrt{3} \times V_P$

Using the star connected load we have access to a 230V supply, which we use in most domestic and low load situations.

Example

A three-phase star connected supply feeds a delta-connected load as shown in the diagram below.

If the star-connected phase voltage is 230 V and the phase current is 20 A, calculate the following:

- The line voltages and line currents in the star connection
- The line and phase voltages and currents in the delta connection.

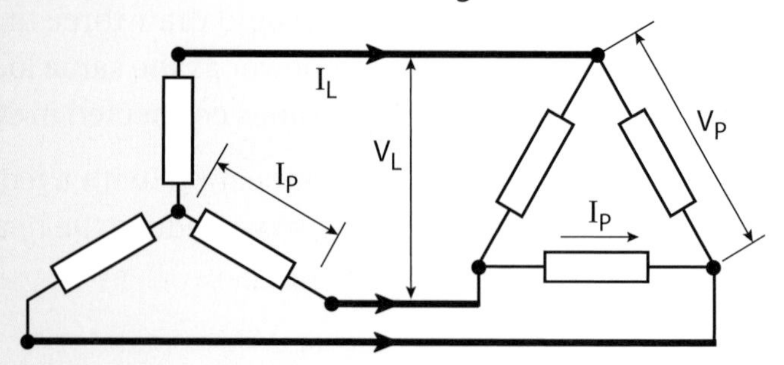

The star connection

In a star system the line current (I_L) is equal to the phase current (I_P). Therefore if we have been given I_P as 20 A, then I_L must also be 20 A.

We find line voltage in a star connection using the formula:

$$V_L = \sqrt{3} \times V_P$$

$\sqrt{3}$, the square root of 3, is a constant having the value 1.732. Therefore if we substitute our values, we get:

$$V_L = 1.732 \times 230 = 398\text{ V}$$

The delta connection

In a delta system, the line current (I_L) is 1.732 times greater than the phase current (I_P). We calculate this using the formula: $I_L = \sqrt{3} \times I_P$

However, we know that I_L is 20 A, so if transpose our formula and substitute our values, we get:

$$I_P = \frac{I_L}{\sqrt{3}} = \frac{20}{1.732} = 11.5\text{ A}$$

We know that for a delta connection line voltage and phase voltage have the same values.

Therefore: $V_L = V_P =$ **398 V**

Example

Three identical loads of 30 Ω resistance are connected to a 400 V three-phase supply. Calculate the phase and line currents if the loads were connected:

(a) in star (b) in delta.

Star connection

First we need to establish the phase voltage.
If $V_L = \sqrt{3} \times V_P$ and $\sqrt{3} = 1.732$ then by transposition:

$$V_P = \frac{V_L}{\sqrt{3}} = \frac{400}{1.732} = 230.9\text{ V}$$

Using Ohm's Law: $I_P = \frac{V_P}{Z} = \frac{230.9}{30} = 7.7\text{ A}$

However, in a star connected load, $I_P = I_L$, therefore I_L will also = 7.7 A.

Delta connection

In a delta connection $V_L = V_P$ and therefore we know the phase voltage will be 400 V.

Using Ohm's Law: $I_P = \frac{V_P}{Z} = \frac{400}{30} = 13.33\text{ A}$

However, the line current:

$$I_L = \sqrt{3} \times I_P = 1.732 \times 8 = 23.09\text{ A}$$

As can be seen from this example, the current drawn from a delta connected load (13.33 A) is three times that of a star connected load (7.7 A).

Neutral currents

As we have already discussed, where we have a balanced load, we can have a three-phase system with three wires. However, in truth it is more likely that we will find an unbalanced system and will therefore need to use a three-phase four-wire system.

In such a system, we are saying that each line (Br, Gy, Bl) will have an unequal load and therefore the current in each line can be different. It therefore becomes the job of the neutral conductor to carry the out of balance current. If we used **Kirchhoff's Law** in this situation, we would find that the current in the neutral is normally found by the phasor addition of the currents in the three lines.

Example

For a three-phase four-wire system, the line currents are found to be $I_R = 30\,A$ and in phase with V_{Br}, $I_{Gy} = 20\,A$ and leading V_{Gy} by 20° and $I_{Bl} = 25\,A$ and lagging V_{Bl} by 10°. Calculate the current in the neutral by phasor addition.

The phasor diagram for this example has been provided. However, you should note that in order to establish the current in the neutral (I_N), you would need to draw two parallelograms. The first should represent the resultant currents I_{Br} and I_{Gy}. The second, I_N, should be drawn between this resultant and the current in I_{Bl}.

When this has been done to scale, we should find a current in the neutral of 18 A – look at Figure 3.43.

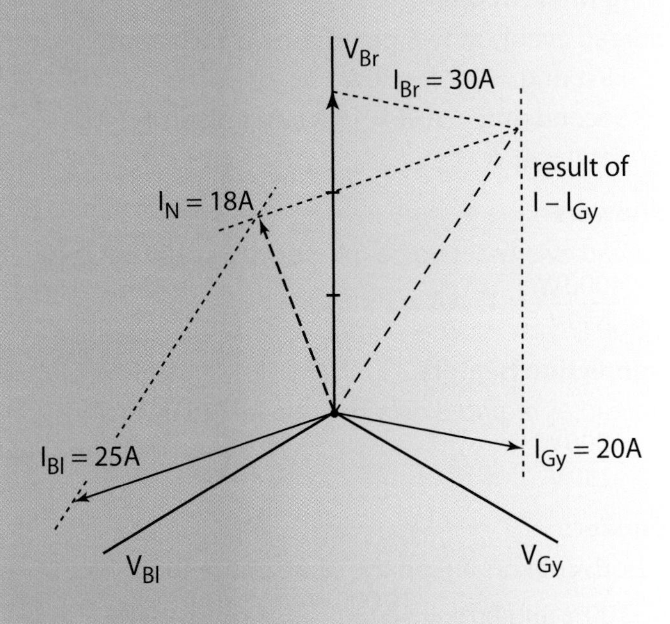

Figure 3.43

Load balancing

The Regional Electricity Companies also require load balancing as a condition of their electricity supply, because it is important to try to achieve balanced currents in the mains distribution system.

In order to design a three-phase, four-wire electrical installation for both efficiency and economy, it needs to be subdivided into load categories. By doing this, the maximum demand can be assessed and items of equipment can be spread over all three phases of the supply to achieve a balanced system. This section looks at load-balancing on three-phase systems. The designer needs to make a careful assessment of the various installed loads, which in turn leads to the proper sizing of the main cable and associated switchgear.

Standard circuit arrangements exist for many final circuits operating at 230 volts. For example, a ring final circuit is rated at 30 A, a lighting circuit at 5 A, and on a cooking appliance is rated at 30–45 A. Where more than one standard circuit arrangement is present, such as three ring final circuits and/or two cooking appliances, then a diversity allowance can be applied.

Once the designer has made these allowances for diversity, the single-phase loads can be evenly spread over all three phases of the supply so that each phase takes approximately the same amount of current. If this is done carefully, minimum current will flow along the neutral conductor; the sizes of cables and switchgear can be kept to a minimum, thus reducing costs, and the system is therefore said to be reasonably well balanced.

Example

A small guesthouse with 10 bedrooms is supplied with a 400/230 volt, three-phase four-wire supply. It has the following installed loads:

- 20 × filament lighting points each rated at 100 watts
- 6 × ring final circuits supplying 13 socket outlets
- 6 × 4 kW showers (instantaneous type)
- 3 × 3 kW immersion heaters
- 2 × 10 kW cookers.

Apply diversity as required by the *On-site Guide* Table 1B and 'spread' the loads evenly over the three-phase supply to produce the most effective load-balanced situation.

Filament lamps

$$\frac{2000\,\text{W}}{230\,\text{V}} = 8.69\,\text{A @ } 75\%$$
$$= 6.5\,\text{A (after diversity)}$$

Ring final circuits

Spread evenly at two per phase on each phase:

First ring = 100% = 30 A

Second ring = 50% = 15 A (after diversity)

Total = 45 A

Showers

Spread evenly at two per phase both at 100 per cent:

$$\frac{4000\,\text{W}}{230\,\text{V}} = 17.4\,\text{A} \times 2 = 34.8\,\text{A}$$

Immersion heaters

Spread as required over the phases (no diversity):

$$\frac{3000\,\text{W}}{230\,\text{V}} = 13\,\text{A}$$

Cookers

If both cookers are on the same phase then:

$$100\% \text{ and } 80\% = \frac{10\,000\,\text{W}}{230\,\text{V}} = 43\,\text{A for } 100\%$$
$$= 35\,\text{A for } 80\% \text{ (after diversity)}$$

Or if both on different phases both are at 100% (this is what will be used).

Example (continued)

Based on these calculations, a suggested load balancing is shown below. Note the immersion heaters have all been allocated to the Grey phase.

	Red phase	Yellow phase	Blue phase
Ring circuits	45 A	45 A	45 A
Shower	34.8 A	34.8 A	34.8 A
Immersion			13 A
			13 A
			13 A
Cooker	43 A	43 A	
Lighting			6.5 A
Totals	122.8 A	122.8 A	125.3 A

Table of value

Also, should one phase be 'lost', only a portion of the lighting will fail. If this is the case then care must be taken, as there could be three-phase voltage values present in multi-gang switches. It is also advisable to position distribution boards in large factories and commercial premises as near to the load centre as possible. This will help reduce voltage drop and make the installation more cost effective.

The load balancing shown above provides satisfactory balancing over three phases, but it may be recommended that if discharge lighting is to be used, then it should be spread over all three phases because of the stroboscopic effect.

For a second example see the Appendix.

Knowledge check

1. The length of a football pitch is 1850 metres. What is this expressed in kilometres?
2. Atoms consist of several smaller particles. Which one of these particles possesses a negative charge?
3. Two resistors, of value 6.2 Ω and 3.8 Ω are connected in series with a 12-volt battery. What is the total resistance and total current flowing?
4. An angle less than 90° is known as what?
5. The output of most alternators is approximately what?
6. A single-phase transformer with 2000 primary turns and 500 secondary turns is fed from a 230 volt a.c. supply. What will the secondary voltage and volts per turn in the secondary be?
7. What is the highest value obtained in the cycle known as?
8. To calculate the average value of a sinusoidal voltage or current we must multiply the maximum value by what?
9. What is the correct formula for inductive reactance?
10. What is the correct formula for calculating impedance in a circuit containing resistance and inductance in series?
11. In a delta connected circuit the line voltage is equal to what?
12. Power factor can be expressed as what?

Craft theory

Overview

In this chapter you will learn about the wide range of practical skills required to become a qualified electrician. The work tasks that a competent electrician will undertake are many and varied. Consequently you need to have practical skills that will allow you to understand different wiring systems, enclosures and equipment and enable you to install them safely, correctly and in accordance with industry guidelines and Regulations. You need to inspect and test the work you have carried out and, if necessary, locate and correct any faults that may occur.

The purpose of this unit, therefore, is to look at those practical skills that electricians should know. This chapter will cover:

- **Tools and equipment**
- **Fixings**
- **Conductors and insulators**
- **Installation techniques**
- **Cable types**
- **MICC cable**
- **FP 200 cable**
- **Steel and PVC conduit installations**
- **Trunking**
- **Cable tray**
- **Switching of lighting circuits**
- **Rings, radials and spurs**

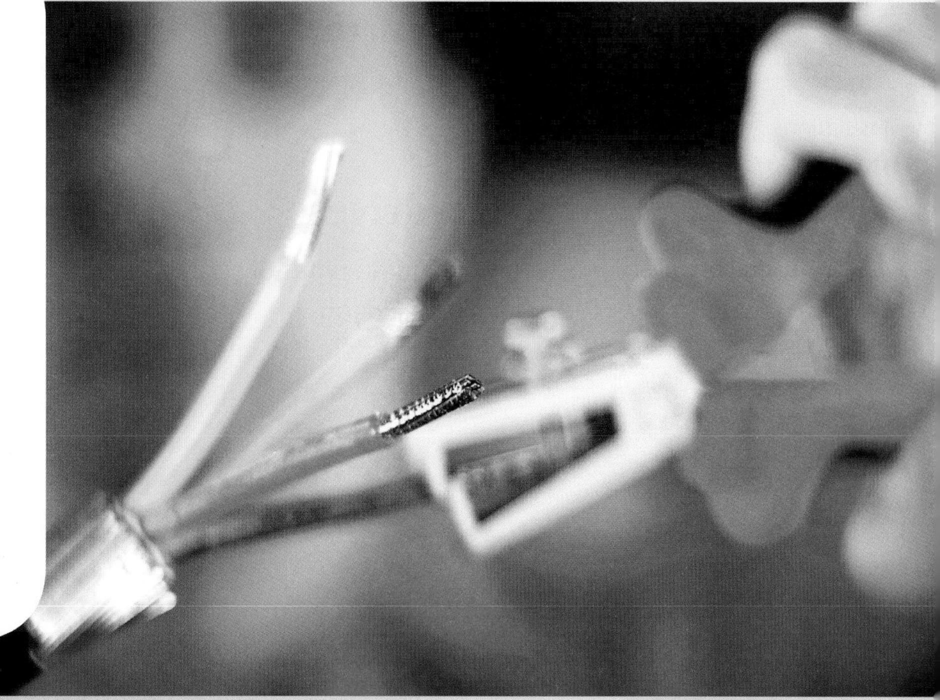

Tools and equipment

At one time or another, all of us have probably started on something and then been frustrated at not having the right tool for the job: 'If I only had...'. Used properly and safely good-quality tools let you work faster and more efficiently, so you need to look after them.

These days in the electrical trade we're using materials that can last for more than 50 years, but only so long as the system is well designed, complies with appropriate safety regulations and is properly installed. As part of an installation team, you could find yourself having to do jobs usually done by other trades, such as building and brickwork. To be a good electrician, you'll need to get to grips with a number of different skills and learn how to use a wide range of tools.

It's not possible to list all the tools you might come across, so this section only describes the most common ones.

Did you know?

You will use several types of pliers, and some are shown here. Other pliers that are useful for special situations include bent-, snipe- and needle-nosed pliers. Round-nosed pliers are used to form loops in wire prepared for termination to an electrical fitting

Pliers and cutters

The main difference between electricians' pliers and any other sort is that, for obvious reasons, they have insulated handles. They have flat serrated jaws for gripping and bending and oval serrated jaws for gripping pipes and cylindrical objects.

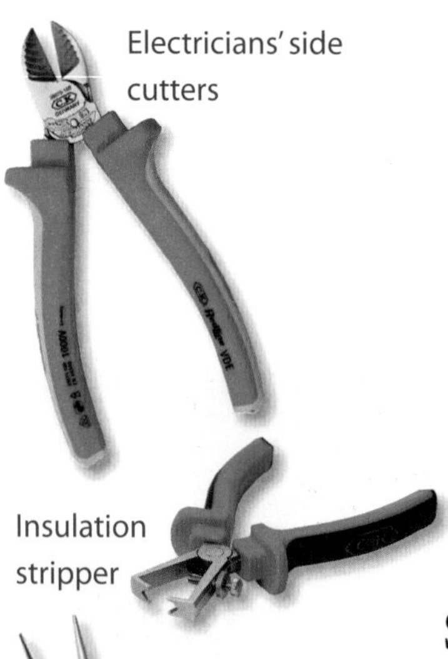

Electricians' side cutters

Insulation stripper

Long-nose pliers

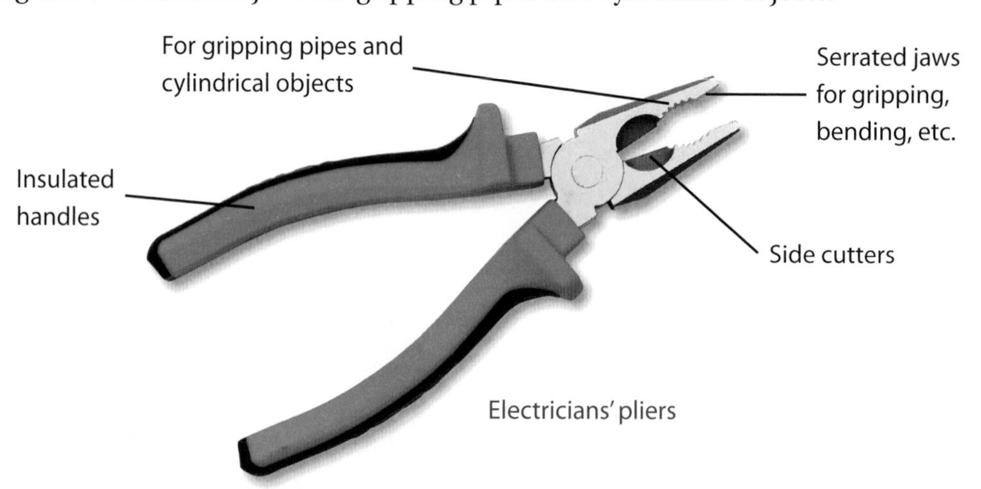

Electricians' pliers

Screwdrivers

The three main types of screwdriver tip are shown here. There are lots of other types for specialised jobs.

Flared slotted
For general use

Parallel slotted
Head size same as shaft

Cross-head (Phillips, Pozidrive)
Gives better 'purchase'

Hammers

Hammers are used to do three main jobs:

- to drive a fixing (e.g. a nail)
- to provide impact on another tool (e.g. a cold chisel)
- to alter the shape of a work-piece (e.g. bend a piece of metal).

It's important to use the right hammer for the job, and to make sure that the shaft (handle) is firmly fitted into the head. Steel hammers are one-piece, so there's usually no problem. With wooden handles, which have wedges to tighten the shaft in the eye, always check that these aren't loose or missing and that the shaft is tight. These are the three types of hammer you're most likely to use:

Ball-pein

Heavier ones are used with punches, cold chisels etc. The rounded end (ball) is used to shape metal and rivets. Its weight can be up to 2 kg.

Ball-pein hammer

Cross-pein

These are often used by carpenters. The tapered end (pein) is used to start small nails held in fingers. Its weight can be up to $\frac{1}{2}$ kg.

Cross-pein hammer

Claw hammer

This is a general-purpose hammer. The claw end is used to lever out nails – but place a piece of hardboard between the work-piece and hammer to protect the work.

Saws

Electricians are most likely to use hacksaws and tenon saws in their work.

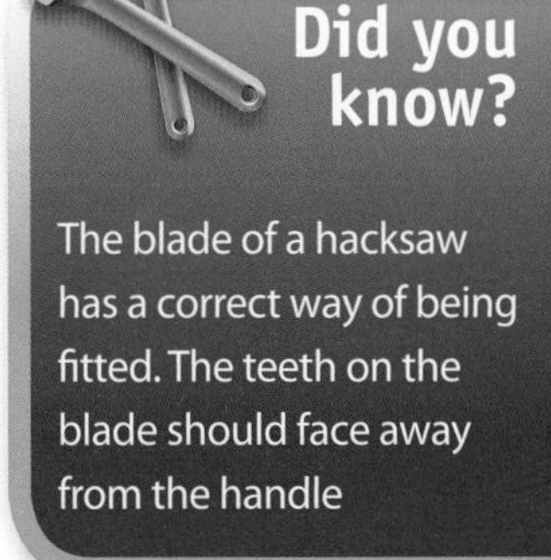

Hacksaw

This is used for basic metal cutting:

- cutting tubes or sheets to length or size
- making thin cuts to help shape the metal.

It consists of a frame, handle and blade. The blade is held in a handle, and tightened by a wing nut. In small 'junior' hacksaws the saw frame itself gives the tension and there is no wing nut.

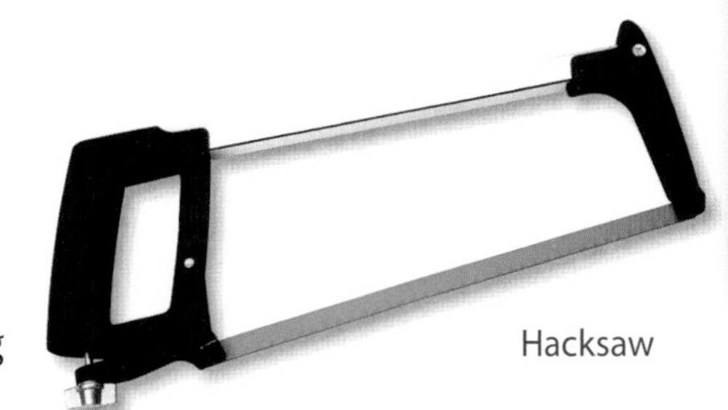

Hacksaw

Tenon saw

This is used mostly for cutting and making joints in timber. The metal strip along the top gives rigidity to the blade. To start a cut, angle the saw so that it cuts into an edge of the wood, then lower the angle for a straighter cut.

Tenon saw

Did you know?

A padsaw.

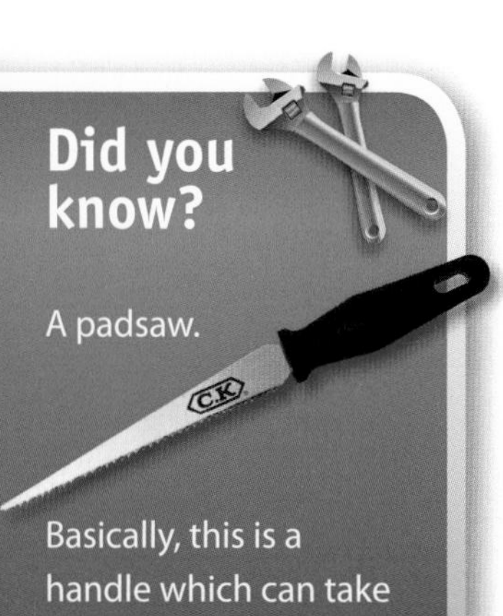

Basically, this is a handle which can take differently sized narrow, tapered blades. The blade shape means you can saw in very tight spaces. It will also cut tight curves and closed shapes (by drilling a starter hole, then putting the blade through the hole)

Flooring saw

The flooring saw is used for cutting the tongues of tongue-and-groove floorboards so they can be lifted. The saw has a curved blade that will slot into the gap between the boards.

Flooring saw

Drills

A drill makes a hole when the cutting edge at its tip is rotated with pressure applied. Drill bits must be chosen to match the hole size, the material being drilled and the tool used to rotate it. There are various manual and power tools for holding and turning drill bits.

Hand drill

This works by rotating the handle on the wheel.

Hand drill

Breast drill and carpenter's brace

Neither of these are used much these days now that cordless electric drills are available. The breast drill is like the 'big brother' to the hand drill, allowing you to apply pressure by using the weight of your body. The carpenter's brace has a cranked frame which you rotate while applying pressure to the domed head.

Power drills

The two most common are the hand-held electric drill and the drill press or bench-mounted drill. Both need to be handled carefully and safely to avoid accidents (see later in this chapter).

Hand-held electric drill

Safety tip

All electrical power tools on site should be the reduced voltage type, operating at 110 volts

Wrenches, spanners and supports

These are used for tightening nuts, bolts and setscrews. Spanners are usually made for one size of nut, marked on the spanner. (One exception is BSF spanners, which are designed to fit a nut one size larger than marked.) Double-ended spanners usually fit two different sizes.

The head of a ring or box spanner may be square, hexagonal (with 6 points) or bi-hexagonal (12 points). They grip all sides of the nut, reducing the possibility of damaging it. However they have to be placed over the nut, and so can't be used when you can't get to the end of its bolt or rod. Box spanners (which are cylindrical) can be used on deeply recessed nuts that are out of reach of normal spanners. The jaws of open-ended spanners close on four sides of a hexagonal nut (or three sides of a square one), giving easier access to nuts on a long bolt or rod, but providing a less secure grip.

Did you know?

Adjustable spanners or wrenches will fit a range of nut sizes

Adjustable spanner

Adjustable spanners or wrenches allow you to change the jaw opening to the size that is needed by adjusting the screw. They will fit a range of sizes of nut, but they do not give such good grip as a spanner and can slip, damaging the work-piece. Use them only when you do not have the proper sized spanner, or need to grip something round.

Adjustable spanner

Footprint wrench

This has serrated jaws that adjust to the size of the nut.

Footprint wrench

Vice grip

This is adjusted with a screw to preset jaw width and released by a clip. The vice grip is good for holding work-in-progress securely; it is also called a mole wrench.

Stillson wrench

This is similar to the footprint wrench, but is larger and more powerful.

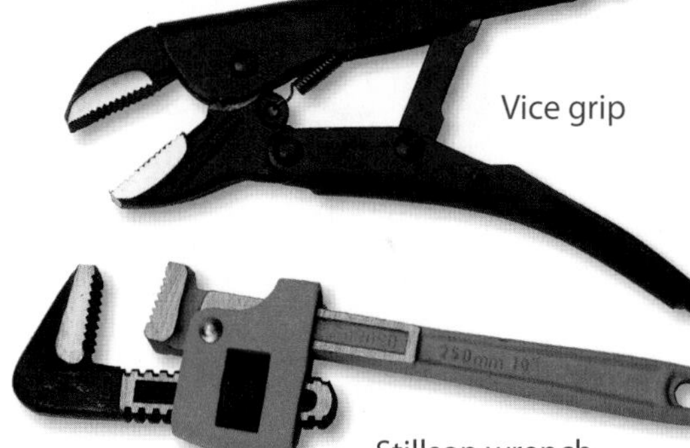

Vice grip

Stillson wrench

Files

Files have a rough face of hardened metal that is pushed across the surface of a work-piece to remove particles of material. They can be used to make an object smoother or smaller, or to change its shape.

Files are classified according to length, cut, grade and shape. The main shapes are flat, square, half-round and round.

The main cuts are bastard, fine and medium. Files of this type must be used with a firmly fitted handle.

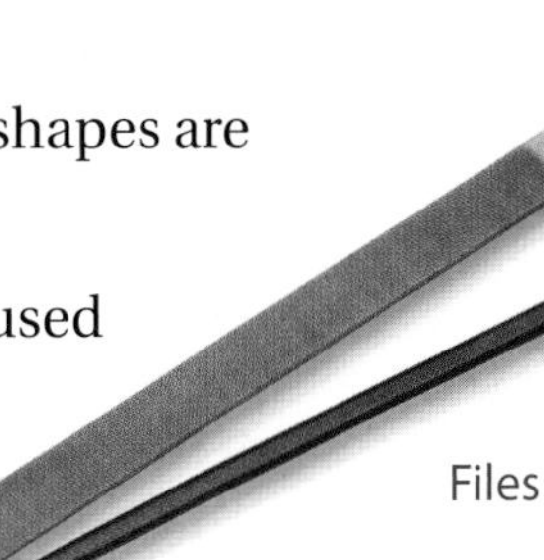

Files

Surform files have a perforated blade (like a cheese grater) held in a frame. They can be used on wood, plastic and mild steel, and are good for removing material quickly without clogging.

Chisels

As with most tools, there is a wide range available, with many used for special purposes. These are the most common ones you are likely to need:

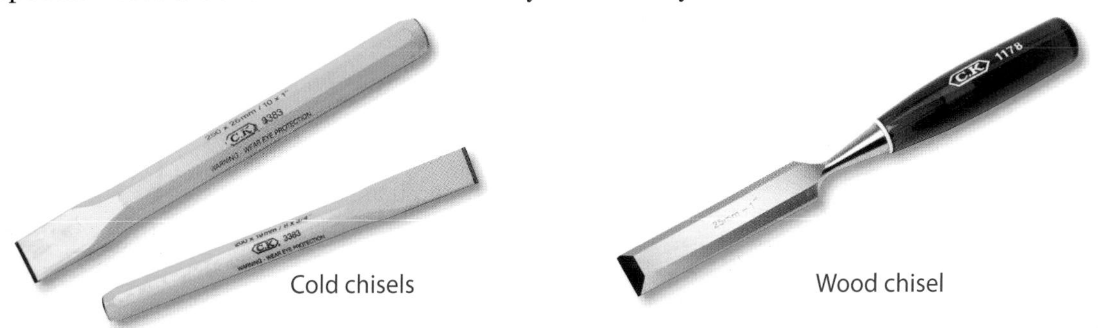

Cold chisels

Wood chisel

Crimping tool

Crimping tools

Many electrical cables are terminated using metal lugs. These are fixed to the conductors either by soldering or crimping, where part of the lug is squeezed tightly onto the conductor using a crimping tool. These can be operated by hand or, for larger sizes, by hydraulics.

Measuring tools

You will often need to check that something is level or measure lengths and distances. Available measuring tools range from a simple rule to a laser levelling device. Although measuring tools are robust, they should be handled carefully to maintain their accuracy. Here are some of the most common:

- Steel rulers are used for general measurement of trunking or tray plates.
- Steel tape measures are used for longer measurements, such as room sizes, trunking and timber.
- Spirit levels are used to check that an object is vertical or horizontal; a bubble in a glass tube shows when this is so.
- Laser rangefinders are used for measurements over longer distances; rooms, corridors etc. (up to 150 m).
- Plumb bobs and chalk lines are used to find or check verticals. They consist simply of a weight on a piece of cord. When the weight stops moving, the string is a true vertical.

Steel tape

Steel ruler

Spirit level

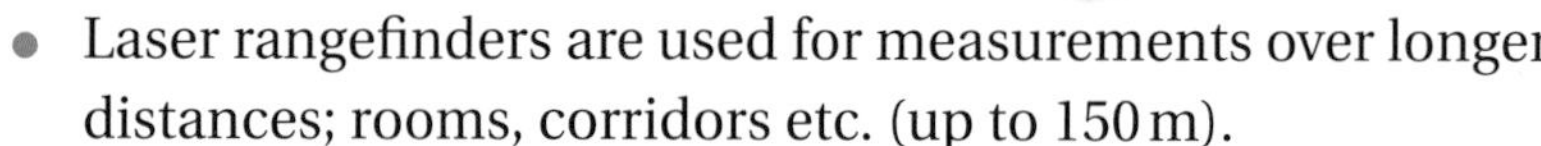

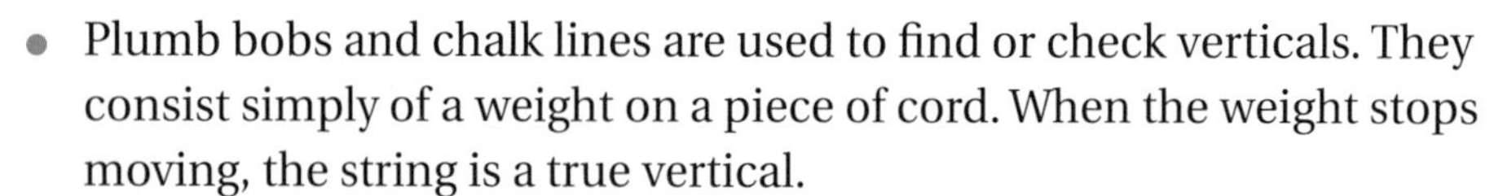

Laser rangefinder

Chalk line

Power tools

So far, almost all the tools we have looked at are hand tools, driven by human effort alone. However, many tasks are done more quickly and easily with tools powered by electricity, batteries or compressed air.

Many of these tools are simply a powered version of hand tools, but some can do things that hand tools cannot. For example, electric drills can have a 'hammer' (or percussion) action that makes it much easier to drill into brick or concrete. Smaller electric tools used on site often operate at 110 V to reduce the risks.

Did you know?

Tools powered by rechargeable batteries (cordless) are now very common, and eliminate electrical risks

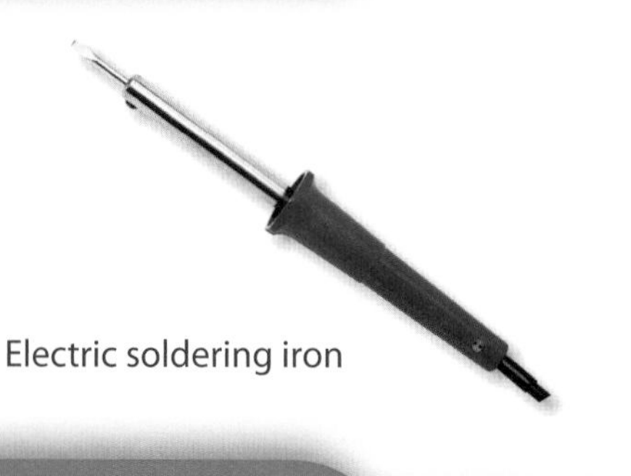

Electric soldering iron

Bench grinder

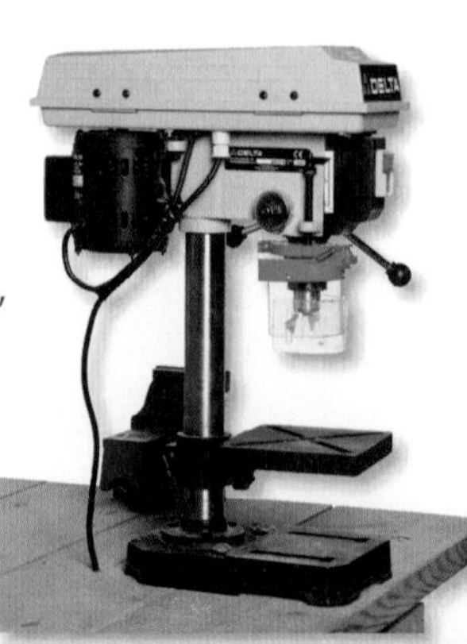

The pillar/bench drill is used in a workshop, usually for repetitive, accurate drilling

Safety awareness

You will use many different tools and equipment in your work. All of them can be dangerous if you don't respect them and use them properly and sensibly. For everyone's safety, you must continue to follow the warnings and advice you get all through your working life. See the Health and Safety chapter for more information on how to use tools and equipment safely, but here's a quick checklist:

Basic rules for hand tools:

- ✓ Always use the right tool for the job – don't make do with the nearest one.
- ✓ Keep tools clean and sharp – blunt tools are dangerous!
- ✓ Make sure handles are secure, tight and have no splinters.
- ✓ Never hit a wooden handle with a hammer.

Basic rules for electrical tools (drills, saws, sanders etc.):

- ✓ Check that the cable is not frayed or damaged.
- ✓ Check that the plug is not broken or that individual wires are showing.
- ✓ Check that mains tools (110 or 240 V) have been properly tested (P.A.T. tested).
- ✓ If in any doubt, don't use the tool and ask someone competent to check it out.

Cartridge tools, such as nail and staple guns, are widely used today. But they can also be very dangerous – they are not toys. Accidental 'firing' (or discharge) can result in serious injury. Even if the tool is not pointed directly at someone, there is a chance that the nail will bounce off another object and injure them seriously. If you need to use these tools in your work, you will be taught how to use them safely. Always follow the instructions carefully.

High-pressure air lines are frequently found in factories and workshops, and must be used with great care. Compressed air can be dangerous! It can cause explosions. Never:

- point it at yourself – or anyone else
- use it where tools or other items might be blown around
- use it to blow dust or dirt away.

Keep the working area tidy

Always try to keep your working area free from clutter and rubbish. Store away tools and materials that aren't being used. Keep cables and air hoses tidy. Clear up spills, oily rags, paper etc. quickly. This will help you to work better and show that you are a skilled and efficient worker.

FAQs

Q Why shouldn't I use a Stanley knife?

A Stanley knives are very dangerous and you can cut yourself badly. One of the best knives to use is an electricians' knife.

Safety tip

Always follow the instructions carefully

Remember

Keep the working area tidy. Always try to keep your working area free from clutter and rubbish. Store away tools

Remember

If compressed air gets into your blood via a cut or body opening (ears, eyes, nose etc.), it can form a bubble in your bloodstream which may be very painful and can even kill you!

Fixings

The electrical industry uses a large variety of fixing and fastening methods. This can lead to confusion over the terminology used to refer to them and their associated devices. This section will look at the various types of fixings and fastenings and where they are used.

Screws and bolts

Wood screws

As the name suggests, they are primarily used when fixing items to wood. In the electrical field, however, they are more commonly employed in conjunction with rawlplugs where fastenings to masonry are required. When ordering wood screws it is important to give the correct description of the screw required. To do this, four pieces of information must be known:

- size of screw
- length of screw
- type of head
- type of metal finish used.

Size of screw

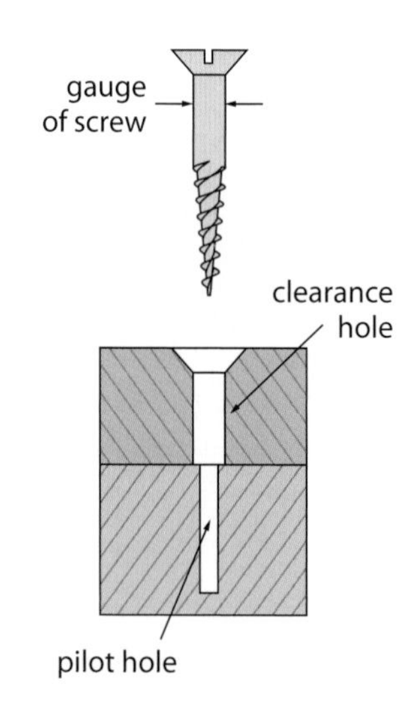

Figure 4.1 Screwing two materials together

The sizes of wood screws are measured across the shank as shown in figure 4.1. This diameter is equal all the way down the screw until the end, which tapers to a point for starting. The sizes or gauges of wood screws are given numbers, starting from no. 2, which is the smallest, and working their way up to no. 24, which is the largest. Wood screws are available in all the intermediate sizes between 2 and 24, although the even sizes (i.e. 2, 4, 6, 8, 10 etc.) are the most commonly preferred. In the electrical field, 6, 8 and 10 are usually used for cable fixing and for mounting boxes and accessories. Larger sizes (12, 14 etc.) may be needed when mounting distribution boards or panels.

You can see opposite the difference between a clearance hole and a threaded hole. The clearance hole is designed to take the screw without the screw gripping its sides, while the pilot hole is made to give the thread of the screw a start. The pilot hole may be necessary if the screw is going to be used in thick or particularly hard wood. These holes are made either with the drill of the correct size or with a bradawl.

Length of screw

The length of a screw is measured in sizes ranging from $\frac{1}{4}$" to 6". They are obtainable in $\frac{1}{8}$" steps for the smaller size screws, $\frac{1}{4}$" steps in the medium screw range, and $\frac{1}{2}$" steps for the larger screw lengths. The most commonly used screw sizes in the electrical field are in the $\frac{1}{2}$" to $2\frac{1}{2}$" range.

Type of head

The various types of screw head are appropriate in different situations, as shown in Figure 4.2.

Screw type	Usage
Countersunk (flat-head)	For general woodwork, fitting miscellaneous hardware, spacer bar saddles etc. This screw must be driven until the head is flush with the work surface or slightly below the surface.
Countersunk (Pozidrive head)	Used with special screwdrivers which will not slip from the cross-slots. Can be carried into confined spaces on the end of the screwdriver. Twinfast type For use in low-density chipboards, blockboard and softwood. They can be driven home in half the time of conventional screws.
Raised head	Used to fix door-handle plates and decorative hardware. They must be countersunk to the rim. They are usually nickel- or chrome-plated.
Round head	Used to fix surface work, fittings, accessory boxes etc. when countersunk screws are not required.
Dome head	A concealed screw for fixing mirrors, bath panels and splash-backs where the head of the screw is covered by a chrome cap that is either screwed into the end of the screw or is push-fitted onto it.
Coach screw	Provides strong fixing in heavy construction and framework. It is turned into the wood with a spanner.

Figure 4.2 Screw-head types and usage

Remember

Screws are available in metric and imperial units. You need to be familiar with both

Remember

If brass screws are tightened too much, the shaft of the screw is likely to break

Type of metal used

The material and finish of the screw must also be considered. The two main materials used are steel and brass. Steel screws may be left with a bare finish or may be sprayed black (black Japanned). They can also be cadmium-coated for rust prevention. Brass screws are either left bare or chrome-plated. The choice of material is dictated by strength requirement. If a load-bearing capability is required then steel screws should be used. The finish of screw is often just a matter of aesthetics.

Self-tapping hardened steel screws

Self-tapping screws are used primarily with sheet steel. A hole slightly smaller than the screw to be used is drilled through the steel, and when the screw is driven into steel it will cut its own thread and become fast. This is particularly useful when joining two pieces of steel together. Self-tapping screws and their head type are ordered as for wood.

They are used in particular on cookers and heaters and on steel boxes.

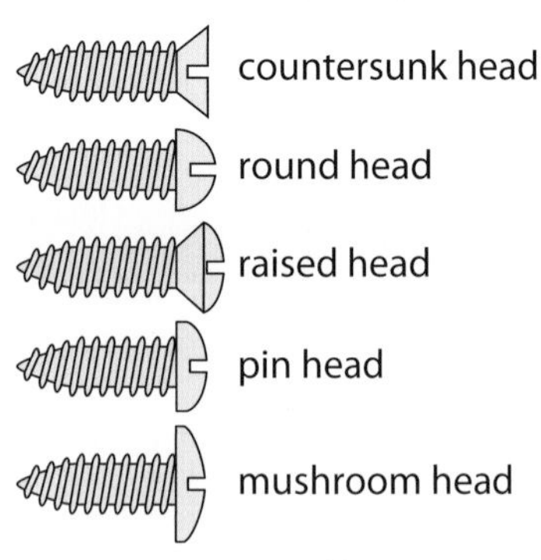

Figure 4.3 Self-tapping screws

Machine screw

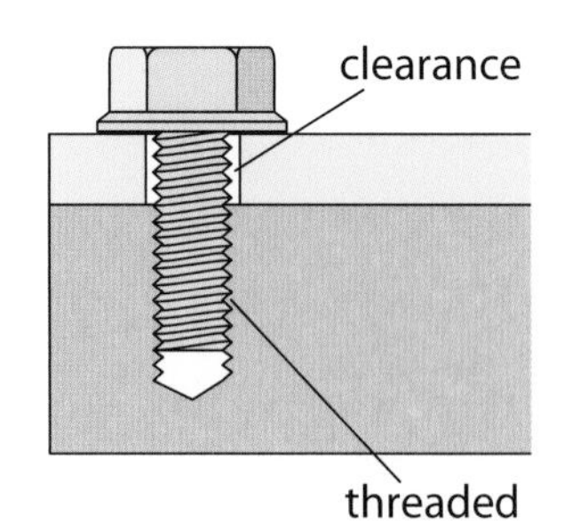

Figure 4.4 Machine-screw fixing

A machine screw, unlike a bolt, is threaded along its whole length. This avoids the need for long bolts when the parts to be joined are very thick, but requires the hole in one of the parts to be threaded, the other hole being a clearance hole.

Machine-screw heads vary in shape depending on their application. Some have a slot for a screwdriver, others a socket for an Allen key. The heads may stand proud or may be sunk below the surface for neater appearance.

The machine screws that are particularly favoured in the electrical trade have in the past been the BA thread ranges. Metric conduit boxes now use 4 mm screws, and socket outlet and switch covers use 3.5 mm metric screws, which are also often found on panels and equipment.

Stud, nut and washer

If a machine screw is frequently removed and replaced there is a tendency for the threads to wear and strip. This is overcome by the use of a stud, which is tightly fastened into the tapped hole and remains in this position when the nut is removed.

Bolts

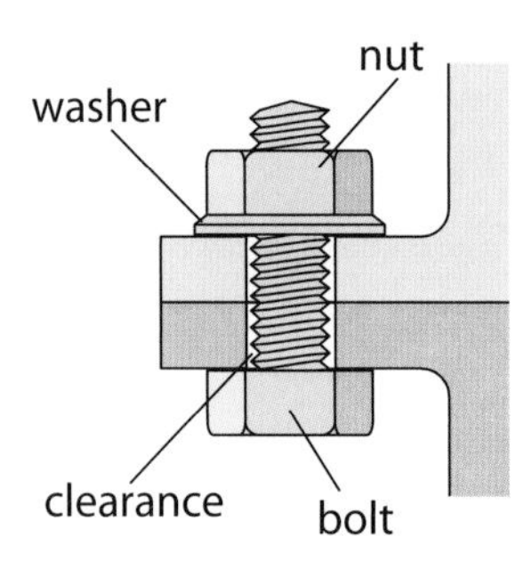

Figure 4.5 Bolt fixing showing clearance

The bolt in conjunction with nut and nut washer are used widely in all branches of engineering. The bolts passes through clearance holes in the parts to be joined. The clearance for general work is 1.5 mm, e.g. for a 16 mm bolt the hole would be drilled 17.5 mm.

Locking devices

Mechanical fastenings, e.g. bolts and studs, are used in order that parts may be removed for overhaul, replacement or to allow access to other parts. When these fastenings are subject to vibrations, such as on machines and engines, there is a tendency for them to work loose. This could result in serious damage, and, to prevent this, locking devices are used. Some of the more common types are shown in Figure 4.6.

Locking device	Features and usage
Spring washer	• Similar to a coil spring. • When the nut is tightened, the washer is compressed, and because the ends of the washer are chisel-edged they dig into the nut and the component, thus preventing the nut from turning loose. • Spring washers may have either a single or double coil. • Depending on the condition, spring washers are used only once.
Locknut	• The bottom nut is tightened with a spanner. • The top nut is then tightened, and friction in the threads and between the nut faces prevents them from rotating. • Locknuts are always bevelled at the corners to ensure good setting of the faces.
Split pin	• This can be used with ordinary nuts or castle nuts (see below). • When used with ordinary nuts, ensure that the split pin is in contact with the nut when tightened. The split pin is opened out after insertion to prevent it falling out. • Split pins can be used only once. • The bolt is left 2 to 3 threads longer for drilling.
Castle nut	• The castle nut has a cylindrical extension with grooves. • The nut is tightened, the stud is drilled opposite a groove, then the split pin is passed through the nut and the stud prevents the nut from turning. The split pin is opened out after insertion to prevent it falling out. • Split pins can be used only once.
Simmonds locknut	• The Simmonds nut has a nylon insert. • When the nut is screwed down the threads on the end of the stud bite into the nylon. • Friction keeps the nut tightened. • This nut can be used only once.
Serrated washer	• When the nut is tightened, the serration is flattened out, causing increased friction between the faces thus preventing rotation. • This type of washer can be used only once.
Tab washer	• This is a more positive type of locking device. • When the nut is tightened, one tab is bent up onto the flat side of the nut and the other tab is bent over the edge of the component. • Tab washers can be used only once – they tend to fracture when straightened out and re-bent.

Figure 4.6 Locking devices

Fixing devices

Whenever it is necessary to fix a piece of apparatus to a wall, ceiling or partition, a fixing device is required to ensure a good hold without causing damage. The drill, screw and device should all be of the correct size with respect to each other. This will depend on the material from which the fixing surface is made and on the weight to be supported.

Fixing holes should be made with an appropriate drill. A plain drill bit should be used for timber while a tungsten carbide-tipped drill should be used for masonry. Select a slow speed for drilling masonry with an electric drill. Hammer or percussion drills are recommended for concrete.

Light fixing devices

As their name suggests, these are used for relatively light fixing jobs and for partition walls and thin sheet materials.

Colour	Length	Screw size
White	¾″	4–6
Yellow	1″	6–8
Red	1½″	8–10
Brown	1¾″	10–12
Blue	2″	12–14

Table 4.1 Colour and size guide for plastic plugs. These sizes are only approximate, and the screw size may have to be altered depending on the type of wall material

Device	Description
Fibre plugs	• General purpose. • Size numbers match the screw numbers, i.e. a number 12 screw should be used with a number 12 plug. • Provide good holding power but may weaken with age. • Supplied by the hundred either of the same size or mixed.
Plastic plugs	• More popular than fibre plugs. • Should not be used when fixing a heating appliance, e.g. a storage radiator, as the heat will cause the plastic to soften and the appliance could become insecure. • Come in strips of 10 or 20 in boxes of 100. • Colour coded to denote size and usage, although colours can vary between manufacturers (see Table 4.1).
Plastic filler-type plugs	• Use loose powdery substance tamped into the hole. • Some are mixed with water first. • Holding strength is not equal to fibre or plastic. • Have the advantage of fitting any hole size.
Gravity toggles	• Only suitable for vertical surfaces. • Intended for use in hollow partition walls (plasterboard), partition thickness 10 mm minimum. When inserted horizontally through the hole the long end falls to a vertical position.
Spring toggles	• Used with partition walls and ceilings (plasterboard). • Wings are spring-activated and automatically open out when inserted vertically through the hole.
Rawl nuts	• Gives a secure fixing in thin, thick, solid or hollow material. • Vibration-proof and waterproof.
Expansion toggles	• Designed to make permanent fixings in thin sheet materials such as plywood, hardwood etc.

Figure 4.7 Light fixing devices

Heavy fixing devices

These are used for heavier jobs such as fixing a large fuseboard or securing a motor to a concrete plinth. Because of the possible dangers associated with these heavier fixings, the following should be seen as only a rough guide to different methods, and further information should be obtained prior to use.

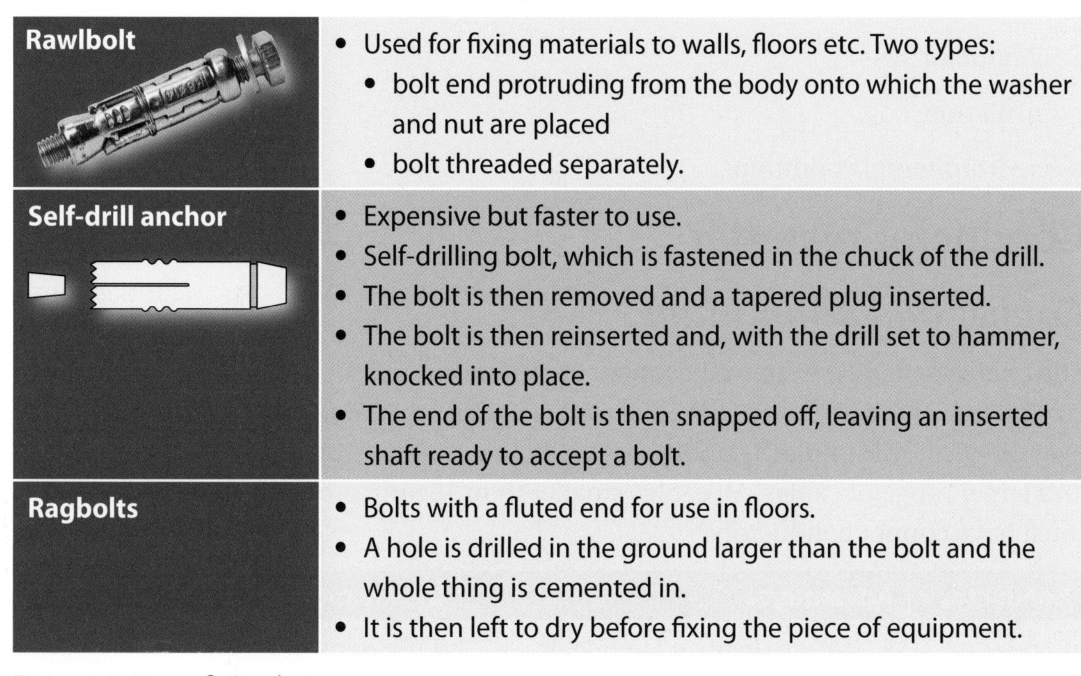

Device	Description
Rawlbolt	• Used for fixing materials to walls, floors etc. Two types: • bolt end protruding from the body onto which the washer and nut are placed • bolt threaded separately.
Self-drill anchor	• Expensive but faster to use. • Self-drilling bolt, which is fastened in the chuck of the drill. • The bolt is then removed and a tapered plug inserted. • The bolt is then reinserted and, with the drill set to hammer, knocked into place. • The end of the bolt is then snapped off, leaving an inserted shaft ready to accept a bolt.
Ragbolts	• Bolts with a fluted end for use in floors. • A hole is drilled in the ground larger than the bolt and the whole thing is cemented in. • It is then left to dry before fixing the piece of equipment.

Figure 4.8 Heavy fixing devices

Safety tip

Any holes through elements of building structures that are intended to take wiring systems must be sealed around the cables in order to prevent the spread of fire

Miscellaneous fixings

- roundhead nail – used for general woodwork
- oval nail – used for general woodwork; prevents splitting of timber, especially thin or heavily grained timber
- brad – used as floorboard fixing; difficult to remove
- galvanised clout nail – handy for fixing channelling over cables prior to plastering
- panel pin – small pin for fastening hardboard or woodsheets, used with buckle clips
- masonry nail – hard nail for use with plastic clips (PVC-sheathed cable)
- rivet – a device for joining together two or more pieces of metal. They should be of the same material as the metal being joined; if this is not possible the rivets should be of a softer metal than the sheets being joined.

The Gripple

The Gripple is a new system for supporting false ceilings, cable basket, or other similar loads. It uses a principle of mini tirfor jacks and can be easily tensioned into the correct position. The Gripple can be released using the small key provided.

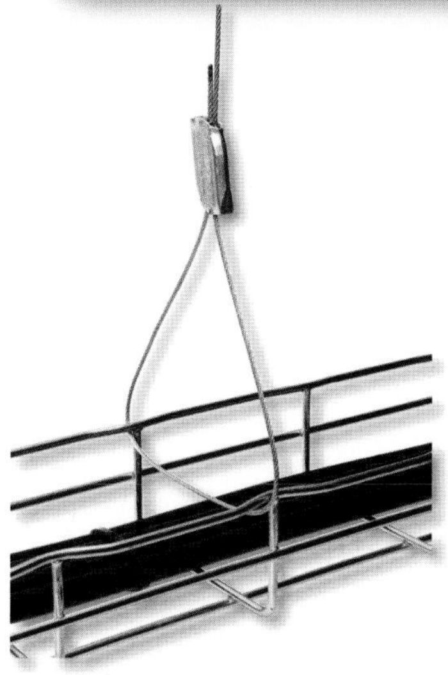

Gripple supporting cable tray

Conductors and insulators

Selecting a cable for an electrical installation is very important; consideration must be given to the following criteria in order to ensure the correct type of cable is chosen:

- conductor material
- conductor size
- insulation
- environmental conditions.

Conductor material

Copper and aluminium

The choice generally is between copper and aluminium. Copper has better conductivity for a given cross-sectional area and is preferable, but its cost has risen over the years. Aluminium conductors are now sometimes preferred for the medium and larger range of cables. All cables smaller than 16 mm^2 cross-sectional area (csa) must have copper conductors.

Conductor	Advantages	Disadvantages
Copper	• easier to joint and terminate • smaller cross-sectional area for given current rating	• more costly • heavier
Aluminium	• cheaper • lighter	• bulkier for given current rating • not recommended for use in hazardous areas

Table 4.2 Copper and aluminium conductors compared

Other conductor materials

Cadmium copper: has a greater tensile strength for use with overhead lines

Steel reinforced aluminium: for very long spans on overhead lines

Silver: used where extremely good conductivity is required. However, it is extremely expensive

Copperclad (copper-sheathed aluminium): cables that have some of the advantages of both copper and aluminium but are difficult to terminate.

Whatever the choice of conductor material the conductors themselves will usually be either stranded or solid. Solid conductors are easier and therefore cheaper to manufacture but the installation of these cables is made more difficult by the fact that they are not very pliable. Stranded conductors are made up of individual strands that are brought together in set numbers. These provide a certain number of strands such as 3, 7, 19, 37 etc. and, with the exception of the 3-strand conductor, all have a central strand surrounded by the other strands within the conductor.

Solid conductors are not as common as stranded; their use has been restricted to either very small or very large conductor sizes

Conductor size

There are many factors that affect the choice of size of conductor:

- load and future development
  The current the cable is expected to carry can be found from the load, taking into account its possible future development, i.e. change in use of premises, extensions or additions.
- ambient temperature
  The hotter the surrounding area, the less current the cable is permitted to carry.
- grouping
  If a cable is run with other cables then its current carrying capacity must be reduced.
- type of protection
  Special factors must be used when BS 3036 (semi-enclosed) fuses are employed.
- whether placed in thermal insulation
  If cables are placed in thermal insulation, de-rating factors must be applied.
- voltage drop
  The length of circuit, the current it carries and the cross-sectional area of the conductor will affect the voltage drop. The length of circuit, the current it carries and the cross sectional area of the conductor will affect the voltage drop. BS 7671 Regulation 525.3 states that the maximum voltage drop must not exceed the values given in Appendix 12, namely 3 per cent for lighting and 5 per cent for other uses.

Insulation

To insulate the conductors of a cable from each other and to insulate the conductors from any surrounding metalwork, materials with extremely good insulating properties must be used. Cables can be installed in a variety of different situations, and you must take care that the type of insulation on the chosen cable is suitable for that particular situation.

Insulation types

Listed below are some of the working properties of the more common types of cable insulation:

- PVC
- synthetic rubbers
- silicon rubber
- magnesium oxide
- phenol-formaldehyde.

PVC This is a good insulator: it is tough, flexible and cheap. It is easy to work with and easy to install. However, thermoplastic polymers such as PVC do not stand up to extremes of heat and cold, and BS 7671 recommends that ordinary PVC cables should not constantly be used in temperatures above 60 °C or below 0 °C. Care should be taken when burning off this type of insulation (to salvage the copper) because the fumes produced are toxic.

Synthetic rubbers These insulators, such as Vulcanised Butyl Rubber, will withstand high temperatures much better than PVC and are therefore used for the connection of such things as immersion heaters, storage heaters and boiler-house equipment.

Silicon rubber FP 200 cable using silicon rubber insulation and with an extruded aluminium over-sheath foil is becoming more popular for wiring such things as fire-alarm systems. This is due largely to the fact that silicon rubber retains its insulation properties after being heated up or burned and is somewhat cheaper than mineral-insulated metal-sheathed cables.

Magnesium oxide This is the white powdered substance used as an insulator in mineral-insulated cables. This form of insulation is hygroscopic (absorbs moisture) by nature and therefore must be protected from damp with special seals. Mineral-insulated cables are able to withstand very high temperatures and, being metal sheathed, are able to withstand a high degree of mechanical damage.

Phenol-formaldehyde This is a thermosetting polymer used in the production of such things as socket outlets, plug tops, switches and consumer units. It is able to withstand temperatures in excess of 100 °C.

Insulation colours

To identify cables the insulation is coloured in accordance with BS 7671, table 51.

Table 51 Identification of conductors	
Function	**Colour**
Protective conductors	Green and yellow
Functional earthing conductor	Cream
a.c. power circuit[1]	
Phase of single-phase circuit	Brown
Neutral of single- or three-phase circuit	Blue
Phase 1 of three-phase a.c. circuit	Brown
Phase 2 of three-phase a.c. circuit	Black
Phase 3 of three-phase a.c. circuit	Grey
Two-wire unearthed d.c. power circuit	
Positive of two-wire circuit	Brown
Negative of two-wire circuit	Grey
Two-wire earthed d.c. power circuit	
Positive (of negative earthed) circuit	Brown
Negative (of negative earthed) circuit	Blue
Positive (of positive earthed) circuit	Blue
Negative (of positive earthed) circuit	Grey
Three-wire d.c. power circuit	
Outer positive of two-wire circuit derived from three-wire system	Brown
Outer negative of two-wire circuit derived from three-wire system	Grey
Positive of three-wire circuit	Brown
Mid-wire of three-wire circuit[2]	Blue
Negative of three-wire circuit	Grey
Control circuits, ELV and other applications	
Phase conductor	Brown, black, red, orange, yellow, violet, grey, white, pink or turquoise
Neutral or mid-wire[3]	Blue

NOTES

(1) Power circuits include lighting circuits.

(2) Only the middle wire of three-wire circuits may be earthed.

(3) An earthed PELV conductor is blue.

Table 4.3 To identify cable insulation in accordance with BS 7671

Safety tip

Great care must be taken when installing and connecting cables, especially where old and new cable colours are present in the same building

You will come across conductors with these colours of insulation for many years to come, but only in existing installations. From April 1 2004 the colours of conductor insulation for new installations changed.

Conductor	Old colour
Phase	Red
Neutral	Black
Protective conductor	Green and yellow
Phase one	Red
Phase two	Yellow
Phase three	Blue
Neutral	Black
Protective conductor	Green and yellow

Table 4.4 Old conductor insulation colours

Environmental conditions

Many factors affect cable selection. Some will be decided by factors previously mentioned and some by the following:

- The risk of excessive ambient temperature
- The effect of any surrounding moisture
- The risk of electrolytic action
- Proximity to corrosive substances
- The risk of damage by animals
- The effect of exposure to direct sunlight
- The risk of mechanical stress
- The risk of mechanical damage.

Ambient temperature

Current-carrying cables produce heat, and the rate at which the heat can be dissipated depends upon the temperature surrounding the cable. If the cable is in a cold situation then the temperature difference is greater and there can be substantial heat loss. If the cable is in a hot situation then the temperature difference of the cable and its surrounding environment will be small, and little, if any, of the heat will be dissipated.

Problem areas are boiler-houses and plant rooms, thermally insulated walls and roof spaces. PVC cables that have been stored in areas where the temperature has dropped to 0°C should be warmed slowly before being installed. However, if cables have been left out in the open and the temperature has been below 0°C (say, a heavy frost has attacked the cables), then you must report this situation to the person in charge of the installation. These low temperatures can damage PVC cables.

Find out

Appendix 4 of BS 7671 gives tables of different types of cable, and shows that the current rating changes in accordance with the conditions under which the cable is installed. Take a look at this and make a copy for your own reference

Moisture

Water and electricity do not mix, and care should be taken at all times to avoid the movement of moisture into any part of an electrical installation by using watertight enclosures where appropriate. Any cable with an outer PVC sheath will resist the penetration of moisture and will not be affected by rot. However, suitable watertight glands should be used for termination of these cables.

Electrolytic action

Two different metals together in the presence of moisture can be affected by electrolytic action, resulting in the deterioration of the metal. Care should be taken to prevent this. An example is where brass glands are used with galvanised steel boxes in the presence of moisture. Metal-sheathed cables can suffer when run across galvanised sheet-steel structures, and if aluminium cables are to be terminated onto copper bus bars then the bars should be tinned.

Corrosive substances

The metal sheaths, armour, glands and fixings of cables can also suffer from corrosion when exposed to certain substances. Examples include:

- magnesium chloride used in the construction of floors
- plaster undercoats containing corrosive salts
- unpainted walls of lime or cement
- oak and other types of acidic wood.

Metalwork should be plated or given a protective covering. In any environment where a corrosive atmosphere exists special materials may be required.

Damage by animals

Cables installed in situations where rodents are prevalent should be given additional protection or installed in conduit or trunking, as these animals will gnaw cables and leave them in a dangerous condition. Installations in farm buildings should receive similar consideration and should, if possible, be placed well out of reach of animals to prevent the effects of rubbing, gnawing and urine.

Many house fires are attributed to rodent damage to cabling

Direct sunlight

Cables sheathed in PVC should not be installed in positions where they are exposed to direct sunlight because this causes them to harden and crack: the ultraviolet rays leach out the plasticiser in the PVC, making it hard and brittle.

1. Road crossing accessible to vehicles

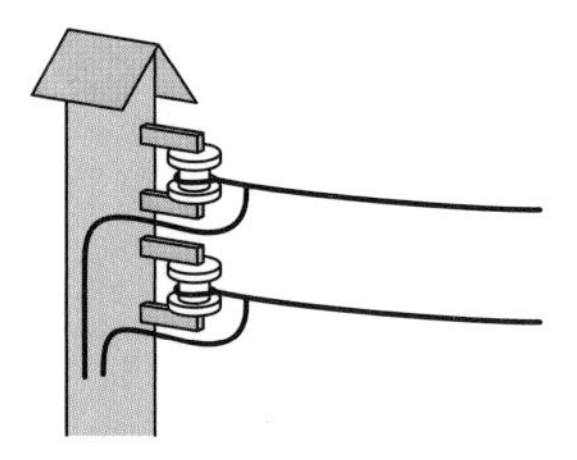

All methods of suspension 5.8 m minimum above ground

2. Accessible to vehicles but not a road crossing

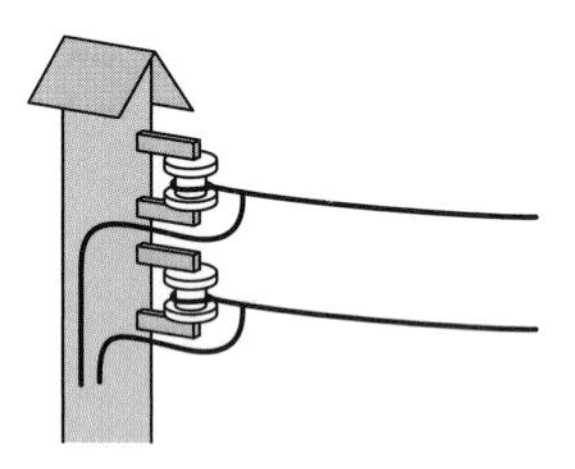

All methods of suspension 5.2 m minimum above ground

3. Inaccessible to vehicles

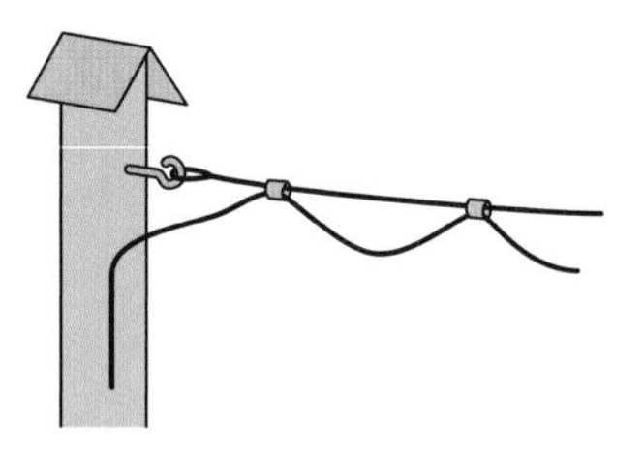

PVC cables supported by a catenary wire. 3.5 m minimum above ground

Figure 4.9 Suspension heights for cables

Mechanical stress

PVC cables, when used for overhead wiring between buildings, can be subjected to mechanical stress if a catenary wire is not used to support them along the way. Figure 4.9 illustrates suspension heights for cables above road crossings. Flexible cables used to suspend heavy luminaries that exceed the recommended weights given in BS 7671 will feel the effects of mechanical stress. See Table 4.5.

Table 4F3A Flexible cords weight support			
Conductor (cross-sectional area mm²)	**Current carrying capacity 1- phase a.c.**	**3- phase a.c.**	**Maximum mass**
0.5	3 A	3 A	2 kg
0.75	6 A	6 A	3 kg
1	10 A	10 A	5 kg
1.25	13 A	10 A	5 kg
1.5	16 A	16 A	5 kg
2.5	25 A	20 A	5 kg
4	32 A	25 A	5 kg

Table 4.5 BS 7671 Table 4F3A Flexible cords non armoured. Current carrying capacity and mass supportable.

These cables can also suffer from stress when subjected to excessive vibration, causing breakdown of the insulation.

Mechanical damage

The main function of the cable sheath is to protect the cable from mechanical damage. All conductors and cables should also have additional protection where they pass through floors and walls or are installed in exposed positions where damage could occur. Cables to be installed underground should have a sheath or armouring resistant to any mechanical damage likely to occur. When cables pass through holes drilled in wooden joists these should be 50 mm from the top or bottom of the joist measured vertically (see Figure 4.10). Numerous other examples of situations where cables can be subjected to mechanical damage can be found in section 521 of BS 7671.

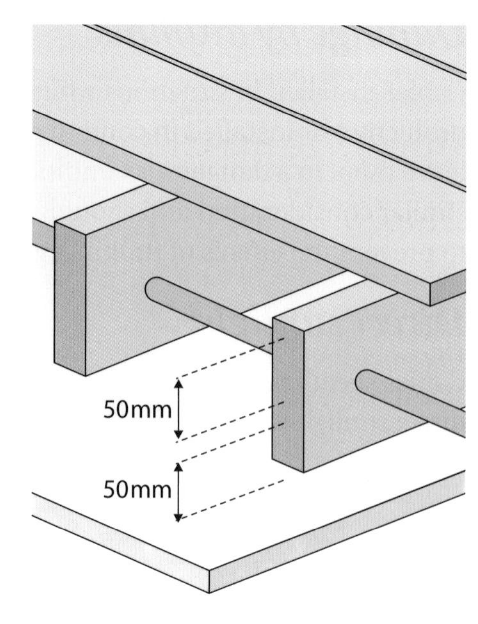

Figure 4.10 Clearance for cables in joists

Installation techniques

Nearly all modern domestic wiring is recessed into the walls, as this leaves a flat surface and a neater appearance. However, most rewiring of houses involves buildings that are at least 25 years old. Before 1956 wiring was usually installed in vulcanised rubber (VRI) insulated with a tough rubber sheath (TRS) or a lead sheath. This type of wiring is greatly affected by temperature changes and eventually the rubber becomes brittle; any interference with the cable usually results in the insulation breaking off. Modern PVC cables have a far greater lifespan. Sometimes wiring was installed in slip-gauge conduit (light-duty conduit with an open seam).

This type of conduit is not to be used as an earth return under any circumstances.

Both of these types of systems are now due for rewiring.

This section will cover some of the techniques required for the installation and rewiring of an existing building and the regulations that apply for the protection of installed cables. Areas that will be covered include the following:

- rewiring an existing building
- floorboards
- cables run into walls
- chasing
- wiring in partition walls
- ceiling fittings
- protection of cables
- protection against heat damage and spread of fire
- cable support and bends
- miscellaneous.

Remember

Heavy-handedness results in unnecessary damage which you or your company will be expected to put right, so taking care will save a lot of additional time and work

Rewiring an existing building

Floorboards

Quite often it is necessary to lift floorboards to be able to install new cables. With some older buildings the boards are of the butt-type finish. This makes lifting the boards comparatively easy. Modern houses tend to have tongue-and-groove floorboards as shown in Figure 4.11.

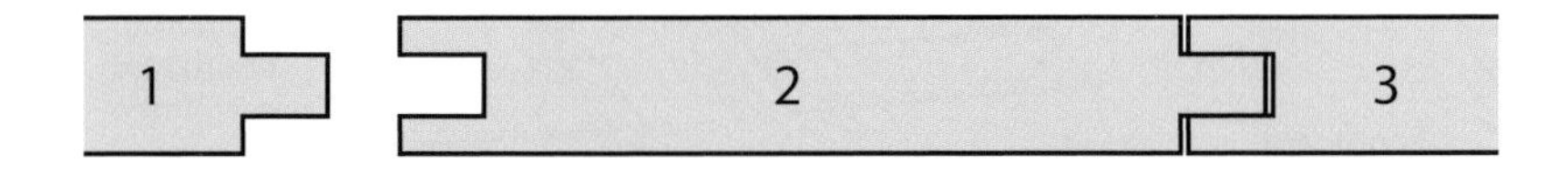

Figure 4.11 Tongue and groove

Remember

Be extra careful lifting floorboards, especially if the boards are old, otherwise they may snap

Most present-day domestic properties have high-density chipboard panels, which interlock with each other and will require a different approach should you be required to install additional points in the future.

Floorboards are normally fitted starting from one wall, i.e. board number 1 placed in position and then board number 2 slotted onto board number 1, then number 3 is slotted onto number 2 etc. until the whole floor is covered. Then they are nailed down. Lifting the boards with minimum damage entails lifting the last board laid and reversing the laying procedure. However, this is a time-consuming operation, especially when only one or two boards need to be lifted in the area you wish to work in. Lifting a middle board means that the two tongues holding the board need to be removed. Ideally a circular saw with a narrow blade or a flooring saw can be used.

Before attempting to lift the floorboard the nails should be punched down to enable the board to be lifted more easily. Very great care must be taken to ensure that any other service pipes, e.g. water or central heating, are not cut or damaged during the work involved with lifting floorboards or chipboards.

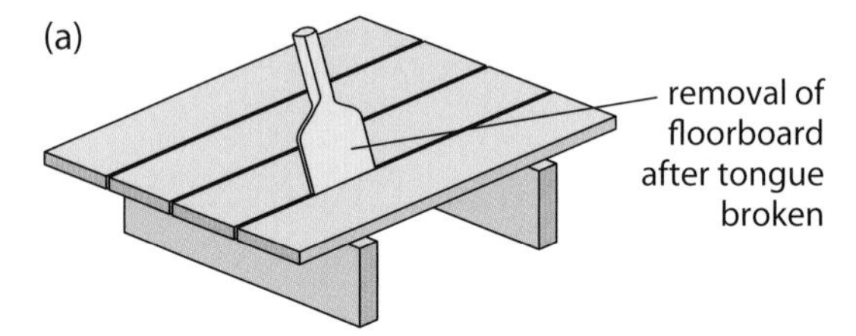

Once the end of the board has been prised free you should endeavour to get the rest up without excess damage.

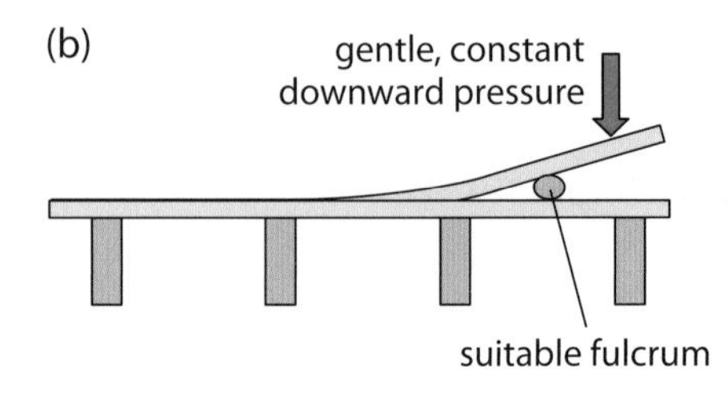

Figure 4.12 (a) and (b) Lifting floorboards

Sometimes it is necessary to cut a floorboard because it disappears under the skirting board or a built-in wardrobe. The place to cut is where the board crosses a joist; this is usually where the board is nailed. If you are unable to lift the floorboard enough to cut it, then the board must be cut at the side of the joist, using a padsaw.

When refitting a board, all that is required is for a small piece of wood known as a fillet to be screwed to the side of the joist. The board can be laid on it and nailed down.

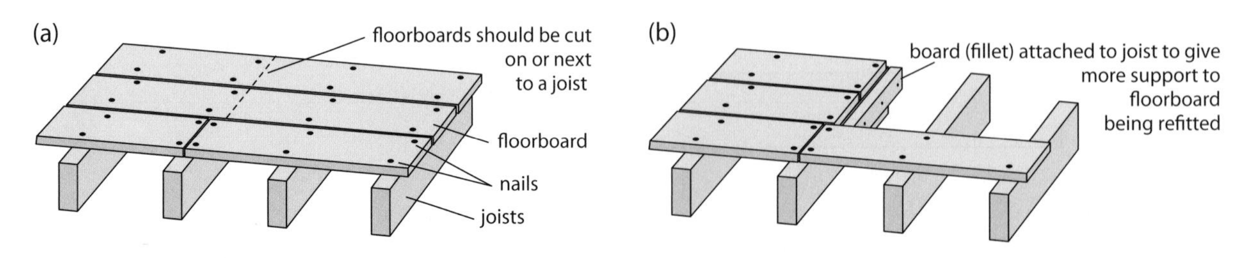

Figure 4.13 (a) and (b) Cutting and refitting floorboards

Cables run into walls

When cables have been installed in a wall (for example to feed a lighting switch or socket outlet) at a depth less than 50 mm from any surface, then in accordance with Regulation 522.6.6 they must fulfil one of the following criteria:

- Incorporate an acceptable earthed metallic covering.
- Be enclosed in acceptable earthed conduit, trunking or ducting.
- Be mechanically protected to prevent penetration by nails, screws etc.
- Be installed in a zone as indicated in Figure 4.14. Please note that a zone formed on one side of a wall of less than 100 mm thickness extends to the reverse side, but only if the location of that accessory can be determined from the reverse side.

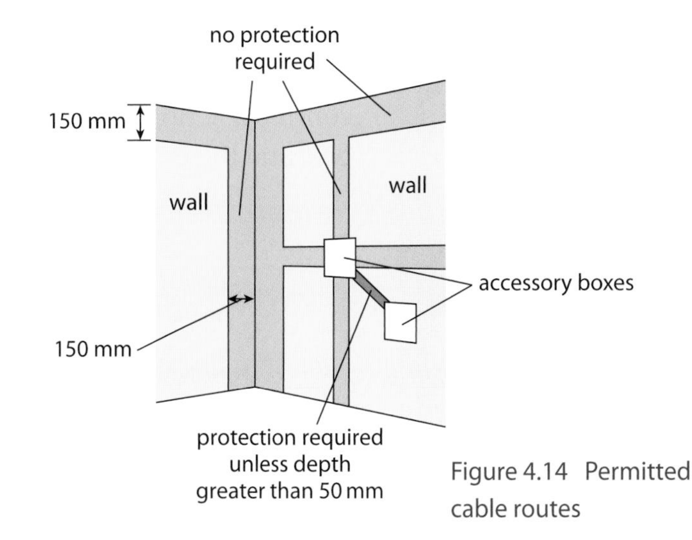

Figure 4.14 Permitted cable routes

Additionally, Regulation 522.6.7 states that where the finished installation is not intended to be under the supervision of a skilled or instructed person, then cables that are not mechanically protected and are installed in a designated zone and in accordance with Regulation 522.6.6 must be provided with additional protection in the form of a suitable RCD.

Regulation 522.6.8 then states that if the finished installation is not under the supervision of a skilled or instructed person then, irrespective of the depth of the cable from the surface of the wall or partition, when that wall is constructed using metallic parts other than nails and screws, cables must do one of the following:

- Incorporate an acceptable earthed metallic covering.
- Be enclosed in acceptable earthed conduit, trunking or ducting.
- Be mechanically protected to prevent penetration by nails, screws etc.
- Be provided with additional protection by means of an RCD.

Additionally, should the cable be installed at a depth of 50 mm or less from the wall or partition surface then the requirements of Regulation 522.6.6 shall also apply.

Chasing

Chasing is the name given to cutting slots into a wall for conduit or cable. This job is made extremely easy by using a chasing tool or a chasing tool attachment (available on most electric drills); all that is required is a line on the wall as a guide to work to. An alternative to this is a bolster chisel and hammer, which takes longer but does the same job. The back boxes for wall-mounted 'flush' fittings must be mounted in a hole cut into the brickwork. The hole is made slightly bigger than the box to allow plaster or filler to be applied around the edge and back of the hole. Once this has been completed the box can be fitted in and secured. The front edge of the box must not protrude from the surface of the wall. The purpose of the filler is to ensure that the box remains firm in the wall.

In older houses the bricks and mortar were not made to today's standards and they tend to be very powdery. Take, for example, a socket outlet recessed into a wall. The continual action of inserting and withdrawing a 3-pin plug puts strain on the back box, and the box may become loose and eventually unsafe.

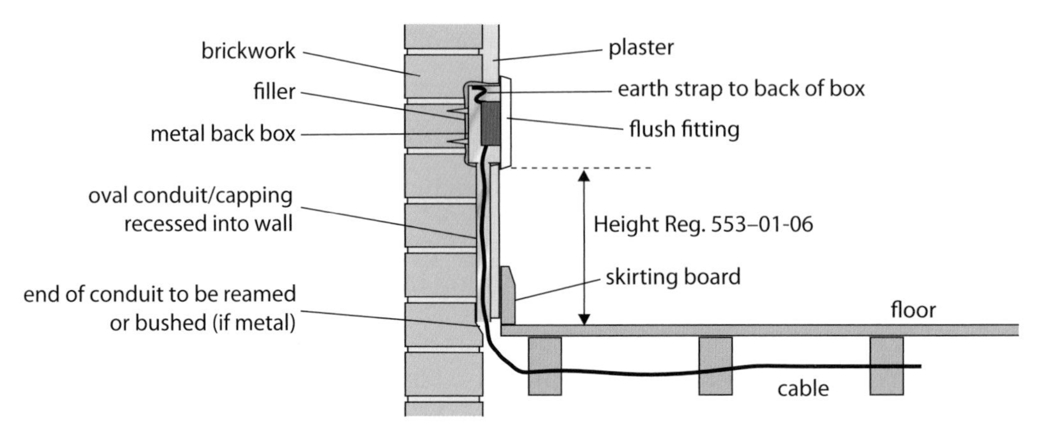

Figure 4.15 Flush fitting fed through the wall via the floor

Did you know?

In new buildings electrical wiring is usually fixed before the plasterers render the walls. In this way chasing is kept to a minimum. The accessories are then usually fixed when the plastering and decorations are completed

Wiring in partitions

Figure 4.16 Wiring in a partition

noggin

Right way

Wrong way

Wiring must be done before any lining (surface) is fixed in position, and if the cables are sheathed they must pass through the centre of the wooden framework and must be a minimum of 50 mm from any finished surface.

However, if the cables are to be run in metal conduit then the conduit can fit into slots in the outer edges of the wooden framework. Regulations for conduit still apply to this form of installation.

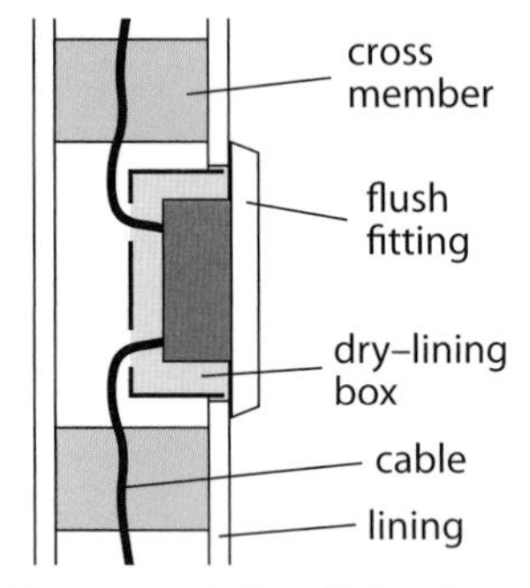

Figure 4.17 Dry-lining box

In either case the back boxes for fitting must be securely fixed, preferably to **noggins** in the structure of the partition. Dry lining boxes can be used if the partition lining is of sufficient strength and thickness.

Definition

Noggin – piece of wood fixed to joists to enable ceiling roses to be fixed

Ceiling fittings

Boxes for ceiling outlets must be securely fixed either to a joist or to a noggin between the joists.

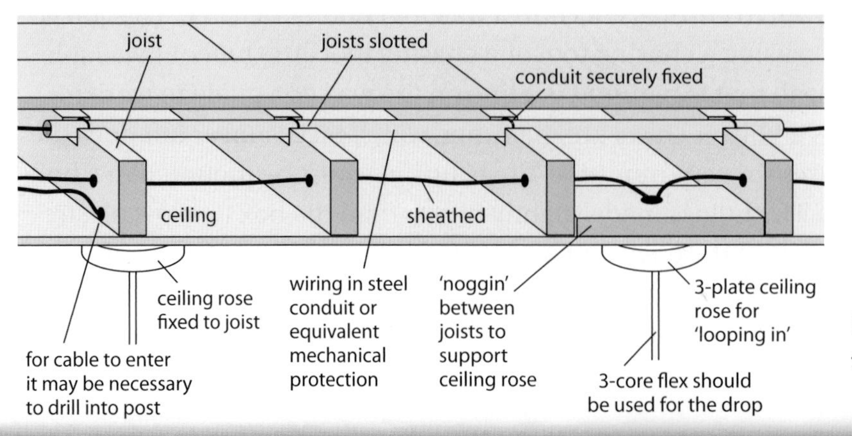

Figure 4.18 Cables through joists

Protection of cables

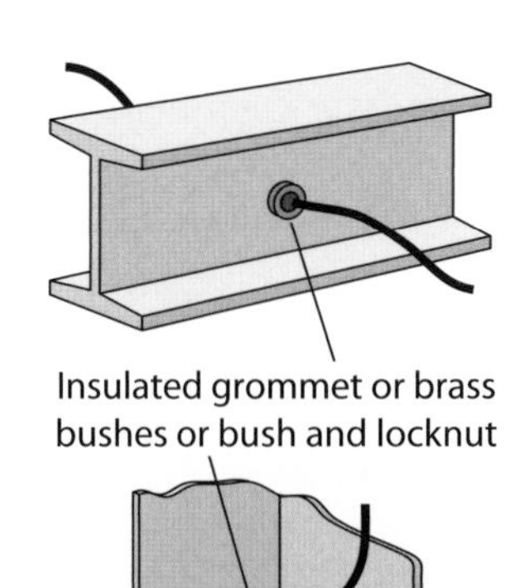

Figure 4.19

Where cables are installed under floors or above ceilings they must be run in such positions that they are not liable to be damaged by contact with the floor or ceiling or their fixings. Unarmoured cables passing through a joist shall be at least 50 mm from the top or bottom as appropriate, or enclosed in steel conduit. Alternatively the cables can be provided with mechanical protection sufficient to prevent penetration of the cable by nails, screws etc.

Where cables pass through holes in metalwork, precautions should be taken to prevent abrasion of the cables on any sharp edges (see Regulation 522.6.1) – for instance by the use of insulated grommets (see Figure 4.19).

Where significant solar radiation or ultraviolet radiation is experienced or expected, select or install a wiring system suitable for the conditions or provide adequate shielding. Special precautions may need to be taken for equipment subject to ionising radiation, such as an X-ray set (Regulations 522.2.1 and 522.11.1).

Protection against heat damage

In order to avoid the effects of heat from external sources (e.g. hot water systems, equipment, manufacturing processes, solar gain) one or more of the following should be used (Regulation 522.2.1):

- shielding
- placing sufficiently far from the heat source
- select a wiring system with regard for the additional temperatures
- local reinforcement or substitution of insulating materials.

Additionally, Regulation 522.2.2 now states that any part of a cable or flex inside an accessory, appliance or luminaire must be capable of withstanding the temperatures determined in Regulation 522.1.1 (i.e. normal use, ambient temperature and fault).

Protection against the spread of fire

Firstly the risk must be minimised by selecting appropriate materials and equipment. However where a wiring system passes through elements of building construction such as floors, walls, roofs, ceilings, partitions or cavity barriers, the openings remaining after passage of the wiring system should be sealed according to the degree of fire resistance prescribed for that element of the building before penetration of the wiring system (Regulation 527.2.1).

Be aware that during installation, temporary sealing arrangements might also be necessary (Regulation 527.2.2). Equally, any disturbances to sealing arrangements made during alterations must be repaired as soon as possible (Regulation 527.2.3).

Regulation 527.2.4 states that when a wiring system such as a conduit, cable ducting or cable trunking, busbar or busbar trunking penetrates elements of building construction having specific fire resistance it should be internally sealed so as to maintain the degree of fire resistance of the respective element as well as being externally sealed (to comply with Regulation 527.2.1).

Cable support and bends

When a conductor or cable is not continuously supported it should be supported by suitable means at appropriate intervals so that it doesn't suffer damage under its own weight (Regulation 522.8.4). The following tables list suitable spacing of supports for cables in accessible positions.

Overall diameter of cable	Maximum clip spacing							
	Non-armoured PVC, XLPE or lead-sheathed cable				Armoured cables		Mineral insulated with copper or aluminium sheath	
	Generally		In caravans					
	Horizontal mm	Vertical mm	Horizontal mm	Vertical mm	Horizontal mm	Vertical mm	Horizontal mm	Vertical mm
up to 9 mm	250	400	250	400	–	–	600	800
9 mm–15 mm	300	400	250	400	350	450	900	1200
15 mm–20 mm	350	450	250	400	400	550	1500	2000
20 mm–40 mm	400	550	250	400	450	600	–	–

Table 4.6 Cable support requirements (Table 4A, *On-site Guide*)

Type of system	Maximum length of span (metres)	Minimum height of span above ground (metres)		
		At road crossings	*In positions accessible to vehicular traffic, other than crossings*	*Positions inaccessible to vehicular traffic*
Cables sheathed with PVC or having an oil-resisting and flame-retardant or HOFR sheath, without intermediate support	3	5.8 for all types	5.8 for all types	3.5
Cables sheathed with PVC or having an oil-resisting and flame-retardant or HOFR sheath in heavy-gauge steel conduit of diameter not less than 20 mm and with no joint in its span	3	5.8	5.8	3
PVC-covered overhead lines on insulators without intermediate support	30	5.8	5.8	3.5
Bare overhead lines on insulators without intermediate support	30	5.8	5.8	5.2
Cables sheathed with PVC or having an oil-resisting and flame-retardant or HOFR sheath without intermediate support	No limit	5.8	5.8	3.5
Aerial cables incorporating a catenary wire	As specified by the manufacturer	5.8	5.8	3.5
Bare or PVC-covered overhead lines on insulators installed in accordance with the overhead line Regs	No limit	5.8	5.8	5.2

Table 4.7 Maximum length of span and minimum height above ground for overhead cables

Normal size of conduit (mm)	Maximum distance between supports (metres)					
	Rigid metal		***Rigid insulating***		***Pliable***	
	Horizontal	***Vertical***	***Horizontal***	***Vertical***	***Horizontal***	***Vertical***
Not exceeding 16	0.75	1	0.75	1	0.3	0.5
Exceeding 16 but not exceeding 25	1.75	2	1.5	1.75	0.4	0.6
Exceeding 25 but not exceeding 40	2	2.25	1.75	2	0.6	0.8
Exceeding 40	2.25	2.5	2	2	0.8	1

Table 4.8 Spacing of supports for conduits

Cross sectional area of trunking (mm^2)	Maximum distance between supports (m)			
	Metal		**Insulating**	
	Horizontal	**Vertical**	**Horizontal**	**Vertical**
Exceeding 300 but not exceeding 700	0.75	1	0.5	0.5
Exceeding 700 but not exceeding 1500	1.25	1.5	0.5	0.5
Exceeding 1500 but not exceeding 2000	1.75	2	1.25	1.25
Exceeding 2000 but not exceeding 5000	3	3	1.5	2
Exceeding 5000	3	3	1.75	2

Table 4.9 Spacing of supports for trunking

Insulation	Finish	Overall diameter*	Factor to be applied to overall diameter to determine minimum internal radius of bend
XLPE, or PVC (with circular, or circular stranded copper of aluminium conductors)	Non-armoured	Not exceeding 10 mm	3 (2) <>
		Exceeding 10 mm but not exceeding 25 mm	4 (3) <>
		Exceeding 25 mm	6
	Armoured	Any	6
XLPE, or PVC (with solid aluminium or shaped copper conductors)	Armoured or non-armoured	Any	8
Mineral (with copper sheath)	With or without PVC covering	Any	6 #

* Denotes for flat cables the diameter refers to the major axis

<> Figures in brackets relate to single core circular conductors of stranded construction installed in conduit, ducting or trucking

Mineral insulated cables may be bent to a radius not less than 3 times the cable diameter over the copper sheath, provided that the bend is not re-worked, i.e. straightened and re-bent

Table 4.10 Minimum internal radii of bends in cable for fixed wiring

Miscellaneous

Other factors that affect the choice of a wiring system and its associated enclosures and equipment include:

- ambient temperature (this affects the insulation properties)
- moisture (can the enclosures safely resist the ingress of water? IP ratings to be checked)
- corrosive substances (chemicals can destroy insulation, enclosures and equipment)
- UV rays (sunlight affects cable insulation)
- damage by animals (by chewing cables or the corrosive effect of animal urine)
- mechanical stress and vibration (use flexible conduits and helix loops at connections to motors to absorb vibration and make sure the cable is adequately supported throughout its length)
- aesthetic conditions (what looks right and appropriate for the building and customer).

Remember that when you are installing cables, someone has to repair the building fabric once the job is finished. While carrying out electrical installation work it will invariably be necessary to work closely with other trades (to coordinate the process). This will ensure a satisfactory completion of each stage of the work process and help reduce damage to a building's structure and fabric.

Patching up afterwards is more than likely going to be the job of the main contractor (builder). However, on smaller jobs you may have this responsibility. Therefore the less damage caused, the easier the repairs are likely to be.

The repair needed will obviously depend on the work done, and certainly in domestic installations you may only have to consider using substances such as Polyfilla to repair damage around ceiling outlets or switch boxes. However, in a domestic rewire also think about flooring, and replace floorboards with care, screwing them back in place after you work to prevent them from 'squeaking'.

On larger sites, you may have to think about replacing drop-in grid ceiling tiles or replacing fire barriers.

Always clean up after yourself and dispose of waste materials in the correct manner.

A successful installation will involve good relationships with all concerned, plus:

- compliance with requirements, specifications and drawings
- fixing of wiring systems and components
- ensuring electrical continuity and system integrity
- avoiding damage to property, components and systems.

Do these things and you'll have a happy client.

Remember

Try to keep all damage to property to a minimum. If a hole is needed in a wall for a cable to pass through, use a drilling machine with an appropriately sized masonry bit; don't use a hand grenade!

Remember

Take care with repair work. The more care and attention taken in restoring the work area to its proper condition, the more likely you and your company are to be recommended for further work

Remember

If you are removing old fluorescent lamps, have them removed and disposed of in special 'crushers' or local-authority facilities

Cable types

PVC (polyvinyl chloride)-insulated and sheathed cables are used extensively for lighting and power installations in domestic dwellings, being generally the most economical method of wiring for this class of work. This section will look at different types of cables and cords. Areas that will be looked at include the following:

- types of cable and cords
- installing PVC/PVC cables
- PVC/SWA/PVC and SWB cables
- fibre-optic cable
- Cat-5 cable
- collector columns
- trailing cables
- terminating cables and flexible cords
- connecting to terminals.

Types of cable and cords

Polyvinyl chloride is tough, cheap and easy to work with; it is also easy to install. It is not surprising, therefore, that PVC-insulated/PVC sheath cable is the most popular type of cable in current use. This form of insulation has its limitations, though, in conditions of excessive heat and cold. It can also be subject to mechanical damage unless you apply additional mechanical protection in certain situations. However, provided this and other factors mentioned later in this section are taken into consideration then the PVC/PVC wiring system is probably the most versatile of all the wiring systems.

Single-core PVC-insulated unsheathed cable

Application

Designed for drawing into trunking and conduit, its construction is that of PVC-insulated solid or stranded copper conductor (Ref 6491X), coloured red or black for single-phase systems. Other colours available include blue, green, grey, yellow, white and green/yellow stripes for use as a circuit-protective conductor.

Remember

From 1 April 2004 the colours changed as follows: red changed to brown, black changed to blue; green/yellow remained the same

Single-core PVC-insulated and sheathed cable

Application

This cable (Ref 6181Y) is suitable for surface wiring where there is little risk of mechanical damage. This cable is normally used as 'meter tails' for connecting the consumer unit/distribution board to the PES (Public Electricity Supply) meter. Single core is used for conduit and trunking runs where conditions are environmentally difficult. The construction of this cable is PVC-insulated and PVC-sheathed solid or stranded plain copper conductor (shown above). The core colours are normally blue or brown for single-phase systems. Sheath colours are normally black, red or grey; however, other colours are available.

Single-core PVC-insulated and sheathed cable with a CPC

Application

This cable (Ref 6241Y) is used for domestic and general wiring where a circuit protective conductor (cpc) is required for all circuits. Its construction is that of PVC-insulated copper conductor laid parallel with a plain copper cpc and PVC-sheathed overall. The core colours are brown or blue, the cpc is plain copper, and sheath colour is normally white or grey.

PVC-insulated and sheathed flat-wiring cables

Application

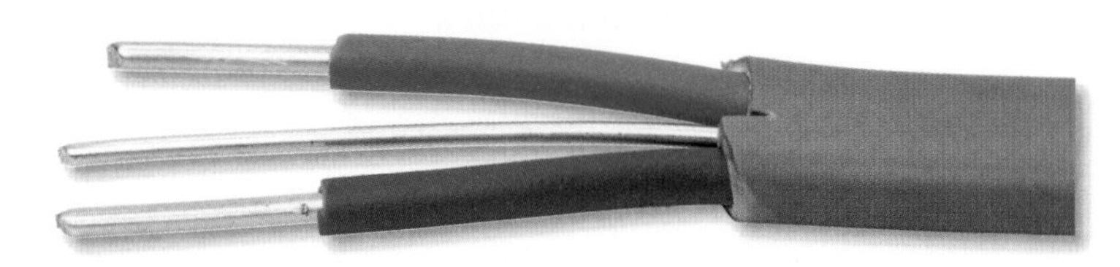

This cable (Ref 6242Y and 6243Y) is used for domestic and industrial wiring. It is suitable for service wiring where there is little risk of mechanical damage. In its construction there are two or three core cables, two or three plain copper, solid or stranded conductors, insulated with PVC and sheathed overall with PVC. The core colours for two cores are brown and blue for single-phase systems. For three cores (used for two-way switching) they are brown, blue and grey. The sheath colours are normally grey or white. The construction of three-core cables is exactly as mentioned above with the inclusion of an uninsulated plain copper circuit protective conductor between the cores of twin cables and between the yellow and blue cores of three core cables.

PVC-insulated and sheathed flexible cords

Application

These flexible cords (Ref 3092Y and 3093Y) are suitable for use in ambient temperatures up to 85°C. They are not suitable for use with heating appliances. Their construction is plain copper flexible conductors insulated with heat-resisting (HR) PVC and (HR) PVC-sheathed. General purpose PVC flexible cords (3182Y, 3183Y etc.) are also available. The core colours are:

- twin-core: brown and blue
- three-core: brown, blue and green/yellow
- four-core: black, grey, brown and green/yellow
- five-core: As above + blue.

Remember

When installing PVC/PVC twin and cpc cables for lighting circuits, twin brown is readily available

PVC-insulated and sheathed flat twin flexible cord

Application

This is intended for light duty in doors for table lamps, radios and TV sets where the cable may lie on the floor. It should not be used with heating appliances. The construction is that of plain copper flexible conductor PVC-insulated two cores laid parallel and sheathed overall with PVC. Core colours are brown and blue; the sheath colour is usually white.

PVC-insulated bell wire

Application

This is used for wiring bells, alarms and other indicators that operate at extra low voltage. The construction is one single-core plain soft copper conductor insulated with PVC. Twin-core wire is produced from two single core wires laid parallel and insulated overall with PVC compound to form a figure 8 section. The standard colour of this wire is white; core identification is by a coloured stripe on the insulation of one core.

Cables with thermosetting insulation

The thermosetting insulation is given the designation XLPE (cross-linked polyethylene). This means that the cable can be used in higher operating temperatures. The maximum continuous operating temperature for XLPE is 90°C compared with 70°C for PVC insulation. The increased temperature permits a reduction in conductor size if XLPE-insulated cables are used in preference to cables having PVC insulation. These cables are used mostly for mains distribution.

PILCSWA cables

These are Paper Insulated Lead Covered Steel Wire Armoured (SWA) cables. They are found on systems at 3.3 kV and over. To work with these cables you need to be specially trained.

LSF cables

These are cables with a low smoke- and fume-giving capacity in the case of fires. Some local authorities may require the installation of this type of cable in special installations.

Installing PVC/PVC cables

Cables are fixed using plastic clips that incorporate a masonry nail. The maximum spacing of clips for fixed wiring, when installed directly on a surface, were given earlier in Table 4.6 on page 248.

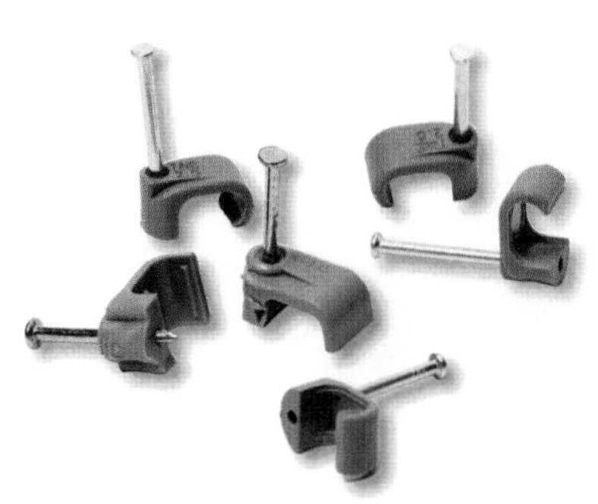

Cable clips

Additionally, BS 7671 Regulation 522.8.3 states that the radius of a bend should be such that conductors or cables do not suffer damage and terminals are not stressed. The bending radii for fixed wiring cables were given in Table 4.10 on page 249. The image opposite illustrates some typical clips used to fasten this type of cable.

Where PVC cables are installed on the surface the cable should be run directly into the electrical accessory, ensuring that the outer sheathing of the cable is taken inside the accessory to a minimum of 10 mm. If the cable is to be concealed a flush box is usually provided at each control or outlet position.

Installing, clipping and terminating PVC/PVC cable

Installing and clipping

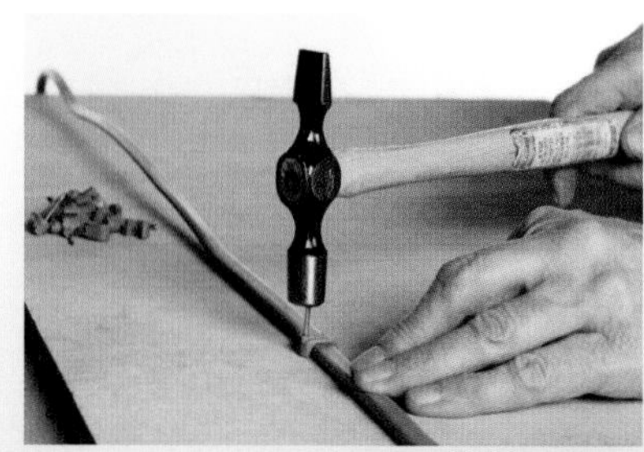

1. Fix the first clip using a small cross pein hammer.

2. In order to ensure a neat appearance PVC cable should be pressed flat against the surface between the cable clips.

3. The end clip should be fixed next.

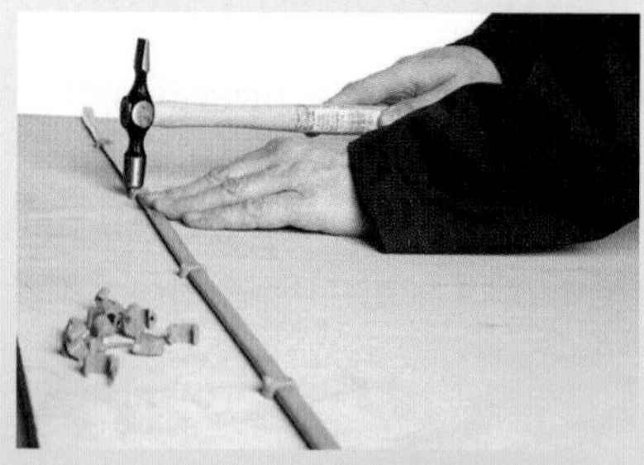

4. Clips should be equally spaced between the first and end clip.

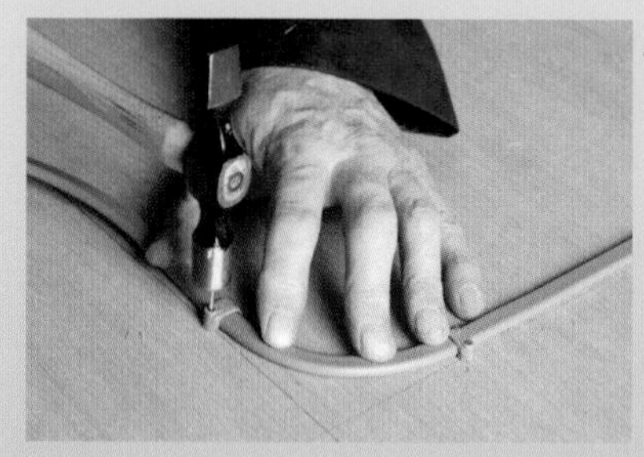

5. When a PVC cable is to be taken around a corner or changes direction the bend should be formed using the thumb and fingers as shown.

6. Final clip after insertion into enclosure.

Terminating

1. Nick the cable at the end with your knife and pull apart as shown.

2. When the required length has being stripped, cut off the surplus sheathing with the knife as shown.

3. The insulation can be stripped from the conductors with the knife or with a pair of purpose-made strippers, as shown. Examine the conductor insulation for damage.

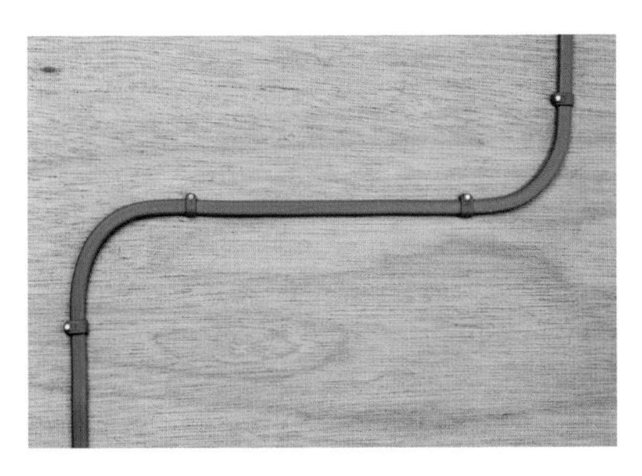

Care must be taken to ensure that the bend does not cause damage to the cable or conductors and that cable clips are spaced at appropriate intervals. The cable can be straightened by running the thumb over it before clipping. The palm of the hand can be used for bigger cables.

Cable suspension and catenary systems

Cables can be run outside between buildings by suspending them from a **catenary wire**. The catenary wire is usually a galvanised steel wire, which should be strained tight. The cables are fastened to this wire using tape or are suspended from it using hide hangers. In order to prevent travel of rainwater along the cable a drip loop should be formed at either end of the catenary wire. Figure 4.20 illustrates the catenary wire and the loop.

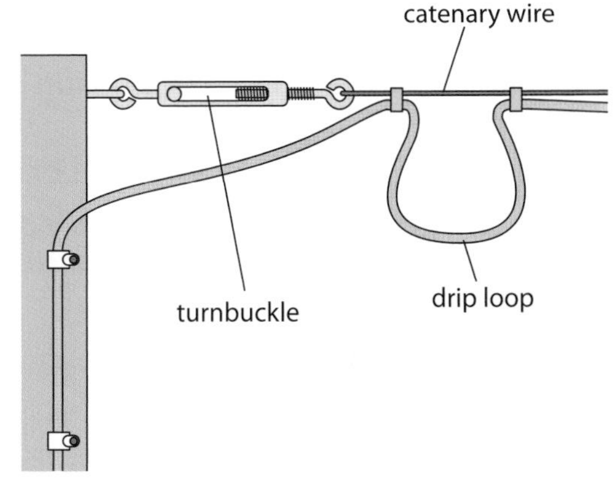

Figure 4.20 Cable suspension

Cable runs

Cable runs should be planned to avoid cables having to cross one another, which would result in an unsightly and unprofessional finish. When cables are to be installed in cement or plaster they should be protected against damage. They should be covered with metal or plastic channel or be installed in oval PVC conduit.

Care must be taken when installing PVC cables to ensure that they are not allowed to come into contact with:

- gas pipes
- water pipes
- any other non-earthed metal work.

PVC-sheathed cables should also not come into contact with polystyrene insulation, as a chemical reaction takes place between the PVC sheath and the polystyrene, resulting in the migration of polymers with the cable known as 'marring'.

FAQ

Q Where would I use single-core PVC-insulated and sheathed cables?

A These are more commonly known as double insulated, and are used extensively in domestic installations to connect the supply to the meter and to the consumer unit. They are used as these cables are not normally given any other protection, being clipped to the surface.

Remember

The disadvantages of PVC/GSWB/PVC cable are that it is more likely to sustain damage when subjected to temperatures in excess of 70 °C for long periods or when it is installed where the temperature is below 1 °C or in conditions where the insulation is likely to be split

PVC/SWA/PVC and PVC/GSWB/PVC cables

PVC/SWA/PVC cable

PVC-insulated SWA (steel wire armoured) and PVC-sheathed cables commonly known as PVC/SWA/PVC cables are used extensively for mains and sub-mains and for wiring circuits in industrial installations. The cable consists of multi-core PVC-insulated conductors made of copper or aluminium with a PVC-sheathed steel wire armour and PVC sheath overall. This type of cable has many advantages: it is more pliable and easier to handle than paper-insulated lead-covered cables, and termination is also relatively easy. This section will cover the installation and termination of various armoured cables.

A typical example of PVC/SWA/PVC cable

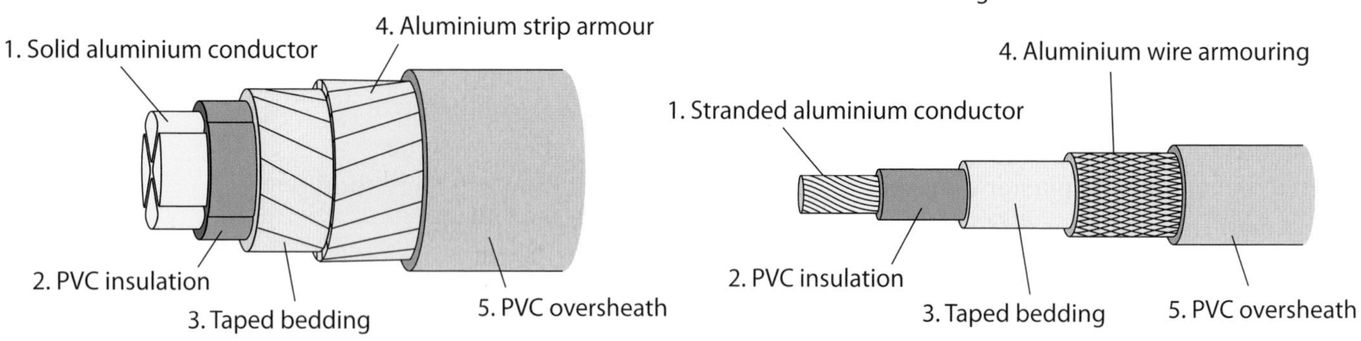

Figure 4.21 Four-core cable aluminium strip armoured and single-core sectoral cable

Terminating PVC/SWA/PVC cable

Remember

From 1 April 2004 the colours for 2-core changed as follows: red changed to brown; black changed to blue. From 1 April 2004 the colours for 3-core changed as follows: red changed to brown; yellow changed to black; blue changed to grey

The cable consists of PVC-insulated conductors with an overall covering of PVC. Between this covering and the outer PVC sheath the galvanised steel wire armouring is embedded. The armouring is used as a circuit protective conductor and special glands are employed to ensure good continuity between this and the metal work of the equipment to which you are connecting.

These glands vary a little from one manufacturer to another and their design also depends on the environment in which they are to be used. An earthing tag (bonding ring) provides earth continuity between the armour of the cable and the box or panel. If PVC/SWA/PVC cable is used to connect directly to an electric motor mounted on slide rails, a loop should be left in the cable adjacent to the motor to permit necessary movement.

Terminations are made by stripping back the PVC sheathing and steel wire armouring then fitting a compression gland, which is terminated into the switchgear or control gear housing.

For correct termination, start by measuring the length of armouring required to fit over the cable clamp; make a note of this measurement. Then establish how long the conductors need to be in order to connect to your equipment (make a note of it), then mark that distance on the cable by measuring from the end of the cable (position A) to position B (as shown in Figure 4.22(a)). This then represents the length of conductor required.

Follow this by marking the length required for the cable clamp from B to C as shown in diagram 4.22(a). At this stage some people will strip off the PVC outer sheathing. However, this is best left on, as it will hold the steel wire armouring in place for you. You must remember to place the shroud over the cable. This is often forgotten – a sign of poor workmanship.

(a)

Next, taking a junior hacksaw, cut through the PVC outer sheathing and partly through the armouring at point B. The PVC outer sheath can now be cut away as shown (b).

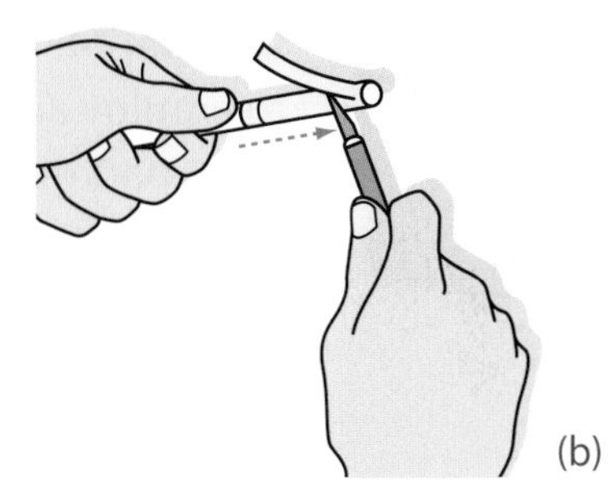

(b)

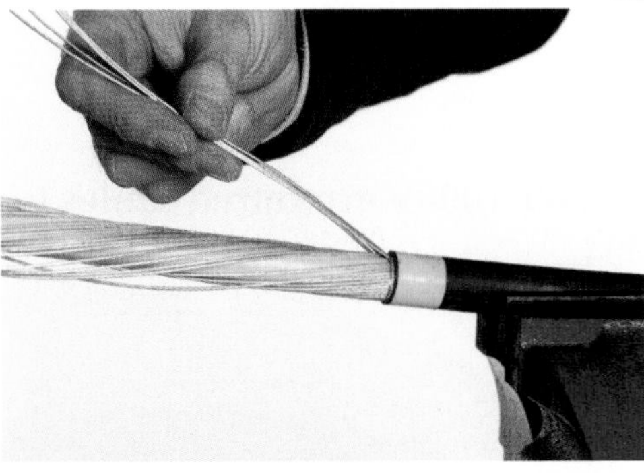

Taking each strand of the armouring in turn, snap them off at the point where they are partly cut through.

Then, using either a hacksaw or a knife, cut neatly around the PVC outer sheath at point C and remove the remaining pieces of outer sheathing; this will leave the cable as shown (c).

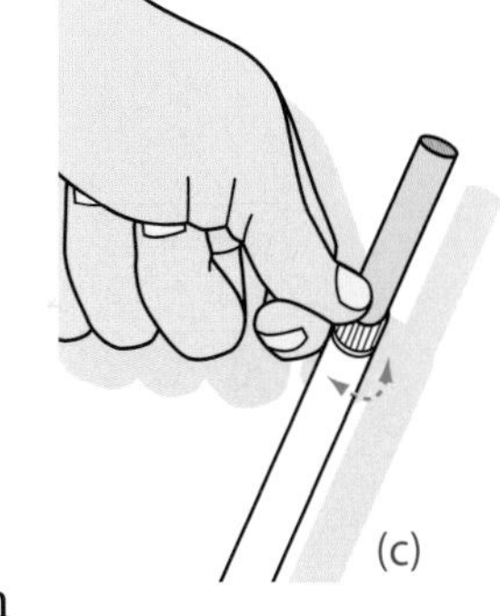

(c)

The gland can now be fitted onto the cable. First slide the backnut and compression rings, if any, on to the cable. Then, taking the gland body, slide this onto the cable, making sure that it fits under all the strands of armouring as shown (d).

(d)

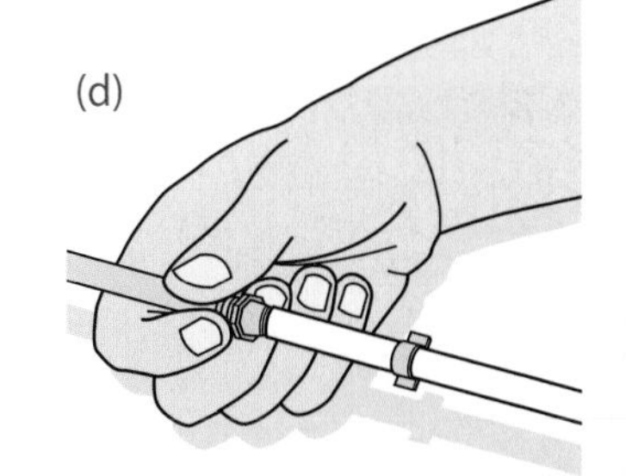

Figure 4.22 Terminating PVC/SWA/PVC cable

Finally, slide up the backnut and screw it onto the gland body, thus clamping the armouring tightly. The inner PVC sheath can be stripped off like any other PVC cable and the gland is then ready for connecting to your equipment.

Safety tip

If, while carrying out an installation of underground cables, you realise that excavations are still taking place, you should inform your supervisor before continuing with the installation

It is important to clean any paintwork from the area of contact before tightening up the locknut and securing the gland. Bonding rings, or earthing tags as they are sometimes called, can be used to provide better contact with surrounding metal work. A cpc (circuit protective conductor) should be fitted between the bolt securing the earthing tag and the earthing terminal of the equipment.

Installation

Cables can be laid directly in the ground, in ducts or fixed directly onto walls using cable cleats. If several cables are to follow the same route they are best supported on cable trays or racks. When several cables are installed in enclosed trenches the current ratings will be reduced due to their disposition. The correction factors for cables run under these conditions are found in BS 7671 Appendix 4.

Cleats

Installation is relatively easy for the smaller size cables but it will become necessary to employ an installation team to handle the bigger sizes or multi-core cables. For the most part one-hole cable cleats constructed of solid PVC will be used. In the case of cables installed on a cable tray, cable ties will be used.

For the bigger cables, cleats made of die-cast aluminium are used. These are often designed to be slotted into steel channels so that, once the channels are in place, multiple runs of cable can be accommodated. Support spacings for this and other cables were given in Table 4.6 on page 248, with bending radii for this and other cables given in Table 4.10 on page 249.

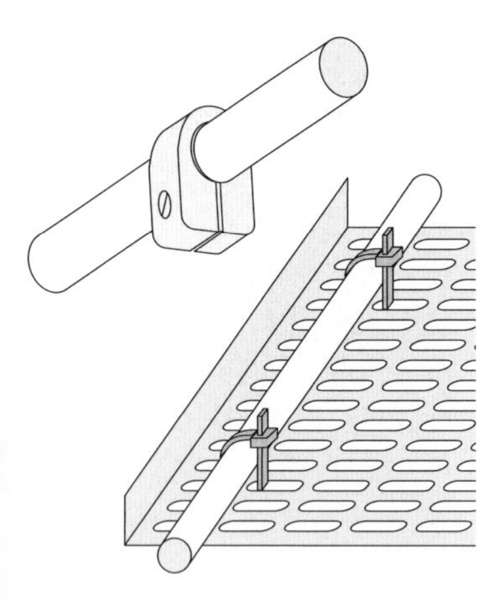

Figure 4.23 Cable tray and cleat

Conductor colour identification for standard 600/1000V armoured cables to BS 6346, BS 5467 or BS 6724 from 1 April 2004

- single core was red or black, is now brown or blue
- two core was red, black, is now brown, blue
- three core was red, yellow, blue is now brown, black, grey
- four core was red, yellow, blue, black is now brown, black, grey, blue
- five core was red, yellow, blue, black, green and yellow is now brown, black, grey, blue, green and yellow.

PVC/GSWB/PVC cable

Find out

This information can be obtained from Appendix 7 of BS 7671

PVC/Galvanised Steel Wire Braided/PVC cable consists of individual conductors within an aluminium screen surrounded by an inner sheath, then by a steel braid, similar to a basket weave, underneath an outer sheath of PVC. The braid makes this cable more flexible than SWA.

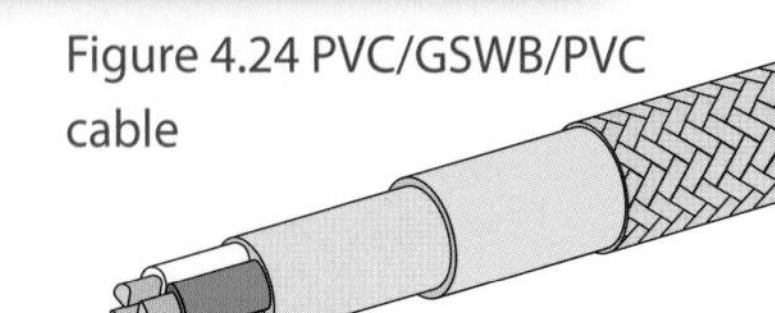

Figure 4.24 PVC/GSWB/PVC cable

This durable cable is used in many instrumentation applications or where shielding is required for signal applications.

Installation techniques are virtually identical to SWA, but cable glands need either to be of the universal type or to have a special **olive** inside them to hold the thin braid.

Fibre-optic/data cables

This cable is used for digital transmissions used by equipment such as telephones and computers. They are made from optical-quality plastic (the same as spectacles) where digital pulses of laser light are passed along the cable from one end to another with no loss or interference from mains cables. They look like SWA cables but of course they are much lighter and contain either one core or many dozens of cores. Tight radius bends in this type of cable should be avoided, as should 'kinks', as the cable will break. Jointing of these cables requires specialist tools and equipment. Never look into the ends of the cable as the laser light could damage your eyes.

Fibre-optic cable

The applications of optical fibre communications have increased at a rapid rate since the first commercial installation of a fibre-optic system in 1977. Telephone companies began early on replacing their old copper-wire systems with optical fibre lines. Today's telephone companies use optical fibre throughout their system as the backbone architecture and as the long-distance connection between city phone systems.

Light rays modulated into digital pulses with a laser or a light-emitting diode move along the core without penetrating the cladding. The light stays confined to the core because the cladding has a lower refractive index (a measure of its ability to bend light).

Cable television companies have also begun integrating fibre-optics into their cable systems. The trunk lines that connect central offices have generally been replaced with optical fibre. Some providers have begun experimenting with fibre to the curb using a fibre/coaxial hybrid. Such a hybrid allows for the integration of fibre and coaxial at a neighbourhood location. This location, called a node, would provide the optical receiver that converts the light impulses back to electronic signals. The signals could then be fed to individual homes via coaxial cable.

Fibre-optic cables are also used in Local Area Networks (LAN). These collective groups of computers, or computer systems, connected to each other, allow for shared program software or databases. Colleges, universities, office buildings and industrial plants, just to name a few, all make use of fibre-optic cables within their LAN systems.

Power companies are emerging as big users of fibre optics in their communication systems. Most power utilities already have fibre-optic communication systems in use for monitoring their power grid systems.

Cat (Category) 5 cable

This cable is used extensively for data transfer in computer networks and telephone systems. It has four pairs of wires that transmit and receive data along them at very high frequencies, typically 350 MHz. Special termination ends are required for these cables.

There are three basic types of cabling used in data systems: coaxial, fibre-optic and Unshielded Twisted Pair (UTP). Coaxial is widely installed in older networks but is not recommended for new network installations. Fibre-optic is used for high-speed networks and to connect networking devices separated by large distances. But UTP is currently the most common and recommended cabling type.

UTP is inexpensive, flexible and can transmit data at high speeds. Most new installations are currently installed with Cat-5 UTP cabling and components.

Collector column

Collector columns

Collector columns supply electrical current to a rotating unit from a fixed source. They can be easily adapted to an endless variety of applications including slewing cranes, sewage treatment plant, test rigs and rotating displays.

The columns usually contain a mounting frame, main body and slipring cover manufactured from aluminium, and are fitted with phosphor-bronze sliprings and dual leg brush gear, incorporating copper graphite brushes. The slipring assembly is mounted on a mild steel shaft rotating in self-lubricating bearings. The slipring cover then incorporates a location pocket to accept the customer's drive or anchor pin.

Trailing cables

Other types of installation might require the use of trailing cables. The main characteristics of a trailing cable are its ability to flex and bend and also to withstand varying levels of mechanical stress. Invariably these cables contain stranded conductors to give them this flexibility.

For example in certain construction sites or in mining, heavy plant may have their operating cables trailed across a site that will see vehicles passing by and have to withstand adverse weather conditions. Such cables are made with heavy inner and outer sheathing and will usually contain wire braid armouring or flexible steel-wound armouring.

Structure of a trailing cable

They will normally be terminated using special couplers and have to withstand regular bending as part of their normal use. However, cables feeding light equipment may only risk being dragged or scraped (e.g. when using a drilling machine), and their construction may therefore only consist of heavy-duty rubber sheathing.

Fibre-optic cables

If you wanted to see down a dark corridor, you might shine a torch down it. But what if the corridor had a bend in it? You could probably put a mirror in just the right place at just the right angle and shine the light round the corner. But what if the corner had lots of bends? Well, what if I made the entire corridor walls out of mirrors, then I wouldn't need to put them in just the right place or angle. The light would be able to bounce around all the mirrors along the walls.

Believe it or not, that's the theory behind fibre-optics, as the glass core is essentially a mirror wound into a thin tube. Some 10 billion digital bits can be transmitted per second along an optical fibre link in a commercial network, enough to carry tens of thousands of telephone calls. The hair-thin fibres consist of two concentric layers of high-purity silica glass, the core and the cladding, which are enclosed by a protective sheath.

These types of cable are also used in factories and manufacturing processes where production machinery is constantly moving, e.g. robotic equipment.

Terminating cables and flexible cords

The entry of the cable end into an accessory is known as a termination. In the case of a stranded conductor the strands should be twisted together with pliers before terminating. Care must be taken not to damage the wires. BS 7671 Chapter 52, Section 526 requires that a cable termination of any kind should securely anchor all the wires of the conductor that may impose any appreciable mechanical stress on the terminal or socket. A termination under mechanical stress is liable to disconnection.

When current is flowing in a conductor a certain amount of heat is developed and the consequent expansion and contraction may be sufficient to allow a conductor under stress, particularly tension, to be pulled out of the terminal or socket. One or more strands or wires left out of the terminal or socket will reduce the effective cross-sectional area of the conductor at that point. This may result in increased resistance and probably overheating. When terminating flexible cords into a conduit box the flex should be gripped with a flex clamp. Any cord grips that are used to secure flex or cable should be clamped down onto the protective outer sheathing.

Remember

Despite the implication, trailing cables are not indestructable and care must be taken to regularly inspect them for damage

FAQ

Q Can PVC twin and earth cable be used unsupported to span between buildings?

A Yes. Table 4.7 on page 248 we can see that the maximum distance should not exceed 3 metres. Though commonly done this is not good practice for two reasons. (1) Over 3 metres the cable is not designed to support its own weight and it would eventually stretch, reducing its csa and therefore current-carrying capacity. Cable used in this way should have a catenary wire fixed to both buildings then the cable secured to the wire. (2) If exposed to direct sunlight the sheath of the cable has to be protected against degradation from solar radiation; this generally requires the sheath to be black in colour.

Q Can PVC twin and earth cable be buried in the ground, such as a feed to a garage?

A Laid in the ground with no additional protection, no. But if additional protection is given, such as concrete tiles being laid directly on top of the cable, as well as the ends where it comes out of the ground being protected, then yes, However, it's not usual to use this type of cable. A SWA, MICC or cable enclosed in conduit are more common.

Connecting to terminals

Types of terminals

There is a wide variety of conductor terminations. Typical methods of securing conductors in accessories are pillar terminals and screwheads with nuts and washers.

Pillar terminals	A pillar terminal is a brass pillar with a hole through its side into which the conductor is inserted and secured with a setscrew. If the conductor is small in relation to the hole it should be doubled back. When two or more conductors are to go into the same terminal they should be twisted together. Care should be taken not to damage the conductor by excessive tightening.
Screwhead, nut and washer terminals 	Using round-nosed pliers form conductor end into an eye, slightly larger than the screw shank but smaller than the outside diameter of the screwhead, nut or washer. The eye should be placed in such a way that rotation of the screwhead or nut tends to close the joint in the eye.
Claw washers 	Claw washers are used to get a better connection. Lay the looped conductor in the pressing. Place a plain washer on top of the loop and squeeze the metal points flat using the correct tool.
Strip connectors 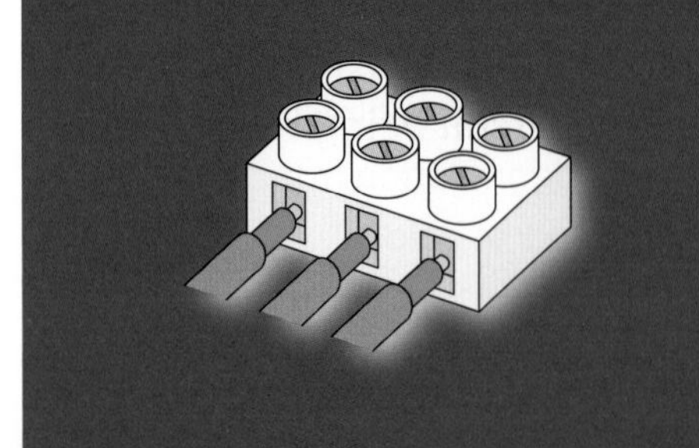	Conductors are clamped by brass grub screws in the connectors mounted in a moulded insulated block. The conductors should be inserted as far as possible into the connector so that the pinch screw clamps the conductor. A good clean, tight termination is essential in order to avoid high-resistance contacts resulting in overheating of the joint.
Solderless lugs 	Lugs are made from tinned solid copper. Used extensively in electrical contracting industry for terminating smaller sized cables. Fastened to cable ends by crimping.

Figure 4.25 Types of terminal

Remember

If the eye is put in the opposite way round, the motion of the screw or nut will tend to untwist the eye and will probably result in an imperfect contact

Remember

Where compression joints are used they should be made with the tools specified by the manufacturer and should be to the approved British Standard

Terminating cable ends to crimp terminals

In order to terminate conductors effectively it is sometimes necessary to use crimp terminals, for instance in the termination of bonding conductors to earth clamps and sink tops. The terminals are usually made of tinned sheet copper with silver braised seams. The crimping tool is made with special adjustable steel jaws so that a range of cable end terminals can be crimped.

Cable joints and connections

Joints and connections should be avoided wherever possible. Where they must be used, they should be mechanically and electrically suitable. They must also be accessible for inspection and testing except for the following:

- A joint designed to be buried in the ground.
- A compound filled of encapsulated joint.
- A connection between a cold tail and the heating element as in ceiling heating, floor heating or a trace heating system.
- A joint made by welding, soldering, brazing or appropriate compression tool.
- A joint forming part of the equipment complying with the appropriate product standard.

The joints in non-flexible cables shall be made by soldering, brazing or welding mechanical clamps, or be of a compression type and be insulated. The devices used should relate to the size of cable and be insulated to the voltage of the system being used. Connectors used must be of the appropriate British Standard, and the temperature of the environment must be considered when choosing the connector. EAW Regulations state that every joint and connection in a system shall be installed so as to prevent corrosion.

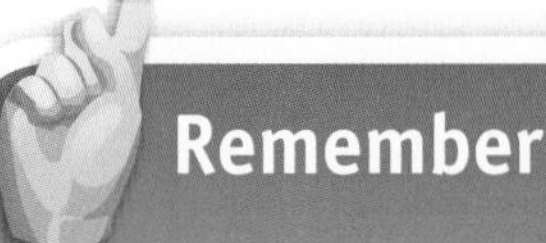

Remember

When a cable is not long enough or alterations are necessary, connectors may have to be used. These are often used in mineral-insulated systems. Where actual switching circuits are wired or where cables are taken from ring final circuits, junction boxes will normally be used

Plastic connector

Among the commonest types of connector, plastic connectors often come in a block of 10 or 12, sometimes nicknamed a 'chocolate block'. The size required is very important and it should relate to the current rating of the circuit being used. They are available in 5, 15, 20, 30 and 50 A ratings.

Porcelain connectors

These are used where a high temperature may be expected. They are often found inside appliances such as water heaters, space heaters and luminaries. They should also be used where fixed wiring has to be connected into a totally enclosed fitting.

Screwits

These are porcelain connectors with internal porcelain threads that twist onto the cable conductors. They are now obsolete, but you may come across them in older installations.

Compression joints

This includes many types of connectors that are fastened onto the conductors, usually by a crimping tool. The connectors may be used for straight-through joints or special end configurations. If the conductors are not clean when making the joint with a crimping tool this may result in a high-resistance joint, which could cause a build-up of heat and eventually lead to a fire risk.

Uninsulated connectors

These are often required inside wiring panels, fuse boards etc. and are used to connect earth cables and protective conductors.

Junction boxes

The two important factors to consider when choosing junction boxes are the current rating and the number of terminals. Junction boxes are usually either for lighting or socket-outlet circuits.

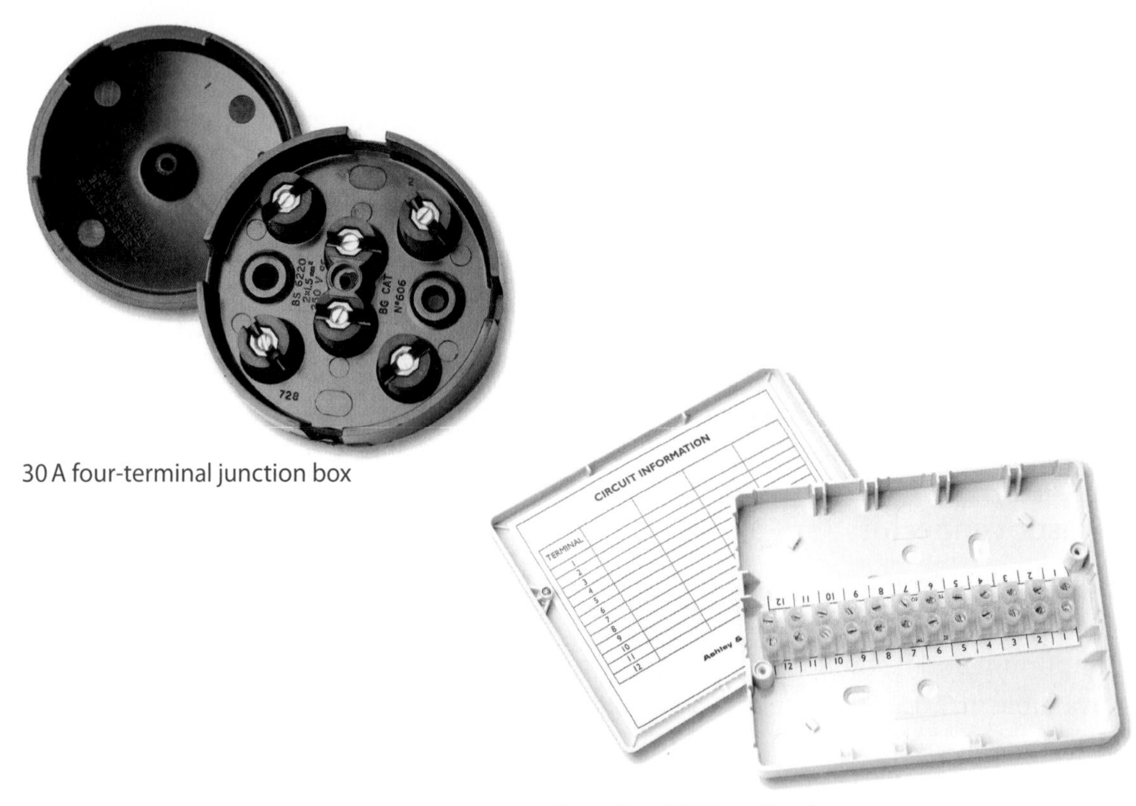

30 A four-terminal junction box

A modern RB4 junction box

Soldered joints

Although soldering of armoured and larger cables is still common, the soldering of small-circuit cables is very rarely carried out for through joints these days. This joint was often referred to as a marriage joint, and involved stripping back the insulation, carefully twisting together conductors, soldering and then finally taping up.

MICC cable

Mineral-insulated copper cables (MICC) consist of high-conductivity copper conductors insulated by a highly compressed white powder (magnesium oxide). A seamless copper sheath encapsulates the conductors and powder.

This type of cable originated in France and was introduced into the UK in 1936. The first company to market these cables in the UK was Pyrotenax and from this name came the term 'pyro', which is still sometimes used when referring to this cable. The cable is made by placing solid copper bars in a hollow copper tube. The magnesium oxide powder is then compacted into the tube and finally the whole tube, powder and copper bars are drawn out by pulling and rolling. This reduces the overall size while further compressing the powder.

In this section we will look at the following:

- properties of MICC cables
- terminating MICC cables
- practical termination of MICC cables
- running of cables
- testing and fault finding

Properties of MICC cables

Mineral-insulated cables have very good fire-resisting properties: copper can

MICC cable

withstand 1000°C and magnesium 2800°C. The limiting factor of the whole cable system is the seal, and where a high working temperature is required special seals must be used. MI cables have the following qualities:

- the cable is very robust and can be bent or twisted within reasonable limits, hence its use in emergency lighting and fire-alarm systems
- for a given cross-sectional area, MI cables have a very high current-carrying capacity
- relative spacing between the conductors and sheath is maintained when the cable is flattened, hence maintaining the cable's insulation properties
- MI cables are non-ageing (many cables installed in the 1930s are still in operation today)
- the cable is completely waterproof, although where it is to be run underground or in ducts a PVC oversheath must be used

- bare copper unsheathed MI cables do not emit smoke or toxic gases in fires
- where PVC oversheath is used, the reduced volume of PVC in comparison with PVC-insulated cables keeps down smoke output. Special oversheaths are also available where a further reduction of flame propagation is required
- the copper sheath can be used for earth continuity, saving the need for a separate protective conductor
- these cables come into their own in areas such as boiler-houses where the ambient temperature can become high and there is moisture present.

Terminating MICC cables

Mineral-insulated cables must be sealed at each end, otherwise the magnesium oxide insulator will absorb moisture, resulting in a low insulation-resistance reading. A complete termination comprises two sub-assemblies, the gland and the seal, each performing a different function.

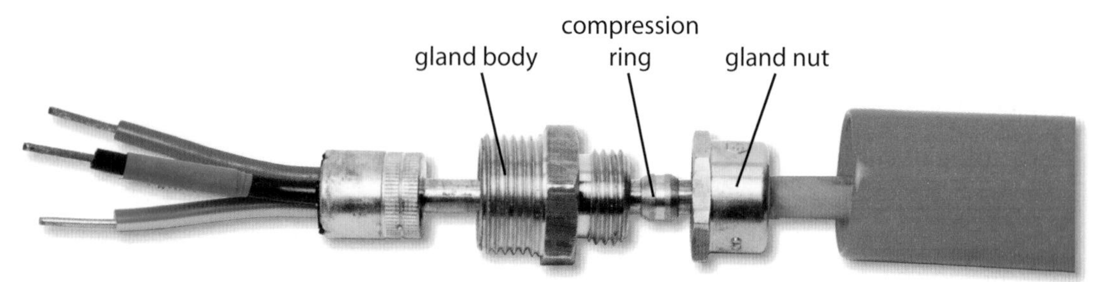

MICC gland and seal

The gland is used to connect or anchor the cable to a piece of equipment and consists of three brass components (gland body, compression ring and gland nut), as shown in the above photograph. The purpose of the seal is to exclude moisture from the cable. The seal consists of a brass pot that is screwed onto the copper sheath of the cable and then filled with a special compound. A plastic disc is then slotted over the conductors and compressed into the 'mouth' of the pot and the neoprene rubber sleeves are fitted to insulate the conductor tails.

Practical termination of MICC cables

The first thing that must be done when terminating MI cables is to strip off the copper sheath. If the cable has a PVC oversheath, then this must be removed before the copper sheath. Some electricians recommend that the cable should be indented with a ringing tool prior to sheath stripping. It may, however, be better to carry out ringing nearer the end of the stripping process. Stripping off the copper sheath can be carried out in several ways, although all of them require practice before the task can be carried out in a reasonable time.

Stripping

Using side cutters

When using side cutters or pliers it is essential that the points of the blades are in good condition, otherwise difficulties can arise. First make a small tear in the sheath, then peel off the sheath with the side cutters, working around the cable in a clockwise direction, at the same time keeping the side cutters at an angle of approximately 45° to the cable.

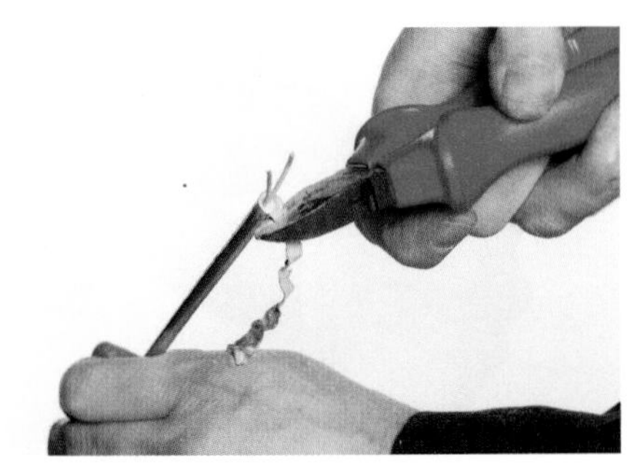

Using side cutters

Using stripping bar (fork-ended stripper)

First the cable sheath is started or broken away with side-cutting pliers as above. Then the torn-off sheath portion is flattened and the slot of the stripping bar pushed into the torn-off sheath. The stripping bar is then rotated while keeping the bar at 45° to the cable.

Using a stripping bar

Using rotary stripping tools

Rotary stripping tools are available in many shapes and forms. They are much easier to use than either side cutters or the stripping bar but can be difficult to set up. This is not a problem if the same size cables are being used all the time. Rotary strippers usually require a good square end to the cable. The stripper is then pushed over the cable, a little pressure applied and then rotated. If the sheath is to be stripped for a longer distance, then the sheath being removed should be cut away at intervals to avoid fouling the tool. Another advantage of the stripping tool is that a ringing tool is not required. When the stripping tool has reached a required position, pliers are applied to the cable sheath and the stripper is further turned.

Rotary stripping tool

Ringing cable sheath

Before the pot can be fixed the sheath must have a good, clean circular end. To achieve this we use a ringing tool, which makes an indent in the sheath. This will be required when the sheath is stripped with either side cutters or the stripping bar, but is not necessary (as previously mentioned) when rotary strippers are used.

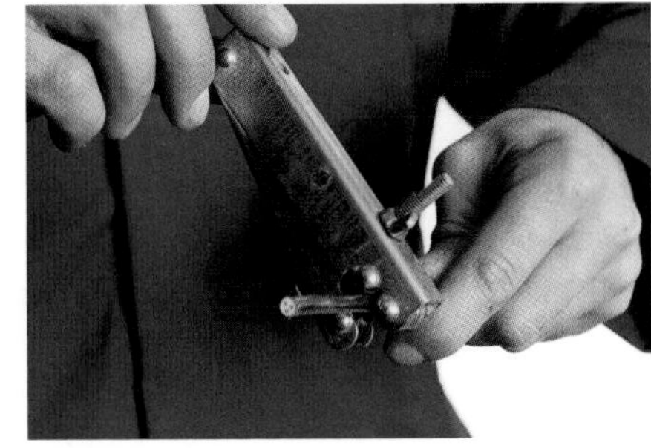

Ringing cable sheath

Actions prior to fitting the pot

Before fitting the pot a check should be made that the gland nut and olive are in position. If the cables have the PVC oversheath then a shroud must be fitted.

Fitting the pot

Fitting the pot

The following describes the fitting of a screw-on pot. Pots are best fitted using a pot wrench. This ensures that the pot goes on square. The pot has an internal thread, which screws on to the copper sheath. The pot should be screwed on to the copper sheath until it just lines up in the inside of the pot.

Finally the pot is tapped to empty the pot of any filings or loose powder. Do not be tempted to blow into the pot as the moisture in your breath could reduce the insulation resistance of the magnesium oxide, causing a short circuit at a later stage.

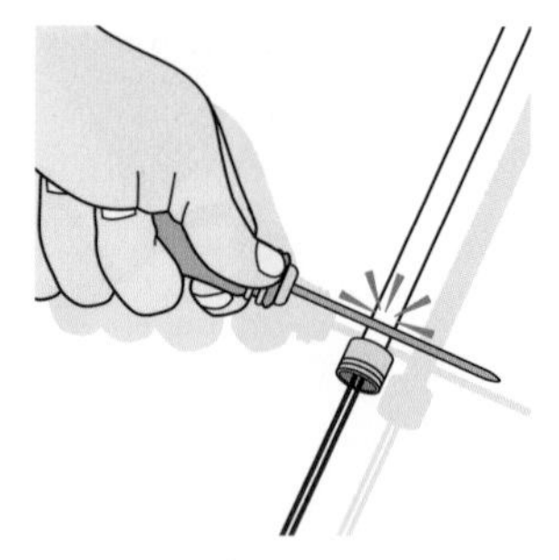
Tapping the pot empty of powder

Actions prior to sealing

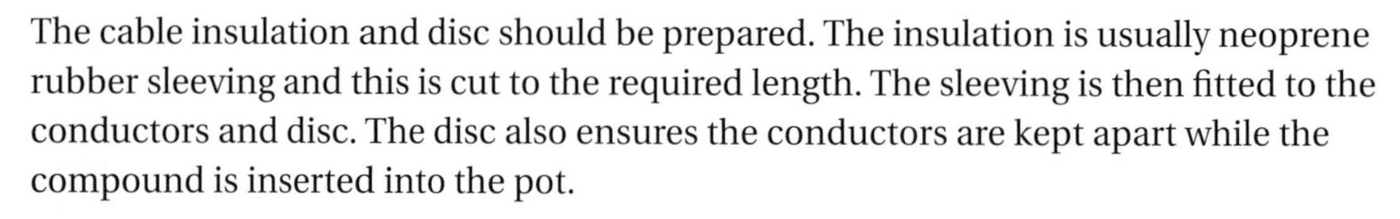
MICC pot and seal

Before sealing, the conductors should be wiped with a clean, dry rag to remove any loose powder. If the cables have become twisted they should be straightened by being firmly pulled with a pair of pliers.

The cable insulation and disc should be prepared. The insulation is usually neoprene rubber sleeving and this is cut to the required length. The sleeving is then fitted to the conductors and disc. The disc also ensures the conductors are kept apart while the compound is inserted into the pot.

Safety tip

Care should be taken when pulling conductors so that they are not stretched, resulting in a reduced cross-sectional area and subsequent reduction in current-carrying capacity

Sealing the pot

Filling of the pot with compound affects the sealing. It is important that the compound is filled from one side only, directed towards and in between the conductors. This prevents air locks, which could lead to condensation problems. Hands should be clean when using the compound and the compound should be kept covered to avoid entry of dirt.

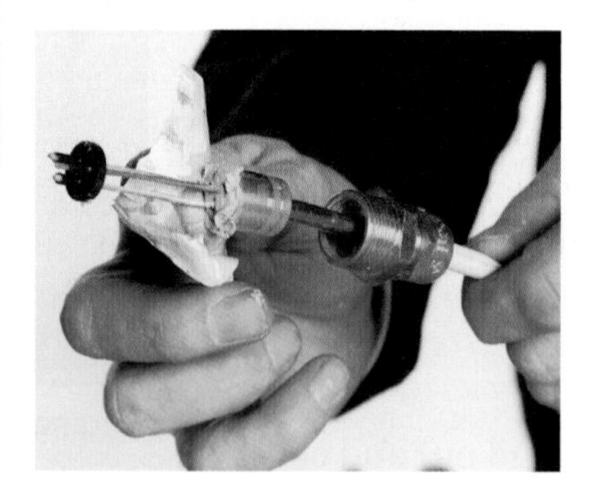

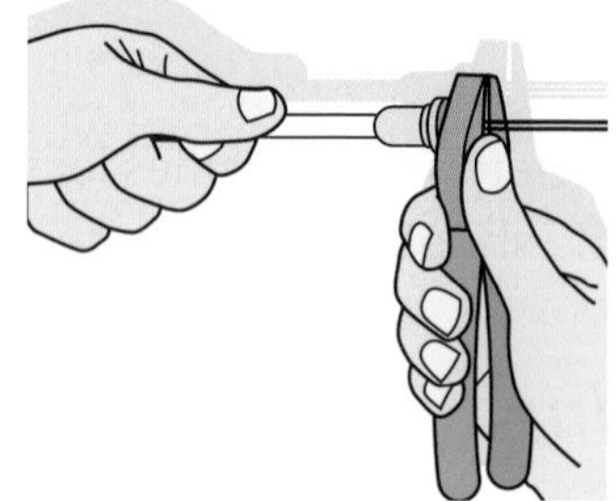
Sealing the pot

Crimping

Before the crimping tool is applied any excess compound should be removed. The crimping tool is applied and gradually tightened while keeping it straight. The operation should be stopped at intervals to allow the compound to seep out.

Crimping the pot

Finished seal and identification

When crimping is completed the seal should be tested to ensure a high value of insulation resistance between cores and from cores to the copper sheath. Having done this, the cable can be wrung out and the other end sealed. The whole cable is then insulation tested and the conductors are given a continuity test to indicate which conductor is which before marking them for identification.

Gland connection

When the cable has been tested and conductors identified, the gland must be tightened up and connected into the equipment or accessory in which it is being terminated. After the conduit thread has been tightened into the accessory the back nut must also be tightened. This will compress the olive ring inside the gland, thereby providing earth continuity. Once this has been tightened it is very difficult to remove.

Types of gland

Standard glands have their sizes stamped on them, e.g. 2LI meaning 2-core, light-duty, 1 mm^2 csa (cross-sectional area) or 4H6 meaning 4-core, heavy-duty 6 mm^2 csa. The size of the conduit thread will depend on the overall size of the cable being used.

Running of cables

One of the advantages of MI cables is that they can be run on the surface as well as being run under plaster. When they are run on the surface, for neat appearance it is imperative that they are run straight. Lines should be run out before fixings are made and care should be taken not to twist the cable unnecessarily before installation. Fixings are made either with one-hole clips or two-hole saddles. Screws are usually brass round-heads.

Where a large number of cables are run together an adjustable saddle is available that will take many cables. Often MI cables are run on cable tray. This is used to provide easier fixing, neat appearance and cleanliness. It also means that a great number of cables can be run together. Where several cables leave or enter an enclosure, e.g. at a distribution board, the cable should be distributed equally. If there are too many cables to fit into one row, two staggered rows of holes should be drilled.

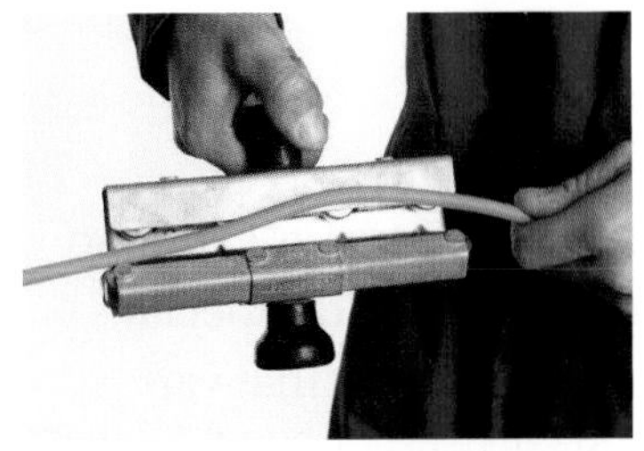

Putting bent cable in roller

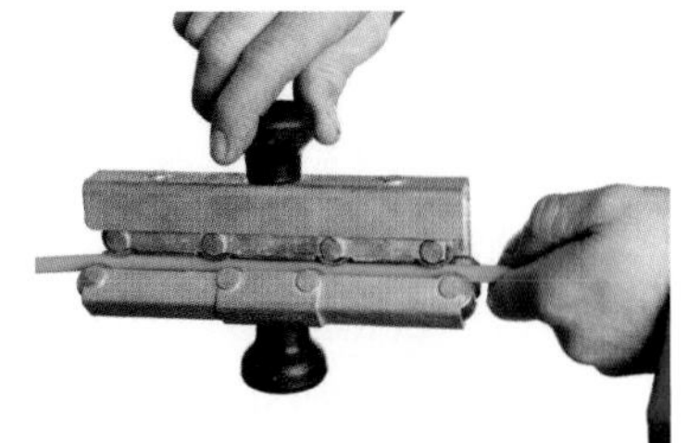

Closing roller

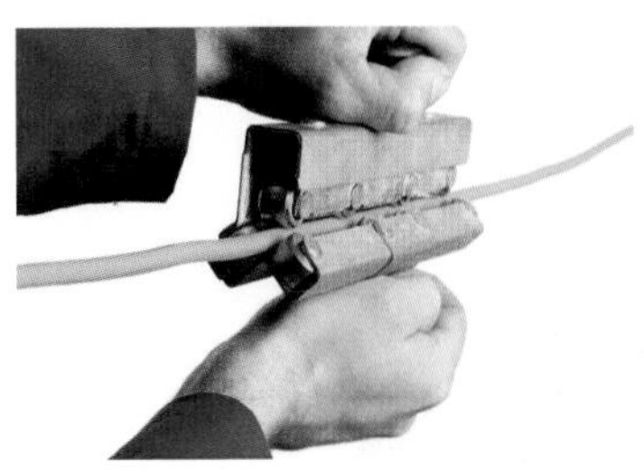

Running roller along cable

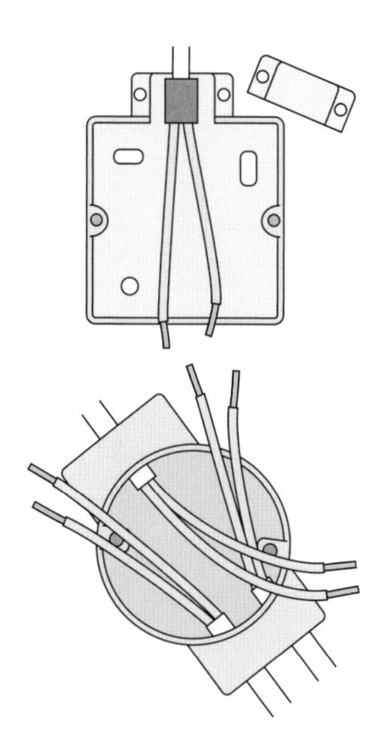

Figure 4.26 Types of plaster-depth termination boxes

MI cables are usually terminated using standard conduit boxes. Where connections have to be made, connectors should be used. Where cables are run buried in plaster, plaster-depth boxes are available. With this type of box no gland is required. It is also important to note that the earth clamp over the pot is secured to give continuity. Earth-tailed pots are also available. These are pots that have an earth conductor manufactured as part of the pot, thus giving good earth continuity. If this cable is to be installed directly into a motor then a loop should be made in the cable to prevent mechanical stress at the termination due to excessive vibration.

Testing and fault-finding

The seal should be visually inspected for obvious defects. If there is a minor fault, e.g. incomplete crimping, it may be practicable just to repeat the operation. However, it may be necessary to remove the seal and re-terminate. After both ends of the cable have been terminated with permanent seals, the cable should be subjected to an insulation-resistance test using a voltage appropriate to the intended operating voltage in accordance with BS7671. The purpose of this initial test is to check for major faults, e.g. short circuit within the pot, in which case the fault should be located and rectified. The insulation resistance should be noted and compared with the value measured at least 24 hours later. The second reading should be at least 100 MΩ and should have risen from the initial value.

FP 200 cable

Construction of FP 200

Normally used for fire alarms and fire detection systems, there are two types of FP 200 cable, FP 200 Gold and FP 200 Flex, the main difference being that FP 200 Gold has solid conductors, whereas FP 200 Flex uses stranded conductors. In this section we only look at FP 200 Gold.

As shown in the photograph below, FP 200 Gold has solid copper conductors covered with a fire- and damage resistant insulation (Insudite). An electrostatic screen is then provided via a laminated aluminium tape screen that is bonded to the sheath of the cable and which is in full contact with a tinned annealed uninsulated cpc. The sheath is a robust thermoplastic low-smoke, zero-halogen sheath, which is an excellent moisture barrier.

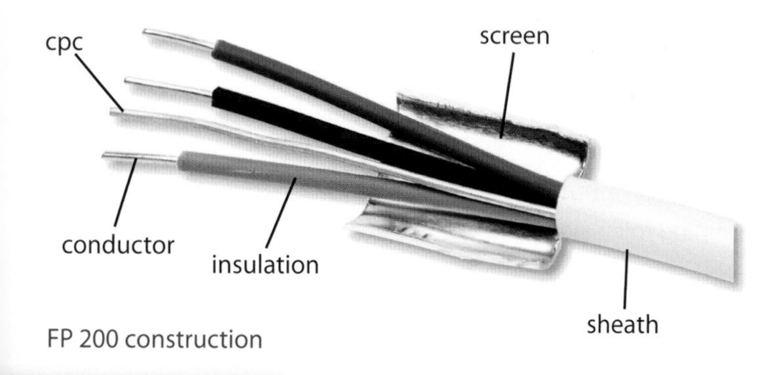

FP 200 construction

Types of FP 200 Gold

FP 200 Gold is available in 2, 3 and 4 cores as standard with 7, 12, and 19 core versions available on special request. All versions have conductor sizes that vary from 1 mm^2 to 4 mm^2. The cable is used in fixed installations in dry or damp premises in walls, on boards, in channels or embedded in plaster for situations in which prolonged operation is required in the event of fire. It is primarily intended for use in fire-alarm and emergency-lighting circuits and has a low voltage rating of 300/500 volts.

The sheath colour is available in red or white, although special colours are available on request. Core colours are:

- two-core: brown and blue
- three-core: brown, black and grey
- four-core: brown, black, grey and blue.

Advantages	Disadvantages
• Easy to handle • Designed to use with conventional terminations • No special tools are required • Age-resistant insulation • Bends easily without the use of bending tools	• Can be damaged if the bending radii are not observed • Cable lacks mechanical strength • Easily damaged if not correctly terminated

Table 4.11 Advantages and disadvantages of FP 200 Gold

Installing FP 200 Gold

When the cable is required to maintain circuit integrity during a fire, it is important that any clips or ties used to support the cable can also withstand that fire. Clips should be copper, steel or copper-coated, or ties suitable for fixing a fire-rated cable; standard plastic or nylon clips or ties should not fix the cable. A new fixing system has been introduced using coated stainless steel clips and gas nailing technology.

There is no need to use glands: the use of glands is entirely a function of the installation requirements. However a range of standard nylon compression glands is available if required. The cable should only be dressed by hand to prevent damage to the cable. The insulation strips easily from the conductors, leaving them in a bright, clean condition, eliminating the risk of any high-resistance terminations. When bending this cable the bending radius should not be less than six times the diameter of the cable.

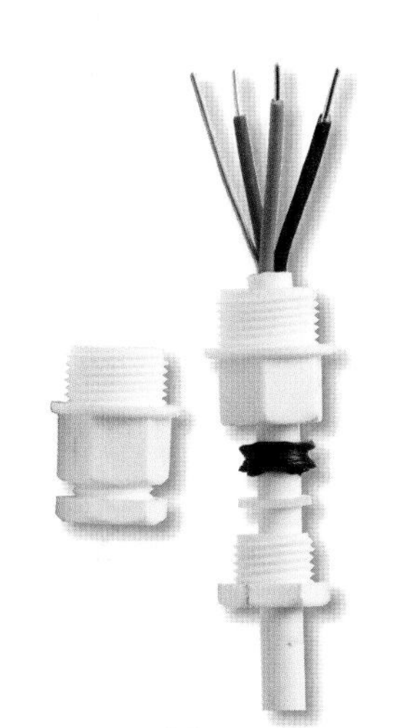

FP 200 Gold termination

Termination

FP 200 does not possess the same mechanical strength as other cables. It is important to terminate the cable according to the manufacturer's instructions. The following sequence explains a recognised method of termination.

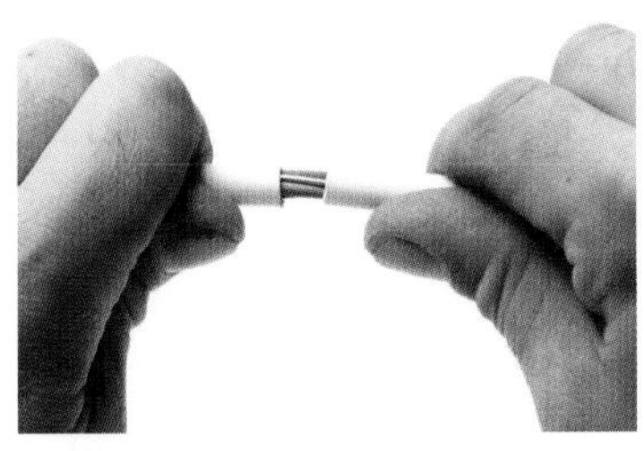

Pulling the sheath

Score around the sheath with a knife or suitable cable-stripping tool, taking care not to cut right through to the aluminium tape. Flex the cable gently at the point of scoring until the sheath yields.

Pull off the sheath, twisting gently to follow the lay of the cores.

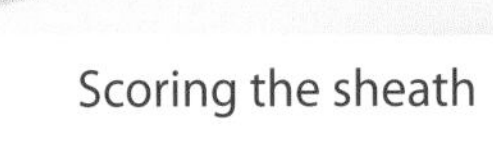

Scoring the sheath

FP 200 Gold does not require a ferrule; the cable may enter wiring accessories or fittings through a simple grommet. When installed in wet conditions or outdoors, a

Remember

When terminating cables in enclosures where space is limited (e.g. plaster-depth boxes), sleeving should be applied over the insulated core and earth conductors to prevent 'pinching' of the insulation and consequent damage

standard waterproof gland incorporating a PCP sealing ring must be used. Earthing of the aluminium screen is achieved automatically by correctly terminating the bare tinned copper circuit-protective conductor. Two-, 3- and 4-core cables have a full size cpc whereas 7-, 12- and 19-core cables have a drain wire which has a short-circuit rating of 75 A for 1 second; suitable protective equipment should therefore be provided.

In order to prevent electrical faults occurring from phase to earth when terminating FP cables, care should be taken to avoid damage to the insulation by not bending the cores sharply over the end of the sheath.

Steel and PVC conduit installations

Annealed mild steel tubing, known as **conduit**, is widely used as a commercial and industrial wiring system. PVC-insulated (non-sheathed) cables are run inside the steel tubing. Conduit can be bent without splitting, breaking or kinking, provided the correct methods are employed. Available with this system is a very extensive range of accessories to enable the installer to carry out whole installations without terminating the conduit. It offers excellent mechanical protection to the wiring and in certain conditions may also provide the means of earth continuity.

The British Standard covering steel conduit and fittings is BS 4568. The two types of commonly used steel conduit are known as black enamel conduit, which is used indoors where there is no likelihood of dampness, and galvanised conduit, which is used in damp situations or outdoors. In this section the following areas will be looked at:

- screwed conduit
- bending machines
- types of bend
- bending conduit
- conduit fixings
- conduit fitting
- running coupling
- termination of conduit
- use of non-inspection elbows and tees
- wiring conduit
- conduit and cable capacities
- plastic conduit
- miscellaneous points.

Screwed conduit

Screwed steel conduit can be either seam-welded or solid-drawn. Solid-drawn is stronger but much more expensive. At one time solid-drawn conduit was the only conduit that could be used in hazardous areas. More recently, because of manufacturing improvements, seam-welded conduit can now also be used in hazardous areas. The thread used on steel conduit is not used on any other pipe, and special conduit dies are therefore required.

Previous to 1970 conduit sizes were imperial, and of course many of these conduits are still around. Since 1970 metric sizes have been used, and where the two must be joined together adapters may have to be used. Adapters may have external imperial threads and internal metric threads or vice versa.

Bending machines

Bending machines used to be considered an expensive item but are easily available these days. They give consistent results every time and require the minimum of practice.

To position the stand as shown in the diagram, swing the rear leg (E) to its maximum. Place the safety pin (D) through the hole beneath where the pin hangs, locking the rear leg in place. The machine should now be standing with the swivel arm (A) hanging downwards. (B) illustrates the conduit guide and (C) is the adjusting arm for the conduit guide.

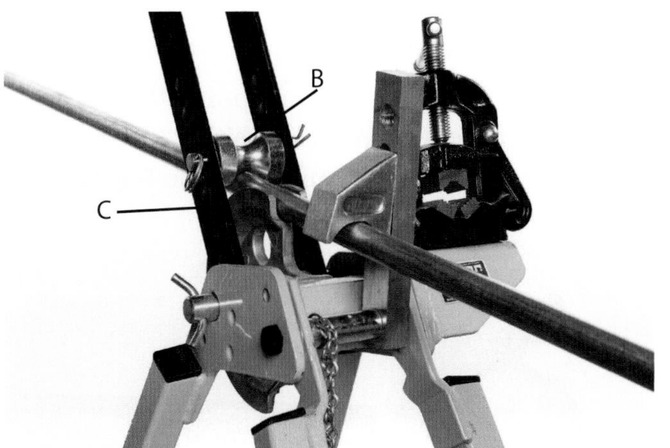

Bending machine with a piece of conduit inserted, which prevents the swivel arm from hanging downwards

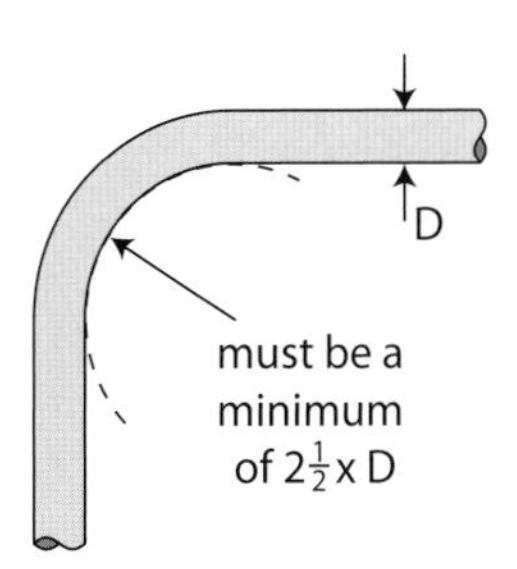

Figure 4.27 Minimum-bending radius allowed

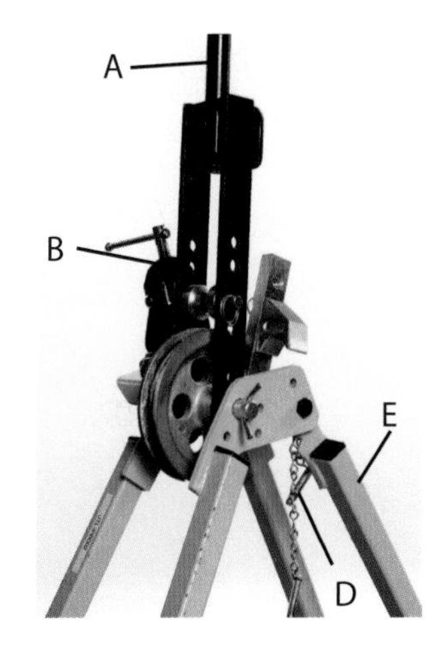

Types of bend

Sharp bends must be avoided. The minimum radius of steel conduit is laid down in BS 7671 as $2\frac{1}{2}$ times the outside diameter of the conduit (Figure 4.27).

Right-angled bend	This is used to go around a corner or change direction by 90°. When bending, measurements may be taken from the back, centre or front of the bend. Allowance should be made for the depth of the fixing saddle bases.	
Set	The set is used when surface levels change or when terminating into a box entry. Sets should be parallel and square, not too long and not too short so that the end cannot be threaded. Where there are numerous sets together all sets must be of the same length. The double set is used when passing girders or obstacles as shown.	Set Double set
Kick	The kick is used when a conduit run changes direction by less than 90°.	
Bubble set or saddle set	The bubble set or saddle set is used when passing obstructions, especially pipes or roof trusses etc. The centre of the obstruction should be central to the set.	

Figure 4.28 Types of bend

Bending conduit

Making a 90° bend from a fixed point

Diagram (a) shows the required bend.

Figure 4.29 a–f Conduit bending sequence

(a)

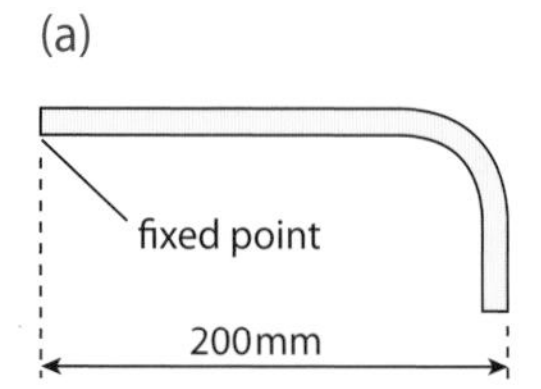

Mark the conduit as shown in diagram (b), 200 mm from the fixed point. If the distance is given to the inside or centre of the tube, simply add on either the diameter or half the diameter respectively to give the back bend measurement and follow the same procedure as for outside measurement.

(b)

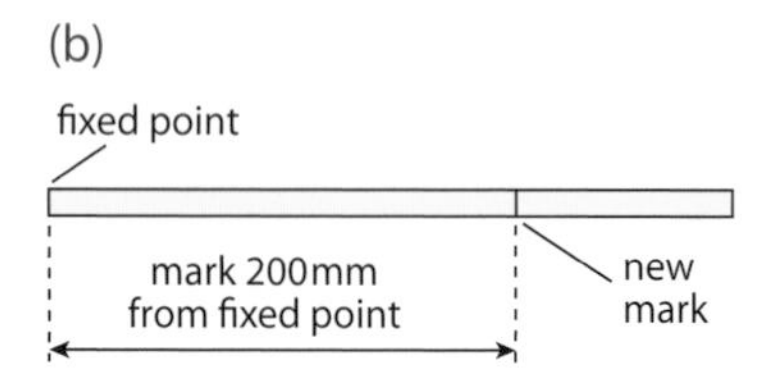

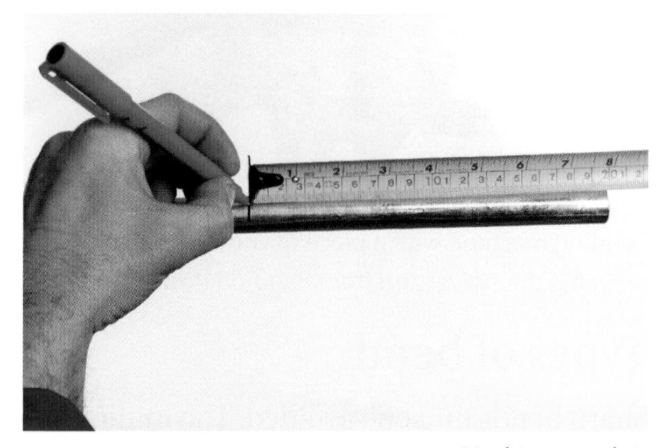

Marking conduit

Place the tube in the 'former' with the fixed point to the rear. Position the tube so that a square held against the tube at the fixed point touches and forms a tangent to the leading edge of the former.

(c)

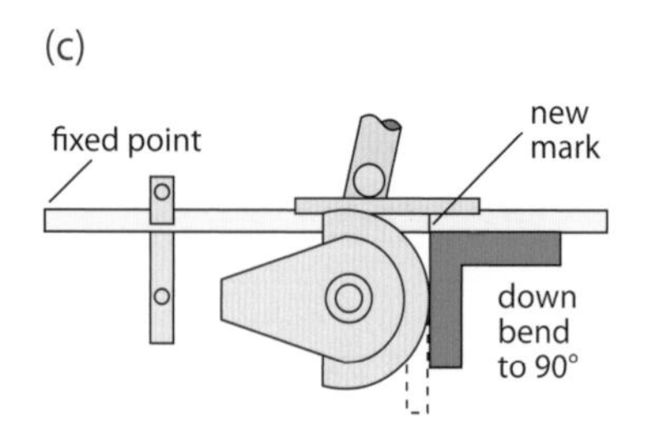

Conduit in former with set square

Where the remaining length of tube from the measured point is too long to down-bend and where it is not convenient or possible to up-bend using the method described, then the problem can be overcome by using the following method.

Figure 4.29 a–f Conduit bending sequence (continued)

(d)

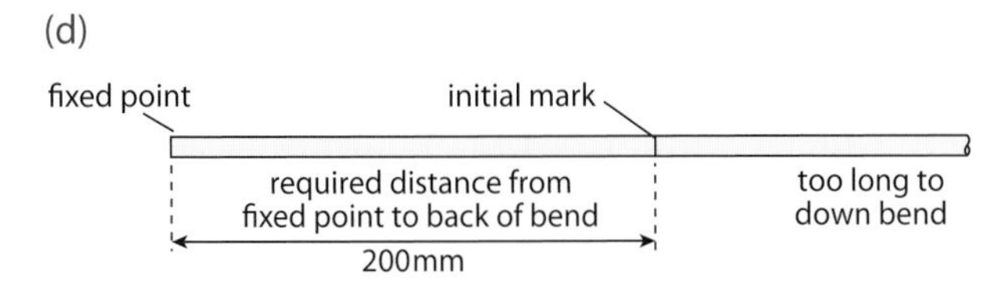

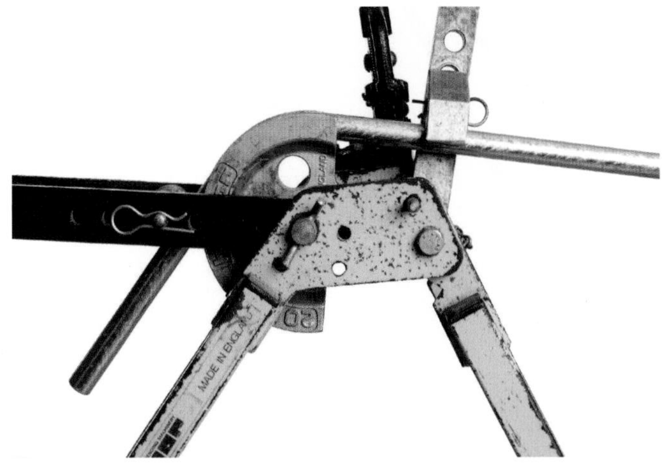

Conduit bent to shape in former

Deduct three times the outside diameter of the tube from the initial mark.

(e)

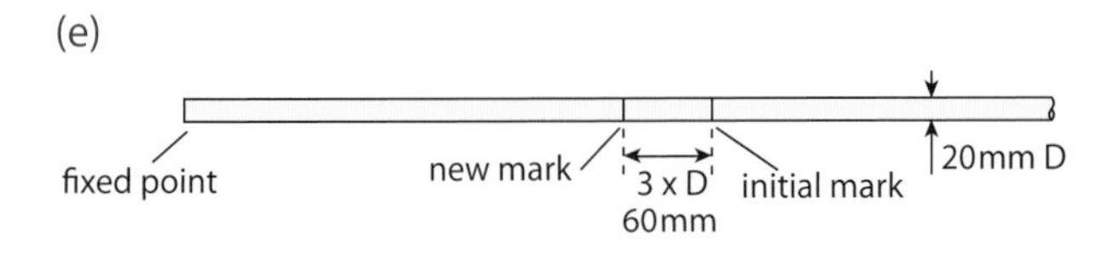

Place the tube in the former with a fixed point to the front with the mark at 90° to the edge of the former. This will give a 90° bend at the required distance from the fixed point to the back of the bend as shown.

(f)

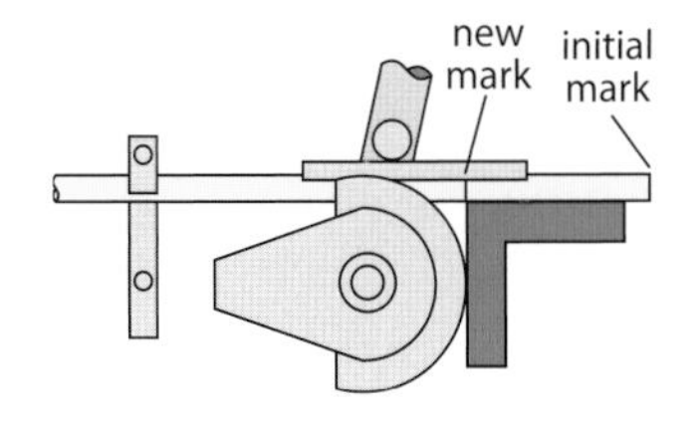

Finished 90° bend

Making double sets

Figure 4.30 shows the required double set.

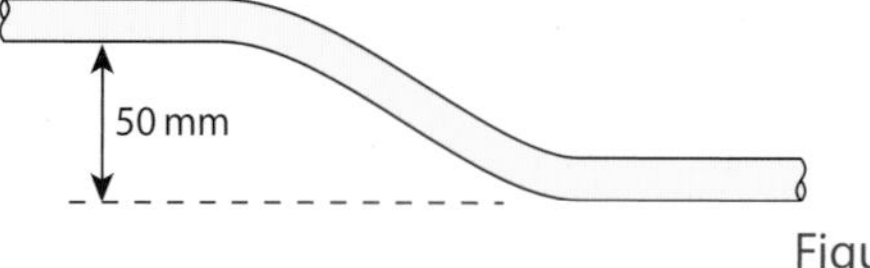

Figure 4.30

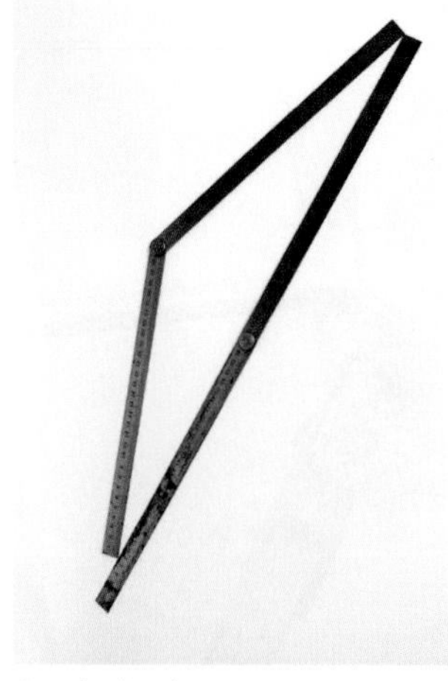

Angled rule

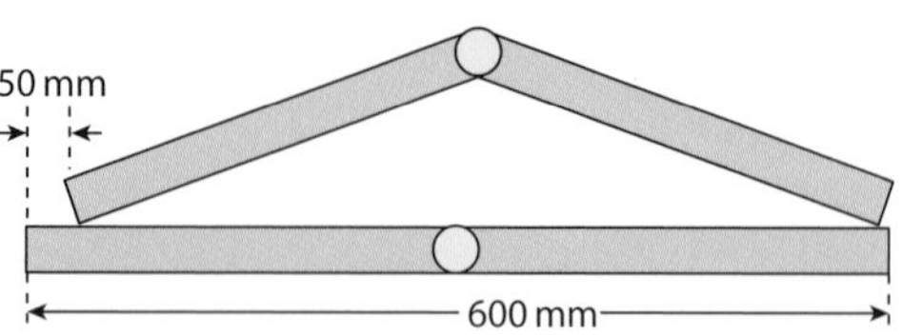
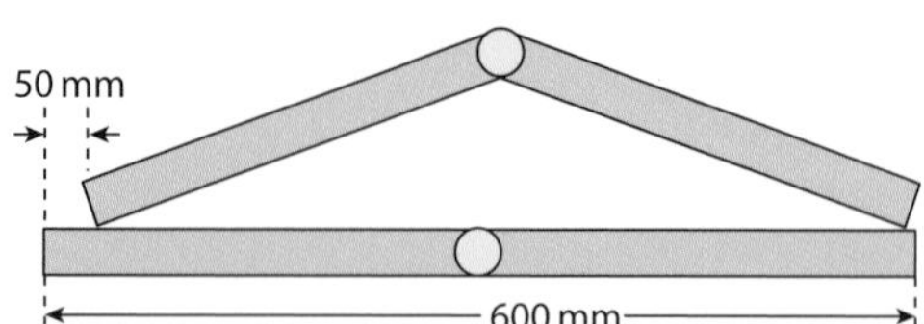

Figure 4.31

To ensure re-entry of the bent tube into the bending machine to complete the return set, a determined angle of initial bend is required. Determine the distance of the set at 50 mm and deduct this distance from a 600 mm rule.

The tube can be bent using the angled rule to indicate the angle of the first set.

Angle of the first set

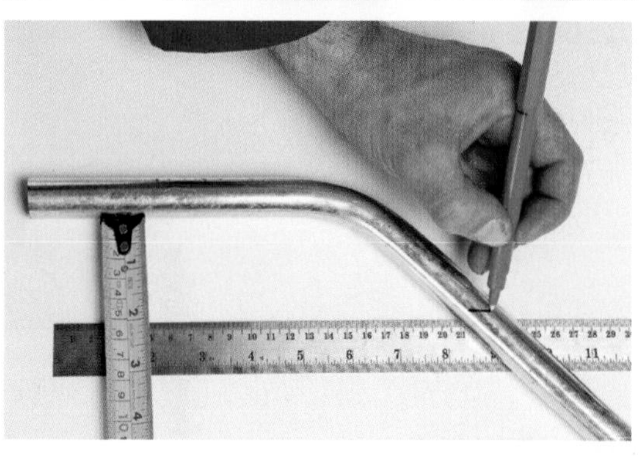

Remove the tube from the bending machine and mark the tube for the return set, making sure to measure the height of the obstacle or accessory from the inside of the tube.

Marking the tube for the return set

Reposition the tube in the machine, ensuring that the mark on the tube forms a tangent to the edge of the former. The final set can be made parallel with the first.

Conduit in former showing marking

Second bend in former

Making bubble sets

Figure 4.32 a–f Making a bubble set

To obtain the correct angle for the first set, multiply the external diameter of the obstacle – say 50 mm tube by 3 (i.e. 50 x 3 = 150 mm) as shown.

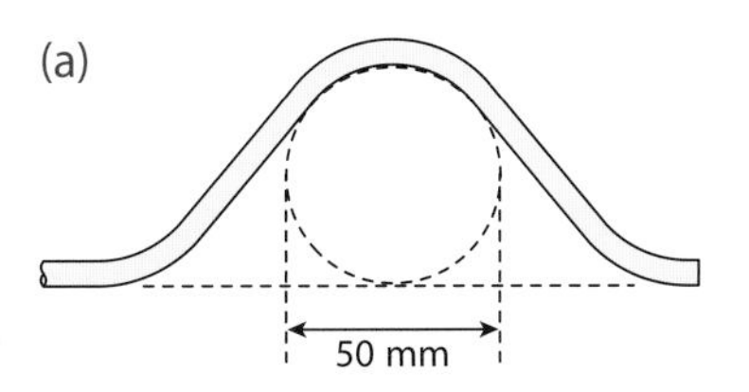

Stagger the legs of a 600 mm folding rule between the 150 mm and 600 mm marks on a second rule.

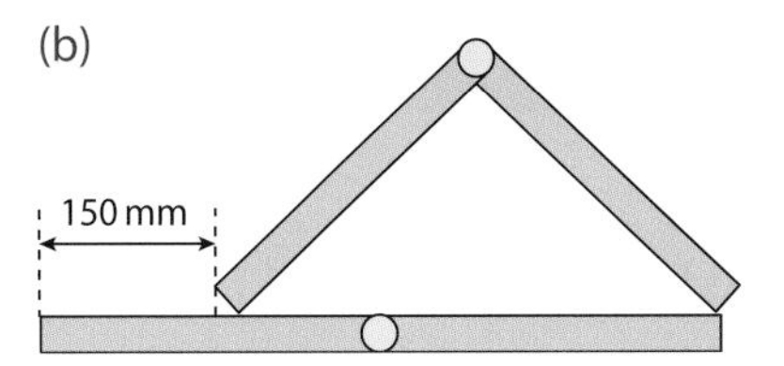

Having marked the centre of the set on the tube, the tube is positioned with the mark vertically above a mark on the former, which is determined by bisecting the angle of the rule when placed as shown.

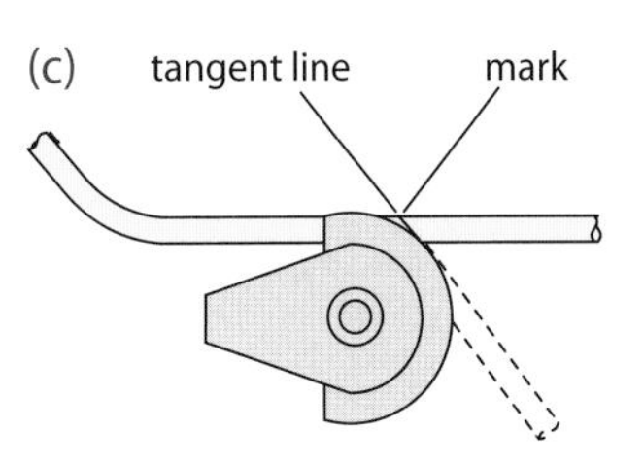

Place the conduit over the obstacle; measure 50 mm from the inside of the first set to a straight edge and mark the tube at A and B as shown.

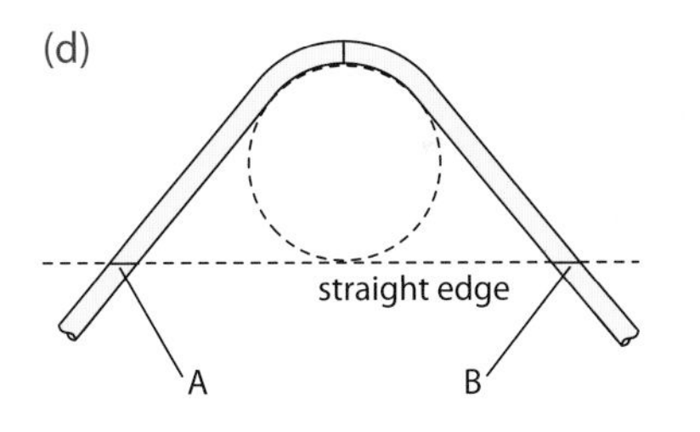

Position the tube in the bending machine so that the mark A forms a tangent to the edge of the former. Bend down until the top edge of the tube is level and in line with mark B.

Reverse the tube in the former and position as for mark A. Down-bend until the top edges of the tube are in line.

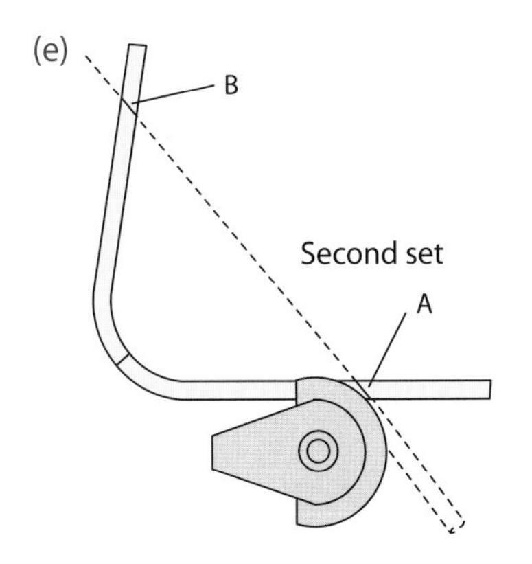

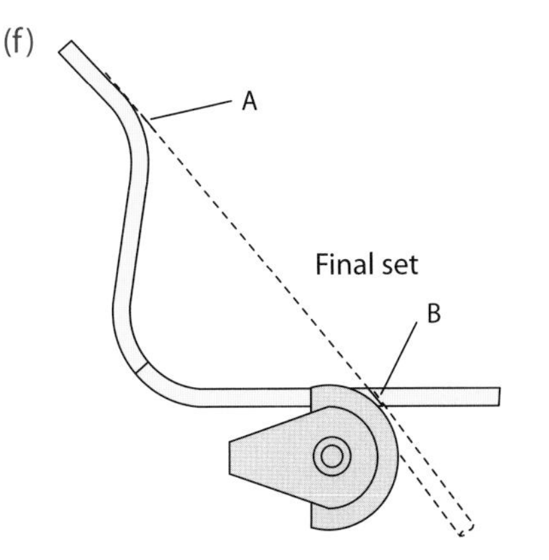

Conduit fixings

Conduits must be securely fixed in accordance with the following distances between supports.

Nominal size of conduit	Maximum distance between supports					
	Rigid metal		Rigid insulating		Pliable	
	Horizontal	*Vertical*	*Horizontal*	*Vertical*	*Horizontal*	*Vertical*
1	*2*	*3*	*4*	*5*	*6*	*7*
m	*m*	*m*	*m*	*m*	*m*	*m*
Not exceeding 16	0.75	1	0.75	1	0.3	0.5
Exceeding 16 but not exceeding 25	1.75	2	1.5	1.75	0.4	0.6
Exceeding 25 but not exceeding 40	2	2.25	1.75	2	0.6	0.8
Exceeding 40	2.25	2.5	2	2	0.8	1

Table 4.12 Conduit fixing parameters (Table 4C, *On-site Guide*)

Methods of supporting conduit

Conduit fixings

- A '**strap saddle**' or '**half saddle**' is used for fixing conduit to cable tray or steel framework.
- The spacer bar saddle is used when fixing to an even surface; it gives a clearance of 2 mm.
- The distance saddle is used if the surface is uneven and where brick on concrete can give rise to heavy condensation.
- The hospital saddle is used where it is necessary to clean around the conduit fixing.
- The multiple saddle strip is used to fasten multiple runs of conduit together.
- The girder clamp will fix conduit to girders and I-beams without having to drill a hole in the girder.
- A pipe hook or crampet is used when conduits are secured to a wall or cast in concrete.

Conduit fitting

Cutting, screwing and terminating conduit

Conduit should be cut with a hacksaw. The cut should be square and the full length of the blade should be used taking steady strokes. Hold the conduit in a pipe vice not a bench vice. The vice should be secured but not so tight that it cuts into the pipe.

Pipe vice and method of cutting

Before threading, the conduit should be chamfered with a file to help the die start. Screwing is carried out using stocks and dies. Another part of the stock and die is the guide, which ensures the screw cut is square. Stocks and dies should be kept clean and any lubricant or steel shavings should be removed after cutting. The cut is made by placing the stock and die on the conduit and then turning clockwise while applying forward pressure; sometimes a great deal of pressure may be required. Once the cut is started the stock and die are removed so that a cutting agent can be applied. Having applied the cutting agent the stock is placed on the conduit again and the threading begins. The stock and die is turned back every turn to clean out the cuttings.

Cutting the thread

When the thread is finished the stock and die is removed and the inside of the conduit is cleaned and reamed. This removes all burrs and sharp edges, which would cut the cables (if not removed) when they were installed. Reaming can be carried out with a reamer or round file. The standard length of thread for a normal joint is half a coupling length.

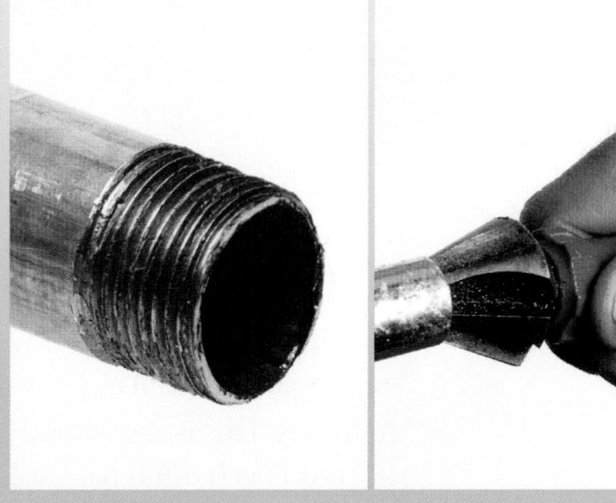

The cut thread

Reaming the thread

All couplings, bushes and conduit boxes must be fully tightened before installation. Where possible couplings, bushes and boxes should be tightened while the conduit is held firmly in a pipe vice.

Tightening the conduit

Running coupling

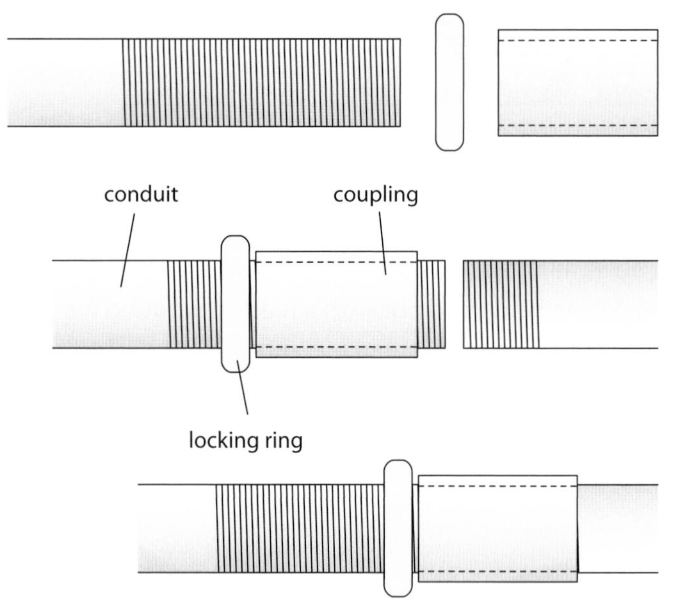

Figure 4.33 The running coupling

Sometimes two conduits must be joined together and neither can be turned. This may be due to one conduit coming through a wall or ceiling or long runs combined with bends making turning impossible. In these cases a running coupling must be used. Running couplings are made by having one thread a normal half-coupling length and the other thread the length of a coupling plus locking ring.

The coupling and locking ring are fixed on the long thread side and the two conduits are then butted together. The coupling is then removed from the long thread to the shorter thread and finally rests across the two sides. After tightening, the coupling is locked. A locking ring must be used because locknuts get caught on the ceiling in tight situations.

Conduit coupling with the locking ring

Because the coupling is traversing two threads simultaneously, the thread must be very clean and well cut. Reversing the dies and running them over the thread can help this. This is particularly important where the running coupling is in an awkward position (as it often is).

Termination of conduit

There are several methods available for terminating conduit, three of which are illustrated.

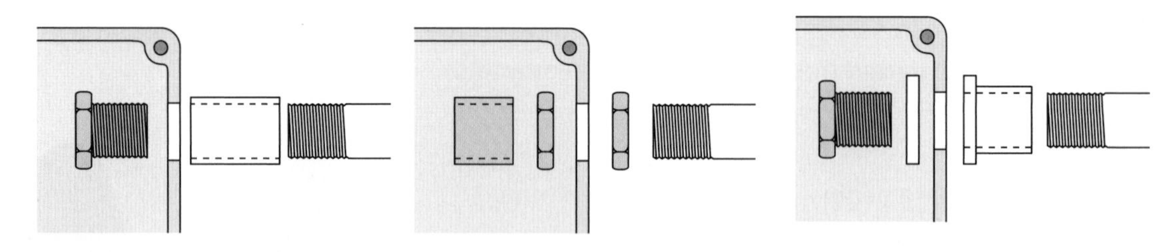

Figure 4.34 (a) Terminating conduit at a box using a conduit coupling and brass male bush

(b) Terminating conduit at a box using locknuts and a brass female bush

(c) Flanged coupling washer and brass male bush method for use with PVC box

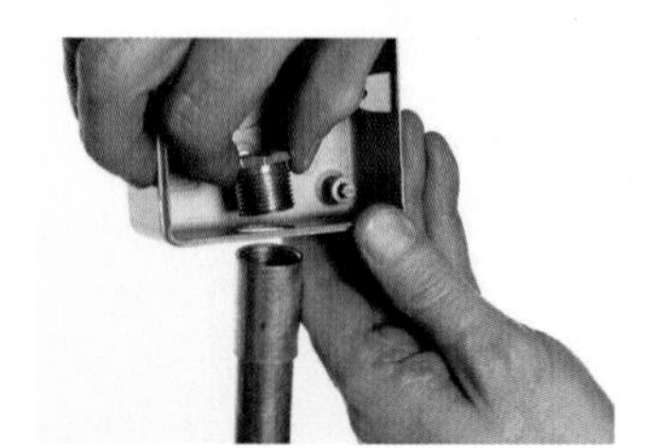

Fitting the bush

Tightening the bush

Use of non-inspection elbows and tees

The main consideration here is that damage to the cables does not occur during installation. Non-inspection elbows are only used adjacent to an outlet box or inspection-type fitting. One solid elbow may be used if positioned less than 500 mm from an accessible outlet, in a conduit run of less than 10 m that has other bends which are not more than the equivalent of one right angle. Figure 4.35 illustrates this use.

solid elbow
one right-angle bend
500mm max
10m max
tee or elbow beside inspection
switchgear

Figure 4.35 Non-inspection elbows and tees

Wiring conduit

Cables must not be drawn into a conduit system until the system is complete. When drawing in cables, you must first run off the reel or drum if there is no supporting mechanism such as a tube to support the reels or they must be allowed to freely revolve.

If a large number of cables are to be drawn into a conduit system at the same time the cable reels should be arranged on a stand or support so they can revolve freely.

In new buildings, cables should not be drawn in until the conduit is dry and free from moisture. If in any doubt, a draw tape with a swab at the end should be drawn through the conduit so as to remove any moisture that may have accumulated.

It is usual to commence drawing in cables from a midpoint in the conduit system so as to minimise the length of cable that has to be drawn in. A steel tape should be used from one draw-in point to another. The draw tape should not be used for drawing in cables as it may become damaged. A steel tape should only be used to pull through a draw wire. The ends of the cables must be paired for a distance of approximately 75 mm and threaded through a loop in the draw wire.

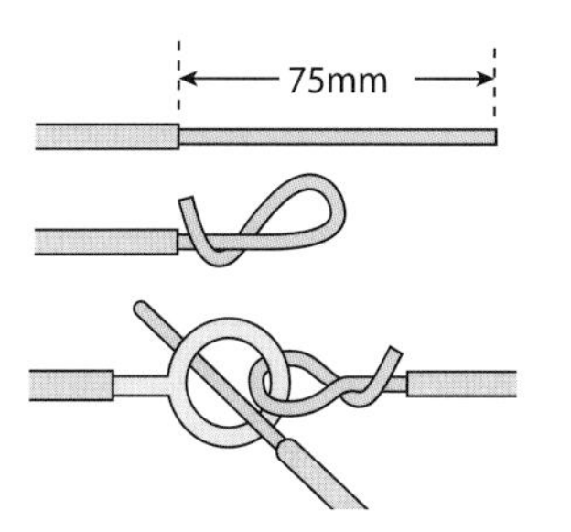

Figure 4.36 Drawing in cables

When drawing in a number of cables, they must be fed in very carefully at the delivery end while someone pulls them at the receiving end. Care should be taken to feed into the conduit in such a manner as to prevent any cables crossing. Always leave some slack cable in all draw-in boxes, and make sure that cables are fed into the conduit so as not to finish up with twisted cables at the draw-in point.

This operation requires care, and there must be synchronisation between the person who is feeding and the person who is pulling. If in sight of each other this can be achieved by some pre-arranged signal, or if within speaking distance by word of command given by the person feeding the cables. If the two people are not within earshot or sight of each other the process is more difficult. A good plan is for the individual feeding the cables to give pre-arranged signals by tapping the conduit.

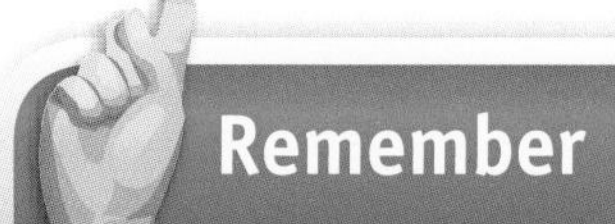

Remember

If the cables are allowed to spiral off the reels they will become twisted and this would cause damage to the insulation

Remember

If cables are not drawn in carefully in this manner they will almost certainly become crossed, and this might result in the cables becoming jammed inside the conduit; this would prevent one or more cables being drawn out of the conduit should this become necessary later

Drawing in cables

1. Pass draw tape through and between outlets.
2. Fasten a draw wire securely to the draw tape.
3. Feed draw wire into the conduit while withdrawing the draw tape. Ensure that the draw wire is long enough and strong enough for the job.
4. Fasten cables to the draw wire. At least 75 mm of insulation should be stripped away and secured as illustrated. Fasten each cable separately.
5. Where possible, cables should be drawn into the conduit directly off the cable drum.
6. First make sure that there is sufficient cable for the job.
7. Feed cables into the conduit, using the fingers of one hand to feed the cables and the other hand to keep cables straight.
8. Ensure that no crossed or kinked cables enter the conduit.
9. Keep hands close to the conduit entry and feed only short lengths of cable at a time.

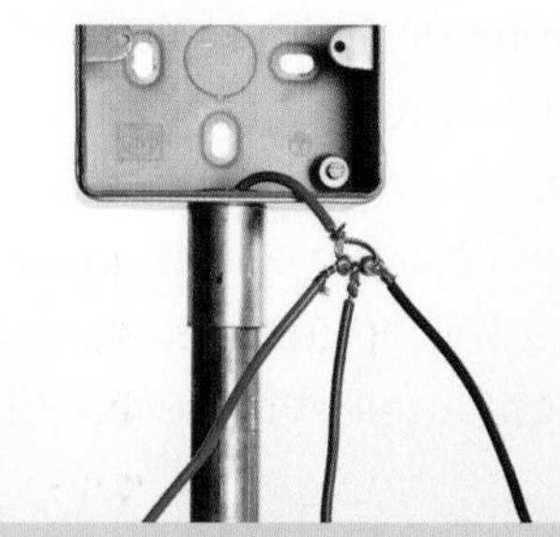

Cables attached to draw wire

Cable drum

Remember

This is only a guide, and larger sizes of conduit may be selected by the designer

Remember

For conduit systems, a bend is classed as 90°, so therefore a double set would be the same as one bend

Conduit and cable capacities

The number of cables that can be drawn into or laid in any enclosure of a wiring system must be such that no damage can occur to the cables or the enclosure during installation. The number of cables that can be used is the overall sum of the cables' cross-sectional area (csa) compared to the overall csa of the conduit. This is expressed as a percentage and should not exceed 45 per cent; this is already taken into account by the standard sizes of cable and enclosures in the *Unite Guide to Good Electrical Practice* Tables. This section looks at the cable capacities of conduit, in other words how many cables can be safely installed into what size conduit.

The tables that will be referred to in this section can be found in Section 5 of the *Unite Guide to Good Electrical Practice*. These tables only give guidance to the maximum number of cables that should be drawn in. The sizes should ensure an easy pull, with low risk of damage to cables and enclosures.

The electrical effects of grouping are not taken into account. Therefore, as the number of circuits increases, the current-carrying capacity of the cables will decrease. Cable sizes would have to be increased, with a consequent increase in cost of cable and conduit. It may therefore be more economical to divide the circuits concerned between two or more enclosures.

The following wiring systems are covered by the appropriate tables in Section 5 of the *Unite Guide to Good Electrical Practice*:

- straight runs of conduit not exceeding 3 m in length (page 67)
- straight runs of conduit exceeding 3 m in length, or in runs of any length incorporating bends or sets (page 68).

Example

A lighting circuit for a school extension requires the installation of a conduit system with a conduit run of 8 m with two right angle bends. The number of cables required is twelve × 1.5 mm stranded, PVC insulated. What size of conduit should be used for this installation?

Step 1 Select the correct table for cable runs over 3 m with bends. (page 68)

Step 2 Obtain the factor for 1.5 mm stranded cable = 22

Step 3 Multiply the number of cables by the factor = 22 × 12 = 264

Step 4 Select the correct table for conduit systems with runs in excess of 3 m with bends. (page 69)

Step 5 Obtain the factor for a length of run which is greater than 264. The table gives a factor of 292 for an 8 metre run in conduit with two bends.

Step 6 The answer is 25 mm conduit.

Plastic conduit (PVC)

Plastic conduit is made from polyvinyl chloride (PVC), which is produced in both flexible and rigid forms. It is impervious to acids, alkalis, oil, aggressive soils, fungi and bacteria and is unaffected by sea, air and atmospheric conditions. It withstands all pests and does not attract rodents. PVC conduit is preferable for use in areas such as farm milking parlours. PVC conduit may be buried in lime, concrete or plaster without harmful effects.

Plastic conduit advantages	Plastic conduit disadvantages
• light in weight • easy to handle • easy to saw, cut and clean • simple to bend • does not require painting • minimum condensation due to low thermal conductivity in walls • quick to install	• care must be taken when applying glue to the joints to avoid forming a barrier across the inside of the conduit • if insufficient adhesive is used the joints may not be waterproof • PVC expands around five times as much as steel and this expansion must be allowed for

Table 4.13 Plastic conduit advantages and disadvantages

Working with PVC conduit

The techniques required for working with PVC conduits differ considerably from those in steel conduit installations. PVC conduit is easily cut using a junior hacksaw. Any roughness of cut and burrs should be removed by simply wiping with a cloth. The most common jointing procedure uses a PVC solvent adhesive. The adhesive should be applied to the female part of the joint and the conduit twisted into it to ensure a total coverage. Generally the joint is solid enough for use after two minutes, although complete adhesion takes several hours. In order to ensure a sound joint, the tube and fittings must be clean and free from dust and oil.

Safety tip

Care must be taken when using these adhesives as they are volatile liquids, and the lid must be replaced on the tin immediately after use. Take care when using in a confined space. Always read the manufacturer's instructions

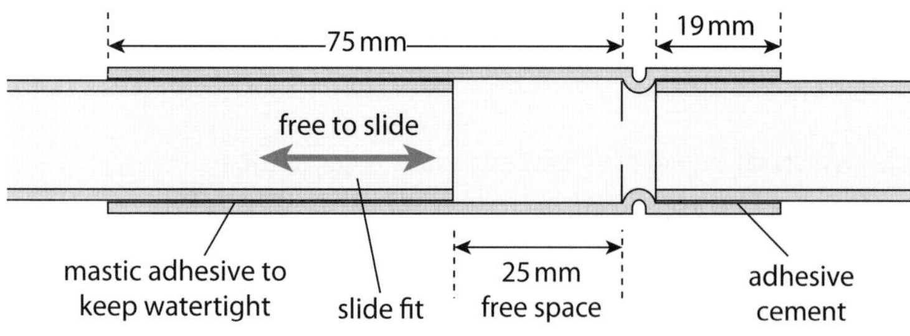

Figure 4.37 Expansion provision in conduits

Where expansion is likely and adjustment is necessary a mastic adhesive should be used. This is a flexible adhesive, which makes a weatherproof joint, ideal for surface installations and in conditions of wide temperature variation. It is also advisable to use mastic adhesive where there are straight runs on the surface exceeding 6 m in length.

PVC conduit expands considerably more than metal conduit with an increase in temperature. The expansion can be ignored where the conduit is buried in concrete or plaster. In surface work precautions must be taken to prevent such expansion from causing the conduit to bow. Usually where bends and sets are close together these take up any expansion. Where longer runs of conduit occur in conditions of varying temperatures, some provision for expansion must be made, using expansion couplers as shown in Figure 4.37.

Remember

A good guide to the use of expansion couplers is one coupler per 6 m in straight runs

PVC conduits not exceeding 25 mm diameter can be bent cold by using a spring. The bend is then made by either the hands or across the knee. In order to achieve the angle required the original bend should be made at twice the angle required and the tube allowed to return to the correct angle. Under no circumstances should an attempt be made to force the bend back with the spring inserted, as this can damage the spring. It is easier to withdraw the spring by twisting it in an anticlockwise direction. This reduces the diameter of the spring, making it easier to withdraw.

In cold weather it may be necessary to warm the conduit slightly at the point where the bend is required. One of the simplest ways is to rub the conduit with the hand or a cloth. The PVC will retain the heat long enough for the bend to be made. In order that the bend is maintained at the correct angle the conduit should be fastened to the surface with a saddle as soon as possible.

Miscellaneous points for steel and plastic conduits

- Ample capacity must be provided at junctions employed for cable connections.
- Where a steel conduit is used as a circuit-protective conductor it must be tested in accordance with BS 7671.
- Where a steel conduit forms the protective conductor, the earthing terminal of each accessory (e.g. socket outlet or switch grid) shall be connected by a separate protective conductor to an earthing terminal incorporated within the associated box or enclosure.
- Cable capacities should be calculated in accordance with the relevant tables found in Section 5 of the *Unite Guide to Good Electrical Practice.*
- Conduits run overhead should be run in accordance with the table on page 72 of the *Unite Guide to Good Electrical Practice.*
- Where conduits pass through walls the hole shall be made good during installation with a fire-resistant material.

Trunking

Trunking is a fabricated casing for cables, normally of rectangular cross-section, one side of which may be removed or hinged back to permit access. It is used where a number of cables follow the same route or in circumstances where it would otherwise be expensive to install a large number of separate conduits or runs of mineral-insulated cable. Trunking is commonly installed, for example in factories where the introduction of new equipment and the relocation of existing equipment may involve frequent modification of the installation.

Surface pattern trunking is available in 3 m lengths and a wide variety of sizes ranging from 38 mm up to 225 mm × 100 mm. This section will look at the following areas:

- square steel trunking with cover and coupling
- floor trunking
- multi-compartment trunking
- flush cable trunking
- overhead trunking
- skirting trunking
- busbar trunking (overhead)
- feed units
- Busbar trunking (rising mains)
- site-built trunking accessories
- regulations concerning trunking
- trunking capacities
- PVC trunking.

Did you know?

Zinc-coated and enamel trunking is generally used internally, whereas outside or in situations where atmospheric conditions are likely to be arduous, a galvanised or electro-zinc plated trunking would be necessary

Square steel trunking with cover and coupling

To facilitate connections, terminations etc. and in order to run the trunking with the contours of the building in which it is installed, a wide number of trunking fittings and accessories are available, a selection of which are illustrated (Figure 4.38).

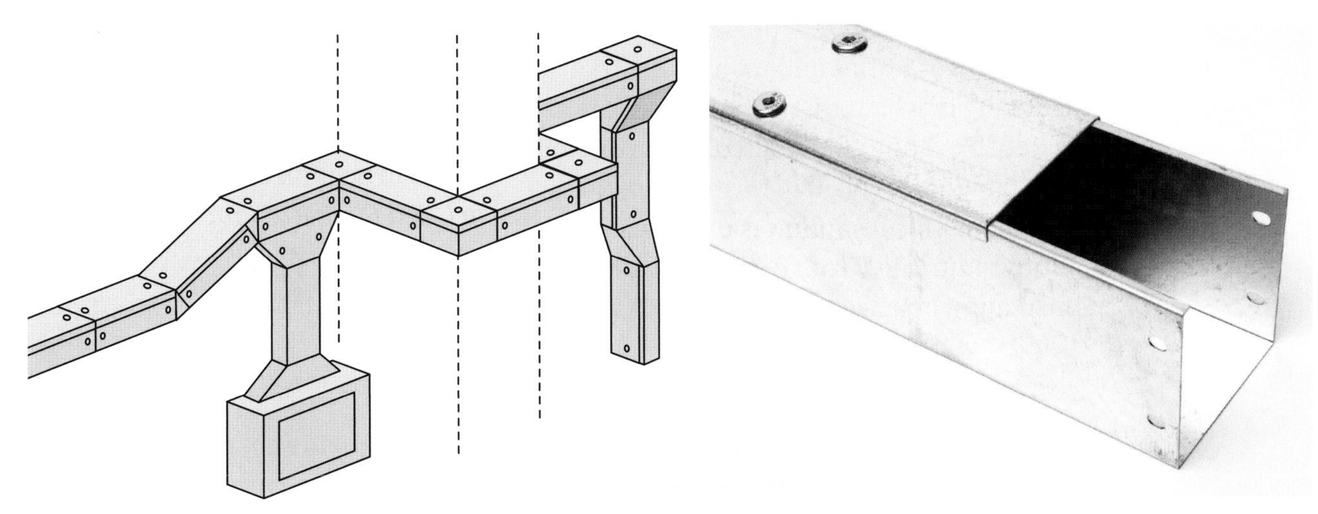

Figure 4.38 Steel trunking

Floor trunking

Floor trunking or ducting is also available in a wide range of sizes and in a wide range of fittings. It is used extensively in schools, hospitals and industrial situations and can either be laid below the floor surface where access is by means of a number of inspection covers or installed just below floor level and covered by a steel chequer plate (Figure 4.39).

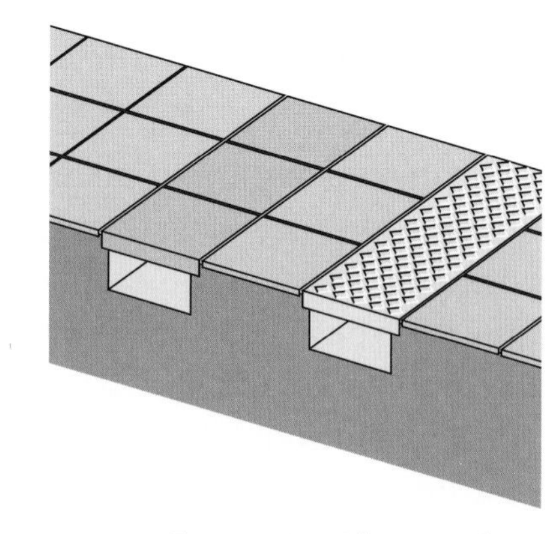

Figure 4.39 Floor trunking

Multi compartment trunking

This is used where segregation is desired. It is normally a broad, flat trunking with internal steel fillets. The fillets are normally spot-welded to the trunking.

Flush cable trunking

Flush cable trunking is used where a neat and unobtrusive cable trunking is desired. It is available in a range of sizes. The covers can be supplied in a number of finishes and have a large overlap on each side.

Skirting trunking

Skirting trunking is used mainly in offices, schools and colleges to provide a large number of socket outlets both for 230V equipment and for telecommunications equipment. It is made with an internal fillet for segregation; it can be installed before or after all plastering; and it eliminates the need for conventional skirting.

Busbar trunking (overhead)

Busbar trunking is a popular means of three-phase power distribution in machine shops, laboratories and many industrial situations. It consists of a broad, flat trunking in which three or four busbars are rigidly fixed onto moulded block insulators. The conductors used in busbar trunking are generally of copper or sometimes aluminium. They can be sleeved with insulating material or left bare. The size of copper will vary with the current-carrying capacity, and its shape can be either round, oval or rectangular. A complete range of fittings is available for right-angle bends, tee pieces and crossovers. These fittings are self-contained assemblies complete with busbars and couplings (Figure 4.40).

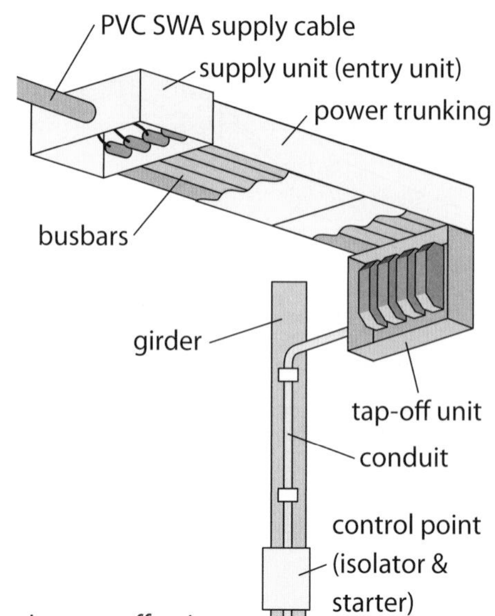

Figure 4.40 Power trunking showing tap-off unit

Feed units

Feeding the busbars can be done in a number of ways. For smaller runs of busbar trunking a screwed conduit or cable trunking could be used with PVC-insulated conductors of appropriate rating. In the case of a large installation or where an armoured cable would be more suitable, a feed unit with the cable-sealing chamber would be used. The busbar trunking can be centre-fed, the centre feed being more popular as this method makes full use of the busbar capacity.

Fused outlet boxes can either be fixed or be of the plug-in type and can be side-mounted or under-slung. Sets of drillings are provided at regular intervals along the busbar trunking to facilitate the tap-off boxes, and blanking-off plates are supplied to cover off the apertures not initially used. HBC fuses must always be used and never replaced with the rewirable fuse links.

Busbar trunking (rising mains)

The increasing number of multi-storey developments for hospitals, offices, flats etc. poses certain problems for electrical power distribution. In some cases a rising main busbar trunking system can provide a compact and economical solution. In this system purpose-made busbar trunking is run vertically through the walls of the building. It is fed and controlled usually from the bottom at the service entry, and has a fixed fuse box mounted at each floor.

The essential difference between horizontal-type busbar trunking and vertical rising main-type busbar trunking is the substantial insulated support rack at the base of each riser. It is designed to carry the full weight of the copper conductors, which are then free to expand upwards. Figure 4.41 illustrates rising mains trunking.

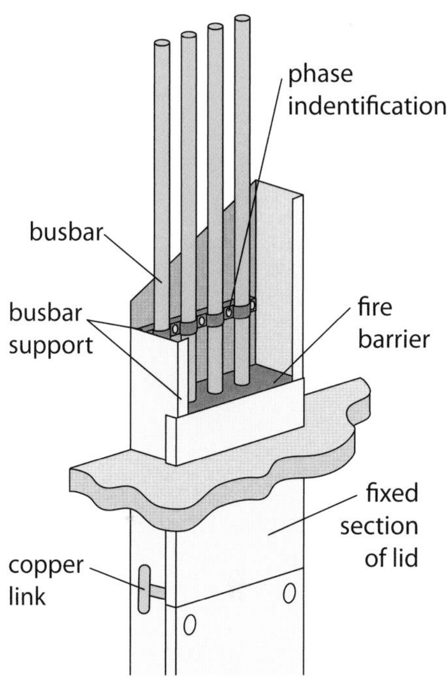

Figure 4.41 Rising mains busbar trunking

Site-built trunking accessories

Prefabricated bends and sets will usually be used to install trunking systems because they are quicker and cheaper. However, there will be times when the prefabricated bends and sets are either not suitable or are not available. In this situation you will need to be able to fabricate your own.

The accessories that are considered here are:

- right-angle internal bend
- right-angle vertical bend
- trunking sets
- tee junction.

Right-angle internal bend

Select a short section of trunking between 900 and 1000 mm in length. Using a soft pencil and a reliable set square, draw a line (called a datum line) around the outside (periphery) of the trunking. This should be done at the mid point positions as shown.

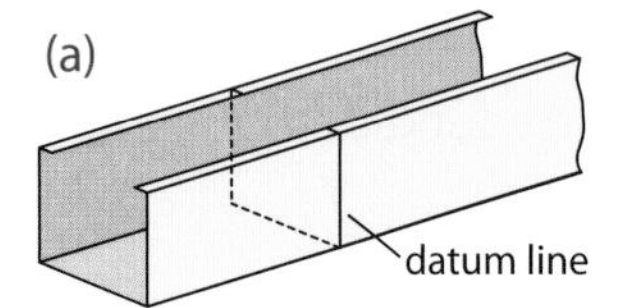

Check the width of the workpiece and transfer this measurement to either the top left or right hand side of the central datum line as shown.

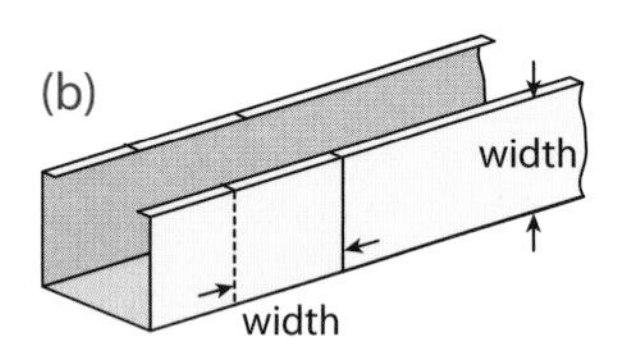

Using an adjustable setsquare as a guide, draw a pencil line from the marked trunking to the bottom of the centre datum line as shown in diagram (c). Repeat this guideline on the opposite vertical side.

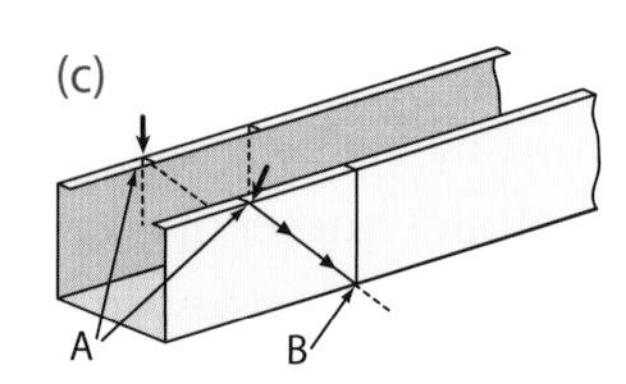

At this stage there will be a right-angled triangle drawn on each outer side of the trunking. Remove the two triangles using a hacksaw with a blade fitted with 25–30 teeth per 25 mm of blade. Once removed, file smooth all rough or jagged edges as these may damage the cables.

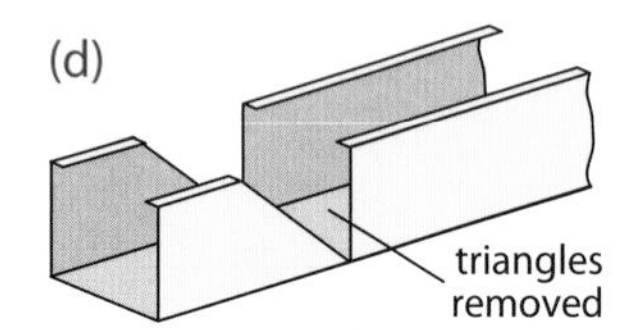

Cut a wooden block with a good square edge on one side able to be fitted comfortably across the internal width of the trunking. Place the wooden block to the vertically cut side as shown in the diagram.

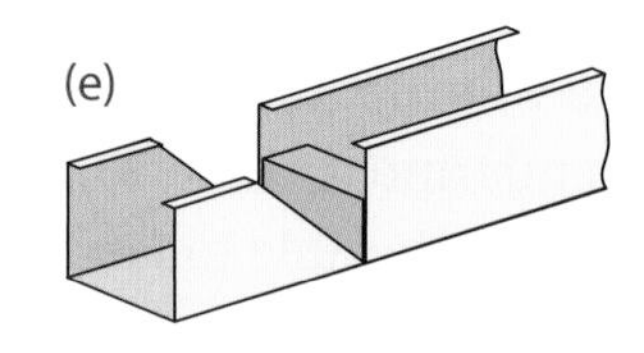

Hold the block firmly in place and with the other hand push up the side of the trunking adjoining the angled cut. Allow the vertical sides to be sandwiched between the angled trunking sides. The wooden block will help provide a sharper edge at the bending point. Once completed, dress the bend with a hammer and remove the wood. Check for squareness and strengthen with pop rivets. Nut and bolt, or spot weld.

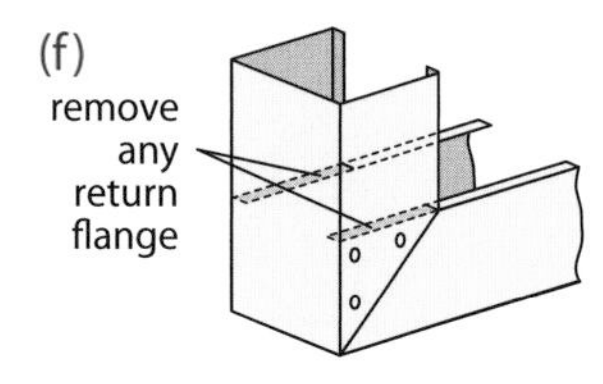

Figure 4.42 a–f Right-angle internal bend

Safety tip

Remove any rough or jagged edges with a medium-cut file in order to prevent damage to the cables

Right-angle vertical bend

Mark out the position of the bend on all sides of the trunking.

(a)

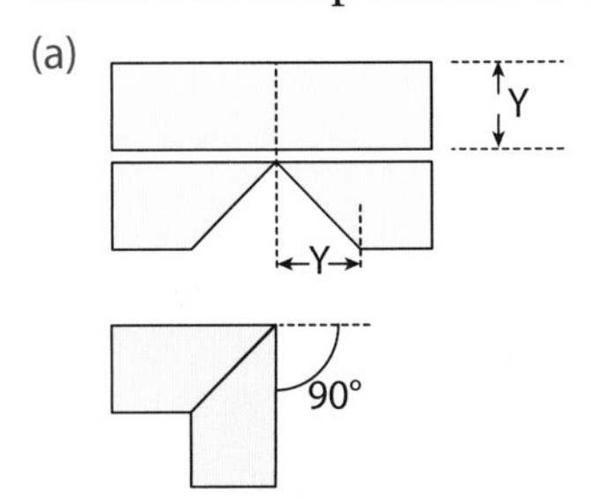

Drill small holes in the corners at the point of the bend to stop the metal from folding. Then place wooden blocks inside the trunking for support. Cut the sides of the trunking with an appropriate hacksaw.

(b)

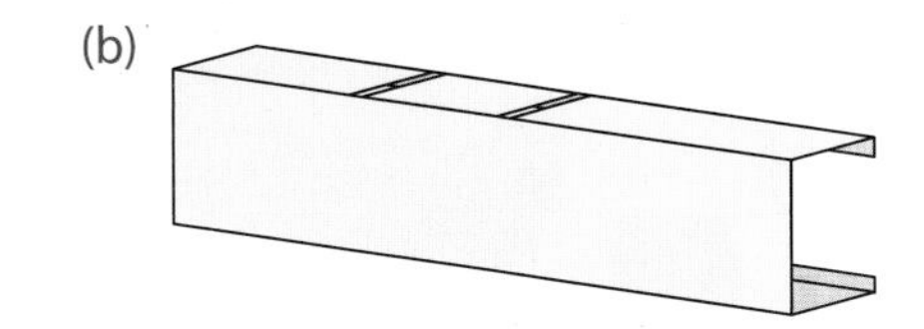

The edge of the trunking can be cut with a file (as shown in (c) opposite) and the waste broken off.

(c)

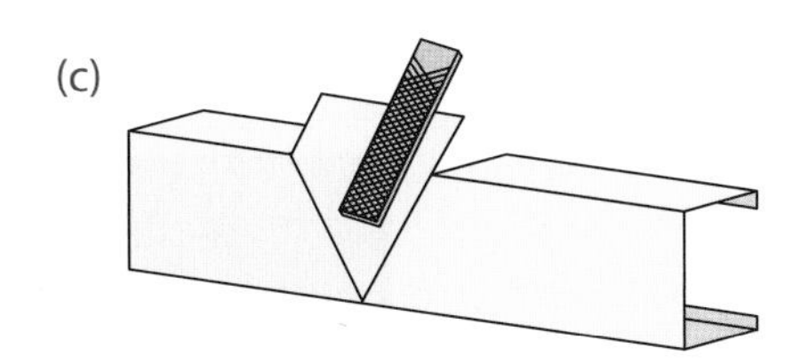

Cut away the back of the trunking using a suitable hacksaw. Then file all the rough and jagged edges and bend the trunking to shape as shown.

(d)

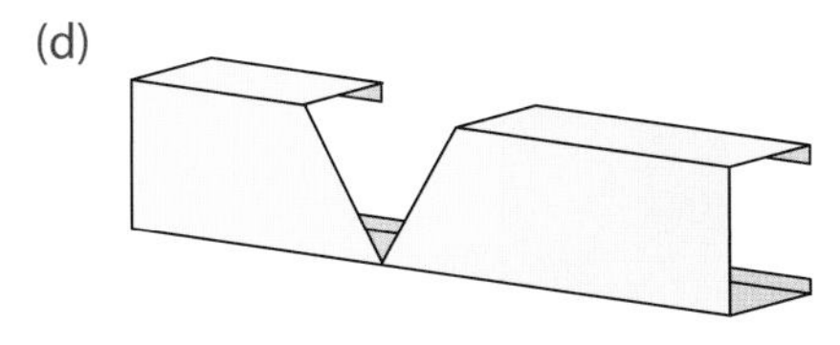

(e)

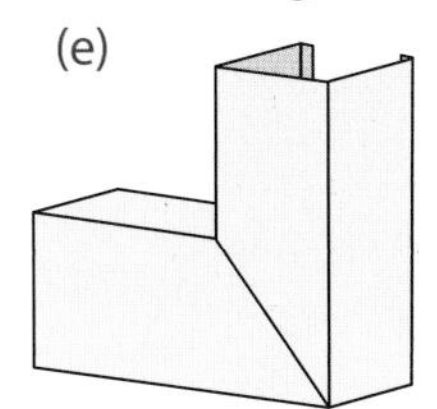

Make a fishplate out of some scrap trunking and drill in some fixing holes.

Finally mark out the trunking from the holes in the fishplate and drill. Secure

(f)

the assembly with nuts and bolts or pop rivets. Alternatively the joint may be spot-welded.

(g)

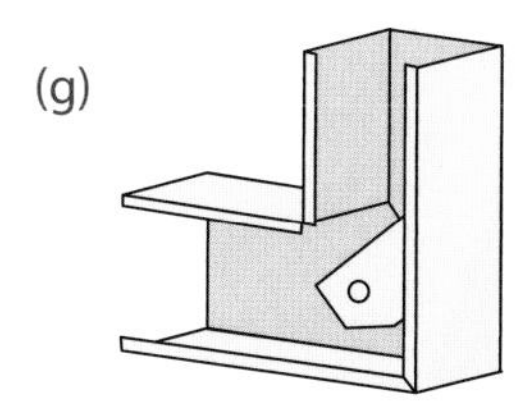

Figure 4.43 a–g Right-angle vertical bend

Making a right-angle vertical bend (alternative method)

1. Mark the trunking for cutting

2. Place wooden block inside trunking to secure vice

3. Use hacksaw to cut trunking

4. Bend trunking into right-angled bend

5. Secure with rivets

6. Final result

Trunking sets

A trunking set and a return set are both constructed in a similar fashion when making a right-angled bend. Figure 4.44(a) illustrates a trunking set (A) and a trunking return set (B).

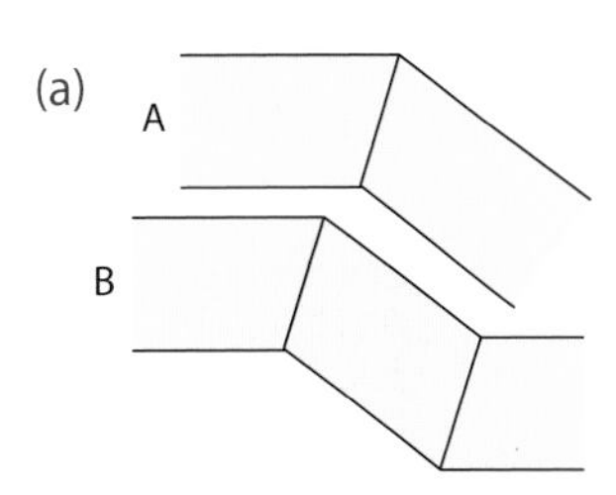

Select a section of trunking that can be worked comfortably yet is long enough to accommodate either a set or return set. Draw a datum line using a soft pencil and set square as a guide.

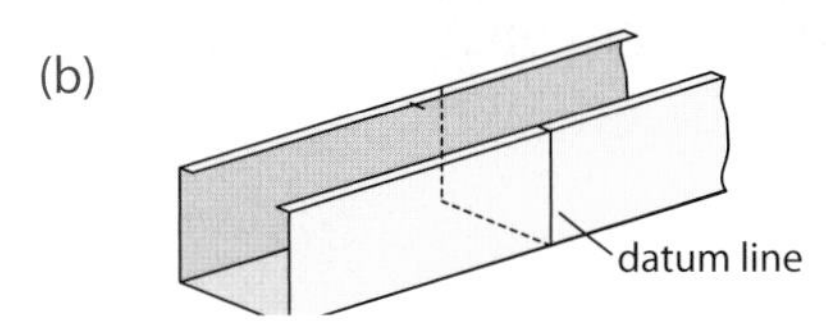

Measure the width of the trunking and transfer half of this measurement to either the top left or top right-hand side of the datum line. Draw a line from this mark to the base of the datum line A.

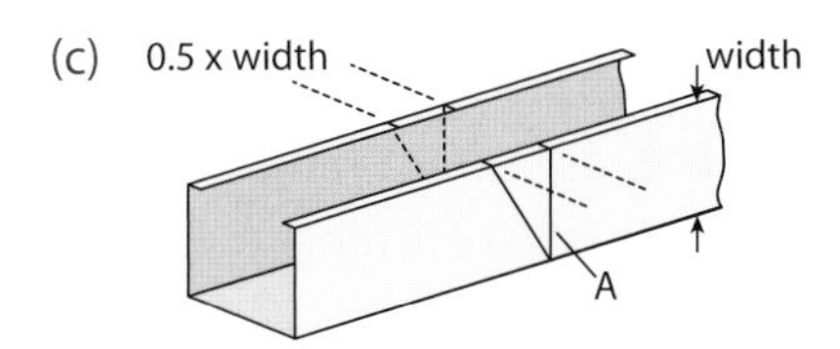

Cut out both triangular shapes from each side of the trunking.

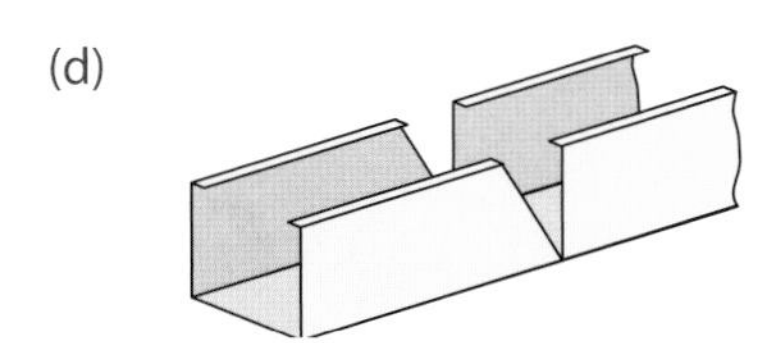

Cut a section of a wooden block which will fit comfortably across the internal diameter of the trunking. Place the piece of wood as shown in the diagram so that it is at the side adjacent to the vertically cut datum line.

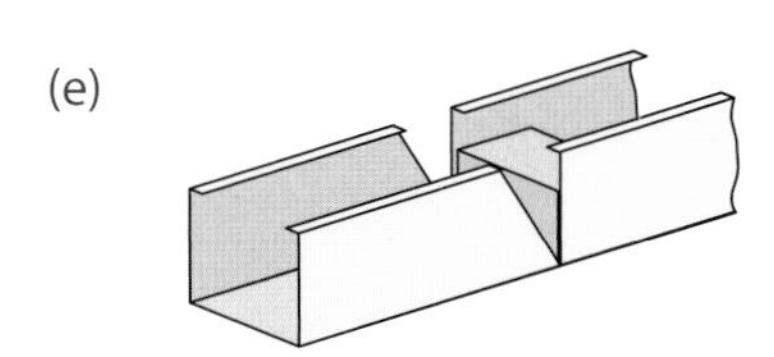

Secure the wooden block with one hand while gently bending up the remaining half of the workpiece until the set is formed. Check that the required angle is correct and secure using pop rivets, nut or bolts or spot-welding. Figure 4.44(f) illustrates the completed set.

If a return set is involved then extra work is required.

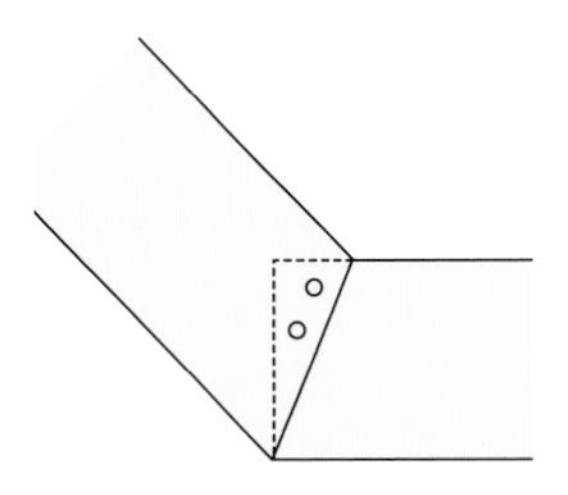

Figure 4.44 a–f Trunking set

Draw a line of reference on a suitable flat surface and offer the shortest section of the set to the line as shown. Measure the depth of the required set (A) and mark on the trunking at (B). (C) represents the shortest leg.

(g)

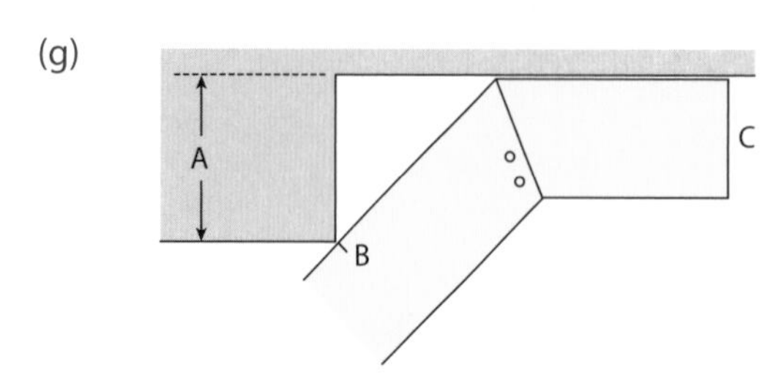

Prepare the trunking as shown in the diagram. Cut out the complete left-hand side of the centre line comprising two side triangles bridged by a rectangular base section. To provide electrical continuity the trunking lip must remain unbroken.

(h)

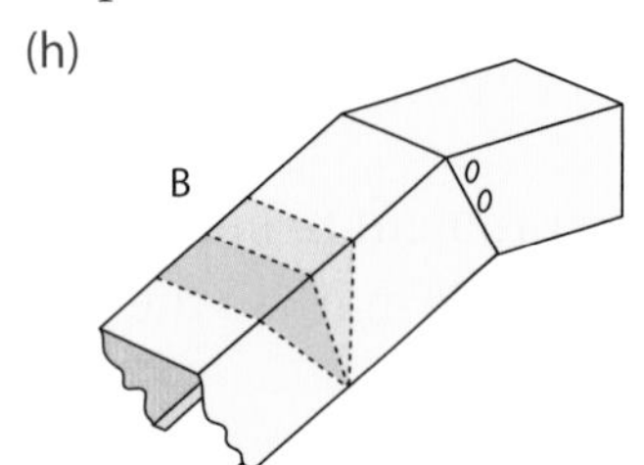

A 2 mm diameter V shape should be cut along both sides of the bottom edge of the trunking from the vertical centre line to the angled dotted line, as shown. This will act as a supplementary lip and help stabilise the return set when it is assembled.

(i)

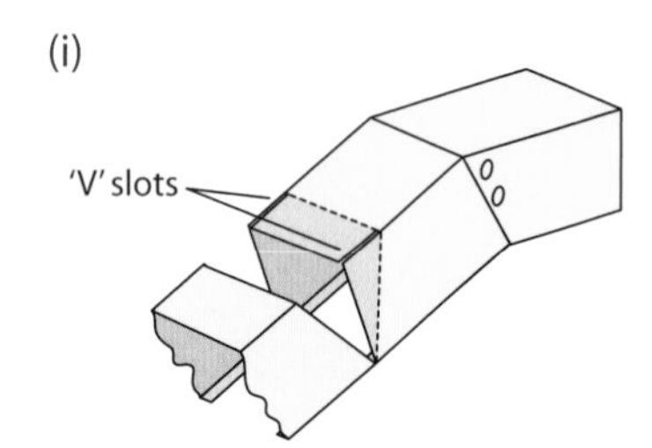

Gently bend the workpiece and unite both sections of the return set, allowing the vertically cut centre edges to be sandwiched between the angled sides of the trunking as shown in the diagram. Check that the set has been worked to meet the required measurement. Secure using pop rivets, nuts and bolts or spot-welding. Dress the supplementary base lip to accommodate the changed angle and secure as necessary.

(j)

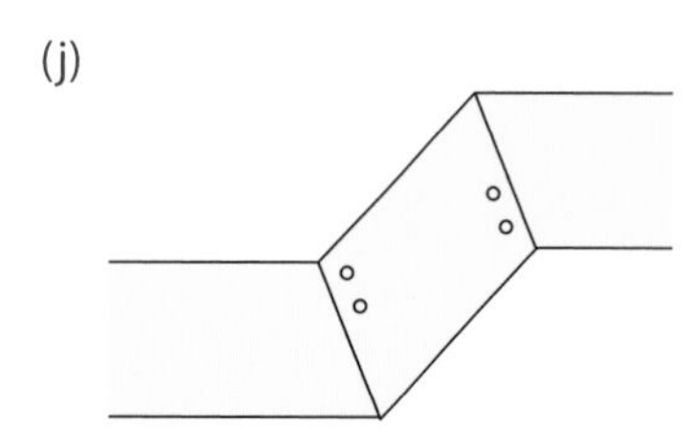

Figure 4.44 g–j Trunking set

Tee junction

Mark out the position of the tee using a second piece of trunking to gauge the width. Cut out a space for the tee as shown in Figure 4.45. Use blocks of wood to support the sections being cut. File all rough edges to protect the cables.

Cut away the section to leave two lugs. These can be bent in a vice using a hammer to give a clean edge. File all edges smooth and check the fit as necessary.

Mark out the holes, drill and secure with nuts and bolts or pop rivet, or alternatively spot-weld.

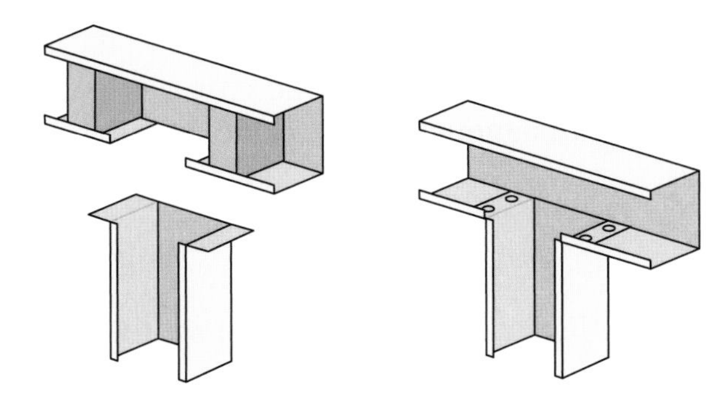

Figure 4.45 Tee junction

Regulations concerning trunking

- **Regulation 521.5.1** – every conductor or cable shall have adequate strength and be so installed as to withstand the electromechanical forces that may be caused by any current (including fault current) it may have to carry in service.
- **Regulation 521.5.2** – the conductors of an a.c. circuit installed in ferromagnetic enclosure shall be arranged so that all line conductors, the neutral conductor (if any) and the appropriate protective conductor are contained in the same enclosure.
- **Regulation 521.6** – cable trunking or ducting shall comply with the appropriate part of the BS EN 50085 series.

The Regulations also lay down the following guidelines:

- Ducts and metallic trunking must be securely fixed and protected from mechanical damage.
- The number of cables installed in trunking shall be such that a space factor of 45 per cent must not be exceeded.
- During installation where conduit, ducts or trunking passes through walls, floors, ceilings or partitions, the surrounding hole must be made good with a non-combustible material, and internal fire barriers are to be installed within trunking at these positions.

- Copper links are used across joints in metallic conduit systems in order to maintain the continuity of the exposed parts.
- Cable entries must be protected with grommets to prevent damage or abrasions to cables.
- Straight runs of trunking are best joined by using a joining section and the nuts and bolts supplied.
- When terminating metal trunking at a distribution board it is essential to ensure that the junction between trunking and distribution board will not cause abrasion to the cables.
- Grommet strips should be fitted over the edges of any holes drilled in trunking to prevent damage to cables.
- To prevent the effects of eddy currents (electromagnetic effects), cables of an a.c. circuit should pass through a single hole.

Trunking capacities

Remember

This is only a guide, and larger sizes of trunking may be selected by the designer

The number of cables that can be drawn into or laid in any enclosure of a wiring system must be such that no damage can occur to the cables or the enclosure during installation. The number of cables that can be used is the overall sum of the cables cross-sectional area (csa) compared to the overall csa of the trunking. This is expressed as a percentage and should not exceed 45 per cent; this is already taken into account by the standard sizes of cable and enclosures in the *Unite Guide to Good Electrical Practice.*

The sizes should ensure an easy pull, with low risk of damage to cables and enclosures. The electrical effects of grouping are not taken into account. Therefore, as the number of circuits increases, the current-carrying capacity of the cables will decrease. Cable sizes would have to be increased with a consequent increase in cost of cable and trunking. It may therefore be more economical to divide the circuits concerned between two or more enclosures.

Trunking example question

Example 1

Calculate the size of cable trunking required for the following conductor numbers and sizes:

25 × 1.5 mm^2 20 × 2.5 mm^2 6 × 4.0 mm^2 2 × 10 mm^2 2 × 16 mm^2

Assume that all the conductors are stranded PVC.

Step 1
Obtain the 'Term' values from the *Unite Guide*.

Conductor size	Term value	Number of conductors	Total term value
1.5 mm	8.6	25	215
2.5 mm	12.6	20	252
4 mm	16.6	6	99.6
10 mm	35.3	2	70.6
16 mm	47.8	2	95.6

Step 2
Multiply the term values by the number of conductors

Step 3
Add the term values together

Total terms = 732.8

Step 4
From the tables, select the nearest trunking term value (equal to or larger than that calculated), which in this case would be:

767 terms, which would give a trunking size of 50 mm × 38 mm

or

738 terms, which would give a trunking size of 75 mm × 25 mm

Whichever is selected will depend on the designer's choice.

Example 2

A steel trunking is to be installed as the wiring system for 80 single-phase circuits, each having a design current of 15A. Single-core PVC stranded cables of 4 mm^2 will be used. 20A BS 88 fuses will be used to protect the circuit, and PVC-insulated copper cables will be installed. Determine the size of trunking required, assuming voltage drop requirements have been met.

Select term values from the *Unite Guide.*

The term value for 4 mm^2 is 16.6.

The number of conductors is 160 (80 circuits each of 2 conductors).

Multiply $16.6 \times 160 = 2656$ terms.

The nearest trunking sizes are:

150 mm × 50 mm which has a term value of 3091

or

150 mm × 38 mm which has a term value of 2999

Whichever is selected will depend on the designer's choice.

PVC trunking

Essentially the same range of accessories and installation techniques are used for the installation of PVC trunking. As well as being lighter and therefore easier to handle, PVC trunking is easier on the eye than metal when installed for data cabling or computer supplies in locations such as shops and offices. Some examples are shown below.

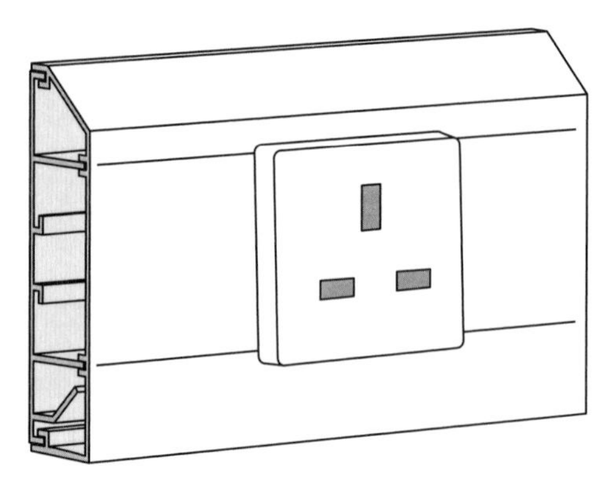

Figure 4.46 PVC trunking

PVC trunking

Cable tray

On large industrial and commercial installations where several cables take the same route, cable trays may be used. A wide range of designs of cable tray and accessories is available to match any cabling requirement, from lightweight instrumentation cable through to the heaviest multicore power cable. In situations where heavy multicore cables are required to cross long, unsupported spans, cable ladders should be used. In this section we will look at the following aspects of cable tray installation:

- types of cable tray
- types of finish
- types of accessories
- installing cable tray
- fabricating on site – the tray bending machine
- bending cable tray by hand
- fabricating a flat 90° bend
- fabricating a flat 90° bend from three pieces of tray
- forming a cable tray reduction
- fixing cable tray
- cable ladders
- cable basket.

Types of cable tray

Standard cable tray

Standard cable tray is suitable for light- and medium-duty installation work. Cables can be quickly dressed and then secured to the numerous perforations. Cable tray is often made to the Admiralty pattern. The flange height is 13 mm for narrow and 19 mm for wider sizes. Cable tray is supplied in standard widths from 51 mm up to 915 mm and in varying lengths. There are a number of patented jointing systems which make for easy assembly without fish plates or additional couplers, but most trays remain compatible with the existing Admiralty-type tray.

Standard cable tray

Heavy-duty tray

The heavy-duty tray remains popular for certain types of installation, although it is now used to a lesser extent than the return-flange designs. The flange height is 38 mm, and the tray is available in standard widths from 152 mm to 610 mm. As with the standard trays the design incorporates a patent jointing system. Heavy-duty trays are manufactured to BS 729 for galvanised trays or BS 2989 for pre-galvanised steel trays.

Return-flange cable tray

The design of return-flange cable tray differs from the standard in that the flange has a turned edge. This makes the tray several times stronger than the standard tray, yet it remains light and easy to install where space is a problem. The flange height is 25 mm with a 6 mm return. This type of tray uses patent snap-on couplers which fit over the outside of the tray flanges and are secured by round-headed bolts through the bed of the tray, eliminating the need for bolts projecting through the tray flange.

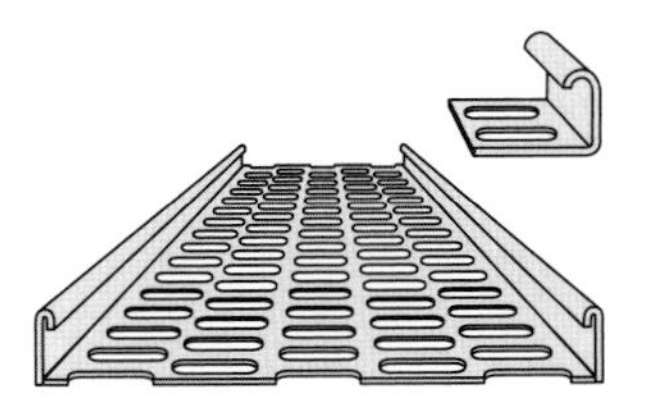

Figure 4.47 Return-flange cable tray

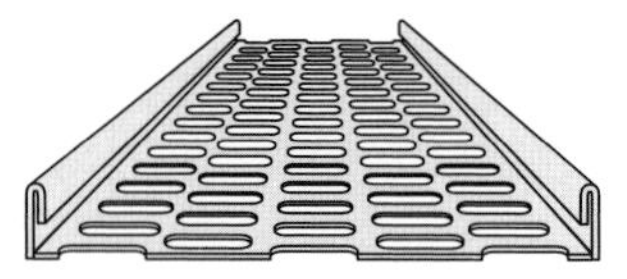

Figure 4.48 Heavy-duty return-flange cable tray

Heavy-duty return-flange cable tray

Heavy-duty flange tray is designed for use in circumstances where a high loading or adverse site conditions are experienced. The heavy-duty perforation pattern combined with a wide flange ensures that the tray is rigid and strong. As with the other types of tray, patent coupling systems, which are convenient to use on site, are usually available.

Wire mesh tray (basket tray)

A modern innovation is the wire mesh tray, a lightweight open tray that is quick and easy to install.

Types of finish

Cable tray is available in a number of standard finishes:

- **Hot-dip galvanised** is the most frequently specified tray finish, being suitable for high or low ambient working temperatures and where normal atmospheric pollutants are likely to be encountered. The galvanising is usually carried out to BS 729.
- **Plastic-coated finishes** are useful finishes for cable trays where protection is needed against chemical contamination or where hygienic cleanable surfaces are required. A wide range of coatings is available; polythene and polyvinylchloride (PVC) coatings are standard finishes, with others being available where specified.
- **Stainless steel** tray work is available for special applications, e.g. food processing and marine applications.

Types of accessories

A wide variety of factory-made accessories is available to suit both standard and return-flange trays of various sizes (see Figure 4.49).

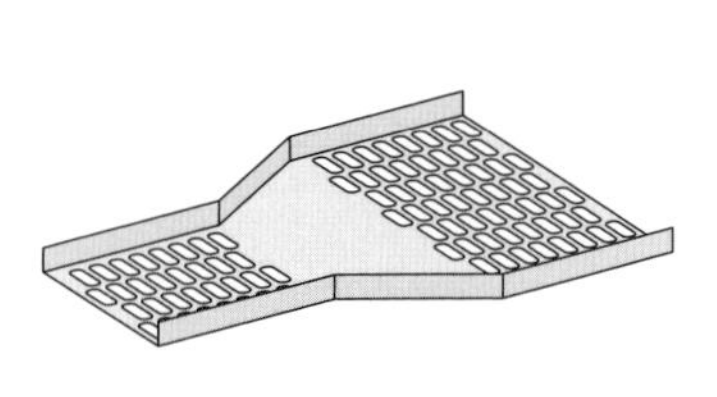

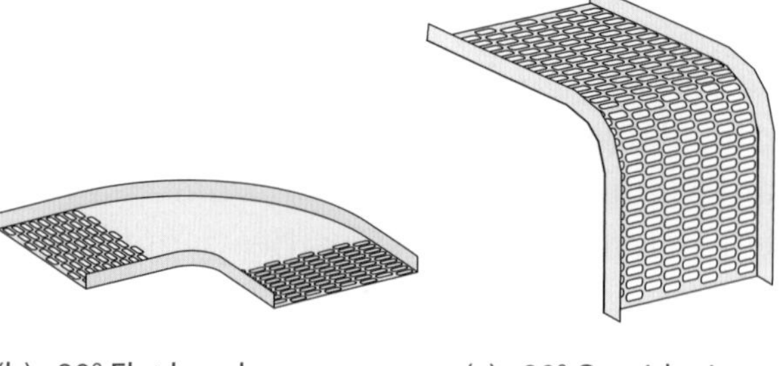

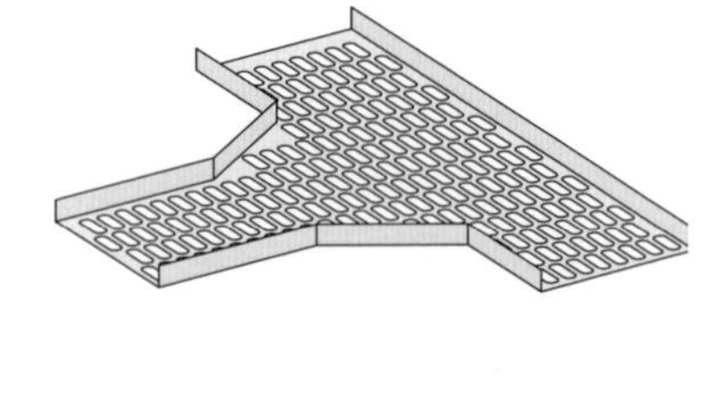

(a) Straight reducer (b) 90° Flat bend (c) 90° Outside riser (d) Equal tee

Figure 4.49

Installing cable tray

When installing cable tray it is essential that it is well supported and that all supports are secured. It is usually possible to complete the installation by making use of the wide range of accessories and fittings generally available, although it may sometimes be necessary to fabricate joints, bends or fittings to meet particular requirements.

Cable trays can be joined in a number of ways. Different manufacturers will supply a variety of patent couplings and fasteners. Links or fish plates are commonly used, and some cable trays are designed with socket joints. In some circumstances a welded joint may be required. If this method is used care must be taken to restore the finish around the weld to prevent corrosion.

Most methods of jointing cable tray involve the use of nuts and bolts. A round-headed or mushroom-headed bolt (roofing bolt) or screw should be used, and this should be installed with the head inside the tray. This reduces the risk of damage to the cables being drawn along the tray.

Remember

When installing a cable tray run, the following points should be considered:

- ease of installation
- economy of time and materials
- facility for extending the system to take additional cables

Fabricating on site (the tray bending machine)

It is sometimes necessary to fabricate joints, bends and fittings to meet particular requirements or where factory-made accessories are not available. Careful measurement and marking out is required. Cable trays can be cut quite easily with a hacksaw but care should be taken to remove any sharp edges or burrs.

Bends can be formed by hand after a number of cuts have been made in the flange to accommodate the bend, although a far better job can be made by using a crimping tool or a tray-bending machine (Figure 4.50). Bending machines are available from various manufacturers. Where a lot of cable tray work is to be installed, machine bending is quicker and more practical. The machines are made to accommodate the various widths and gauges of cable tray. They may also be used to bend and form flat strips of metal, and have a vice to hold the length of cable tray being worked upon. Cable-tray chain vices are also available for this purpose.

Remember

Whenever the tray has been cut the steel edges must be protected as far as possible by sealing with either zinc paint for galvanised tray or an appropriate primer and topcoat for painted tray, or a liquid plastic solution for plastic-coated tray

'Making good' cannot normally match the protection qualities of the original factory-applied finish, but the absence of any protection at all can seriously reduce the effectiveness of the original finish through corrosion spreading from this point. Primed tray should have a proper finishing paint applied over the primer as soon as possible. The purpose of the finish is to protect the cable tray from corrosion.

Figure 4.50 (a) Cable-tray bending machine

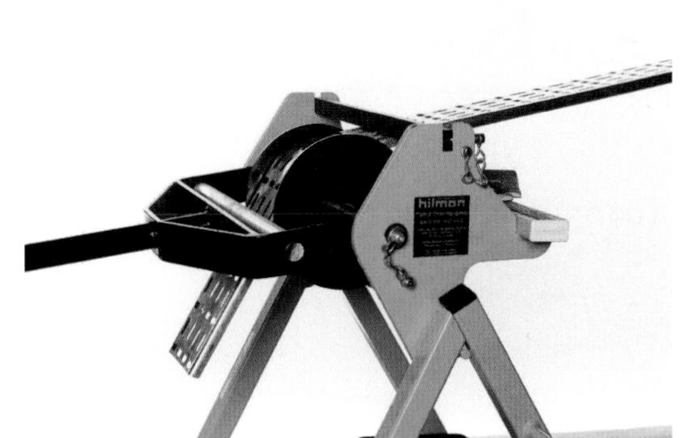

(b) Sharp-radius bend being formed

(c) Large upward-radius bend being formed

Bending cable tray by hand

Light-duty cable tray may be bent by hand with the aid of a crimping tool. This can be made from a piece of 6 mm mild steel bar.

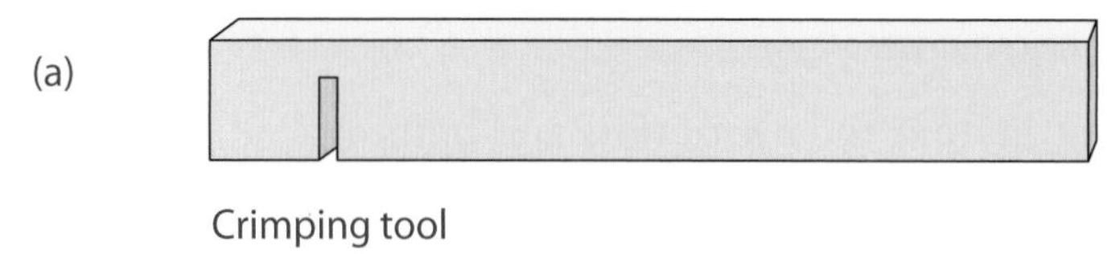

To make an inside bend, first determine the radius of the bend from the diameter of the largest cable to be installed. Mark the points at which the bend will begin and end, and the centre of the bend on the piece of tray to be bent.

Using the crimping tool, crimp evenly on each side of the tray.

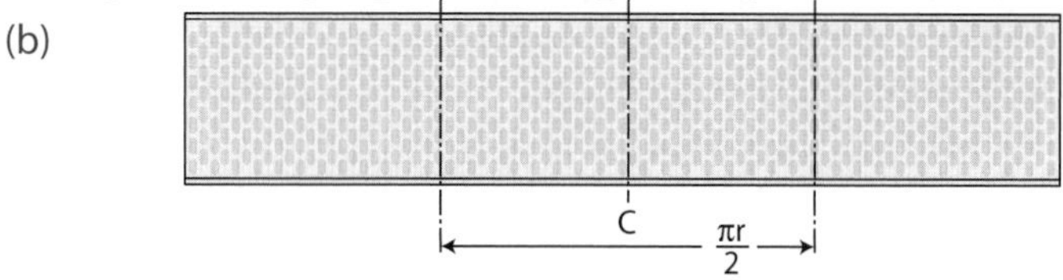

Work slowly, checking the form of the bend.

To make an 'outside' bend, saw through the flange and bend the tray in a vice. Mark

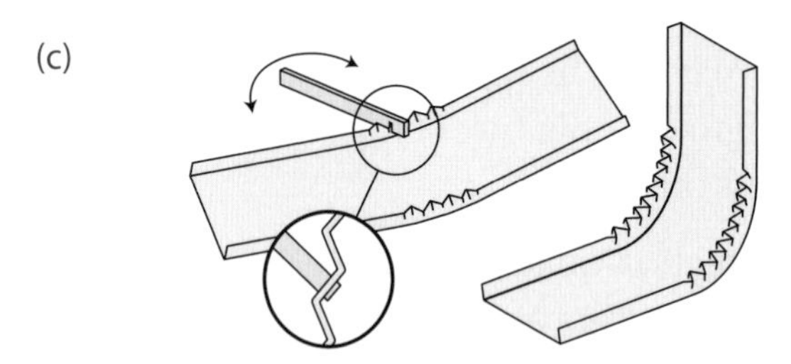

out the bend, as before, taking care that the radius of the bend accepts the largest size cable as stated in Table 4E from the *On-Site Guide*.

Mark off along both flanges on the cable tray a series of equal distance points. Make appropriate hacksaw cuts that are equal to the depth of the tray flange on either side of the centre line as shown.

Grip the tray with wooden blocks in a vice and bend gradually, moving the tray along and checking the bend for evenness. Figure 4.51 (f)

illustrates the completed bend.

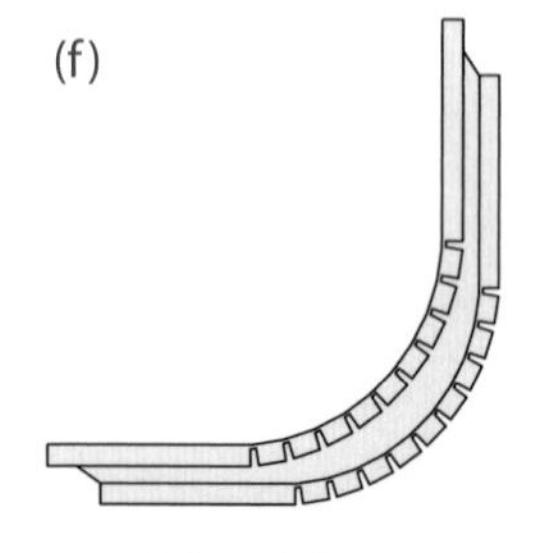

Figure 4.51 a–f Bending cable tray

Fabricating a flat 90° bend

Measure and mark the mid-point of the bend.

Mark off X when $X = \sqrt{2} \times W$, where W = width of tray.

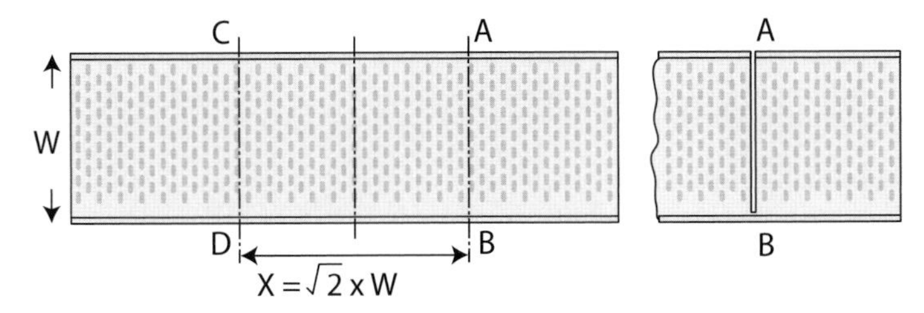

Cut through the flange with a hacksaw at point A and along line A–B but do not cut through the opposite flange. Make a similar cut at point C and along line C–D. Bend the two outer sections of the tray together to form a 90° angle.

(b)

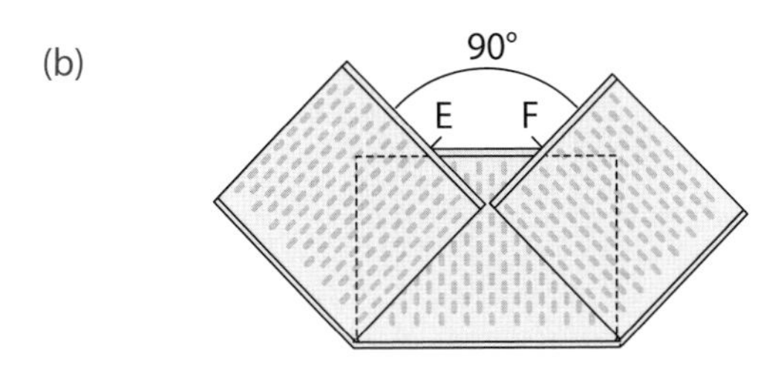

Mark the flange at the points of overlap E, F and E1, F1. Cut through the flanges at these points and bend these flanges flat. Cut away the tongues at both slots A and C as shown.

(c)

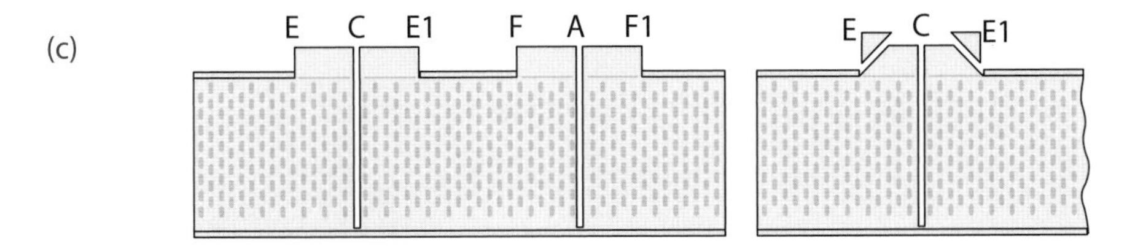

Remove all sharp edges and burrs with a file. Make up the assembly as shown and secure with round-headed bolts. Ensure that the bolt heads are uppermost.

(d)

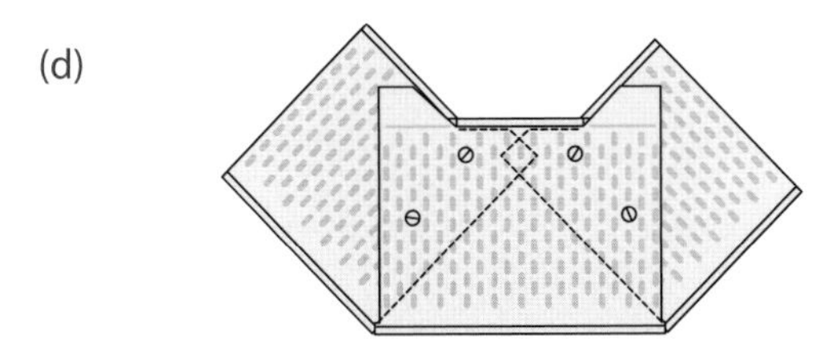

Figure 4.52 a–d Flat 90° bend

Fabricating a flat 90° bend from three pieces of cable tray

Place two pieces of tray together to form a 90° angle at the point of contact marked X.

(a)

X

90°

Place a third piece of tray over these and mark points of contact A, B, C, D, E, F.

(b)

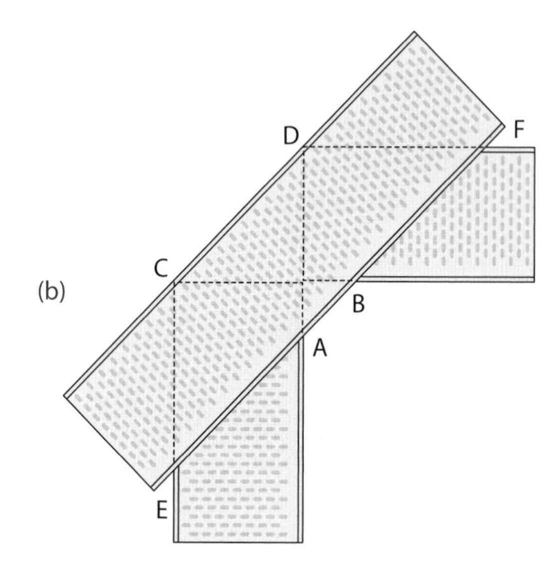

Cut away flanges XA and XB as shown.

(c)

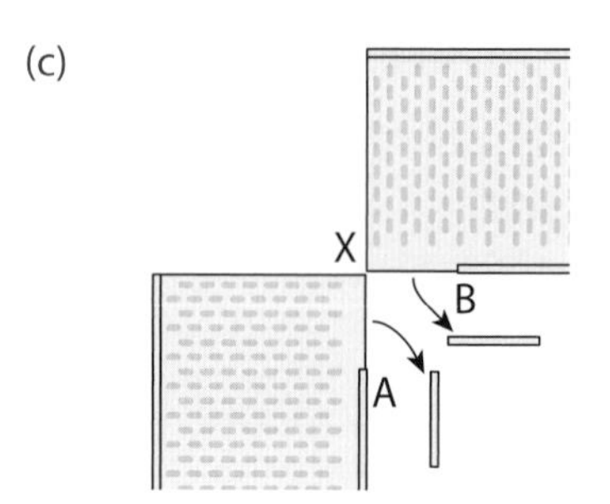

Cut along CE and DF. Cut away inner flange AE, BF.

(d)

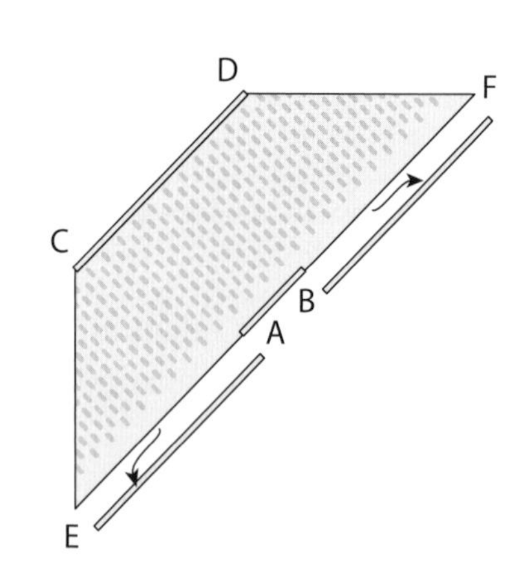

Remove all sharp edges and burrs with a file. Assemble as shown, using round-headed bolts.

(e)

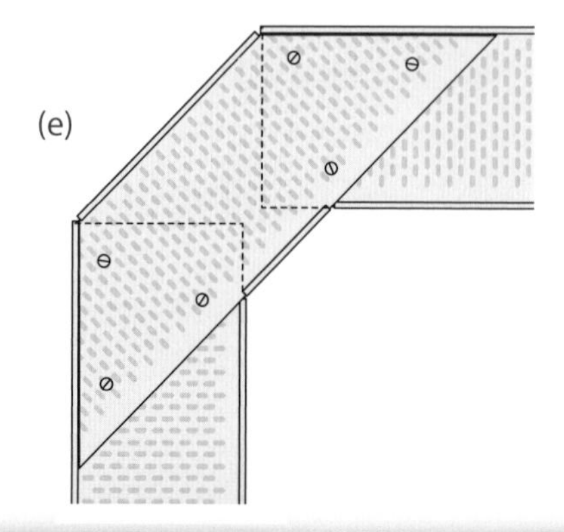

Figure 4.53 a–d Flat 90° bend made from three pieces of cable tray

Forming a cable-tray reduction

Overlap the two segments of cable tray by at least 100 mm. Mark the tray width and reduction angle as shown.

(a)

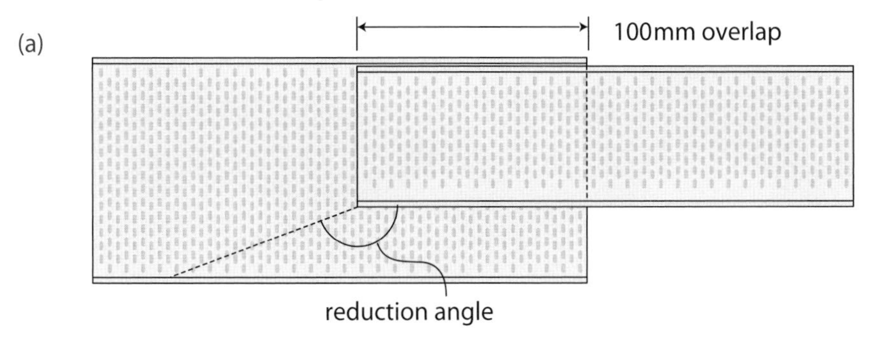

Cut with a hacksaw and cold chisel but do not cut through the flange.

(b)

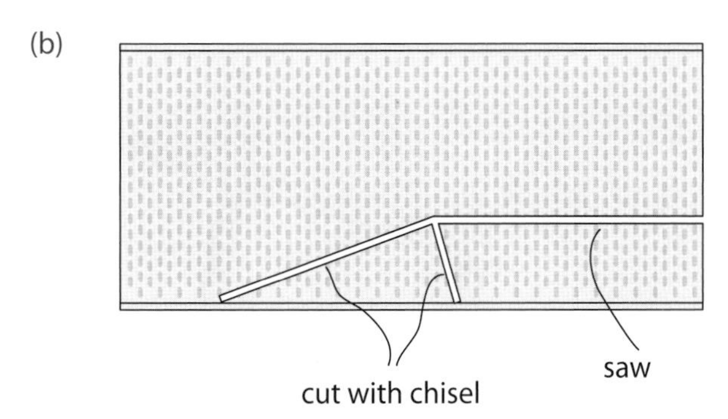

Bend the cable tray into shape. Remove all sharp edges and burrs with a file.

(c)

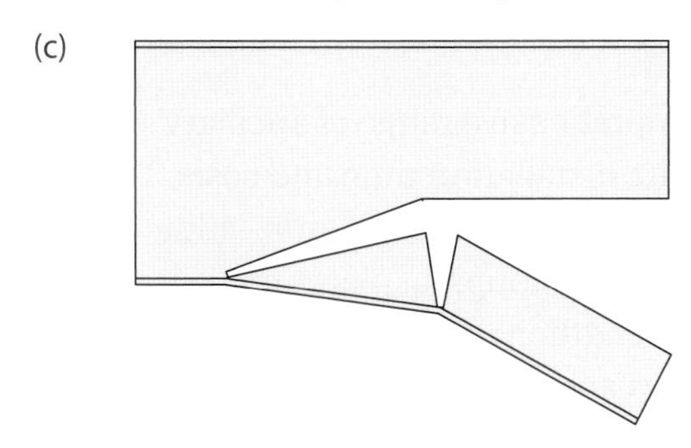

Assemble the joint using round-headed bolts.

(d)

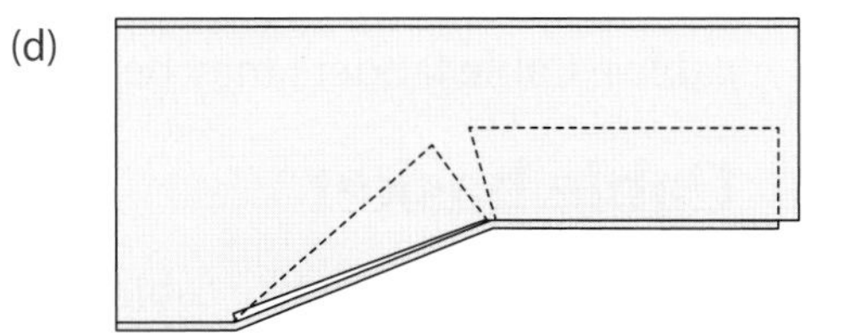

Ensure that the bolt heads are uppermost to prevent any damage to the cables.

(e)

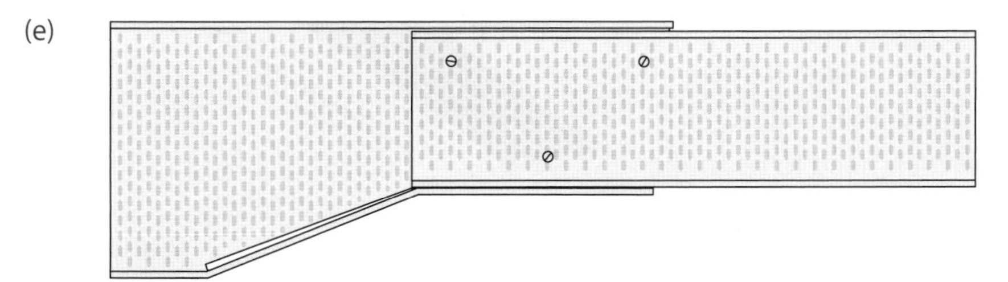

Figure 4.54 a–e Cable-tray reduction

Fixing cable tray

Cable tray is fixed to building surfaces using three basic methods, as illustrated below.

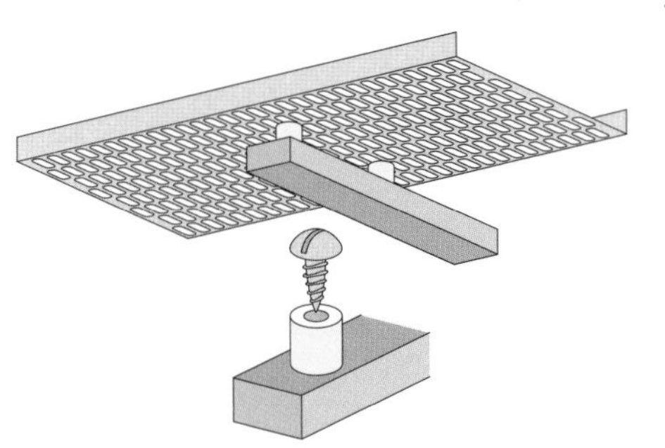

Figure 4.55
(a) Using spacers and round-headed screws

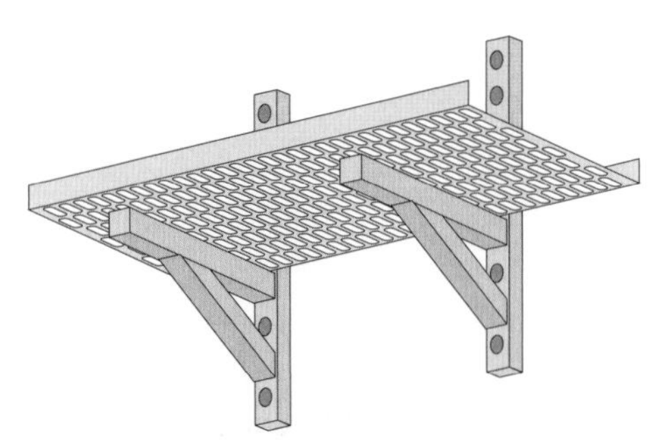

(b) Bolting to brackets

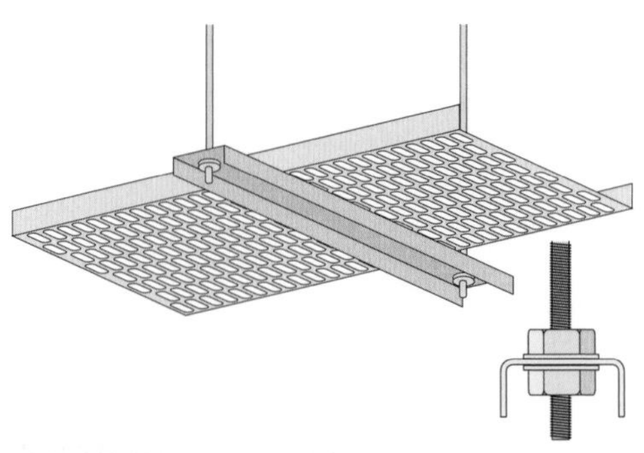

(c) Support channels are fixed to the underside of the tray when suspended from ceilings

Cable ladders

Cable ladders are an effective and convenient method of transporting cables across long unsupported spans or where the number of supports is to be reduced. They may be used in the most adverse site conditions and can withstand high winds, heavy snow, sand or dust settlement, or high humidity.

Cable ladder

Cable ladders comprise a series of prefabricated 'ladder' sections usually available in different widths together with a comprehensive range of ancillary components and accessories which minimise site installation time and costs. The result is a racking network of considerable strength and flexibility. Cable ladders are manufactured throughout in 2 mm mild steel with coupling plates made from 3 mm mild steel. The ladder side channels are strengthened by reinforcing inserts to increase the tortional (lengthways) rigidity. Rungs are slotted to take most available types of cable cleat, cable tie, pipe or conduit clamp. Cable-ladder design allows the maximum airflow around the cables and so prevents possible derating of power cables. Cable ladders may be mounted in virtually any direction as required.

Cable basket

This system is not unlike cable ladder. Made from a steel wire basket, it requires similar installation methods and techniques. Cutting of the basket to form bends or tees is normally achieved using bolt cutters. Any cuts then need to be made smooth as with tray or ladder systems.

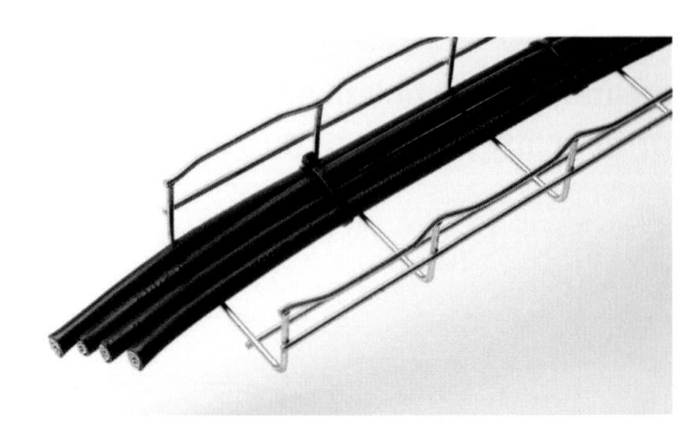

Cable basket

FAQ

Q Why use steel conduit when plastic conduit is cheaper and faster to install?

A It's about protection really. Steel conduit offers a much higher level of mechanical protection than plastic. Remember, cost and speed of installation are not the only design requirements to consider.

Q Where would I use a crampet?

A Crampets or pipe hooks are used as a quick fixing where the conduit is to be hidden, such as in a screed floor or plastered walls. The conduit is laid on the concrete floor, a hole is drilled into the concrete beside the conduit, and then the crampet is driven into the floor to grip the conduit.

Q Why do we need to use multi-compartment trunking?

A BS 7671 requires circuits to be segregated from other circuits. This is to prevent interference and danger if the circuits operate at different voltage bands or are emergency circuits. If these circuits are to be run in a common trunking then there has to be a physical barrier between them, hence the multi-compartment trunking.

Q What advantages are there in using overhead busbar trunking in place of other distribution systems in an engineering workshop installation?

A
- The installation can be completed before the final position of machines is known.
- Machines can be added or removed with minimum of work and disruption – just a few metres of conduit up to the Busbar tap off box.
- Fault-finding is limited to just a few metres of cable and disruption to other machines minimised.

Q Why do I need to know how to make bends and sets in steel trunking and tray when I can buy what I need?

A With good planning, all bends and sets can be bought. However, is it necessary for the job not to come to a standstill when an unexpected wall appears or you have to work round the air-conditioning trunking, and the wholesaler cannot get the required set to you for a few days?

Switching of lighting circuits

There are numerous switching arrangements that make up a lighting circuit. This section will look at the most common ones. It would not be practical to look at all the possible combinations, as there are simply far too many.

Cabling	Switching arrangements
• wiring in conduit and/or trunking • wiring using multicore/composite cables.	• one-way switching • two-way switching • intermediate switching.

Table 4.14 Switching arrangements

Wiring in conduit and trunking

Wiring in this type of installation is carried out using PVC single-core insulated cables (Ref 6491 X). This code number is a manufacturer's code used to denote different types of cable. The line (live) conductor is taken directly to the first switch and looped from switch to switch for all the remaining lights connected to that particular circuit. The neutral conductor is taken directly to the lighting outlet (luminaires) and looped between all the remaining luminaires on that circuit. The switch wire is run between the switch and the luminaire it controls.

Remember

Red sleeving changes to brown sleeving

Remember

Cables supplying switches of any circuit and the cables feeding away from the switch to the light are identified. The cable from the distribution board to the switch is called the switch feed, and the cable from the switch to the light is called the switch wire

Wiring using multicore/composite cables

This type of cable is normally a sheathed multicore twin and earth or a three cores and earth (two-way and intermediate circuits only) (Ref 6242Y and 6243Y respectively). A 'loop in' or 'joint box' method may be employed with this type of installation. In many instances a loop-in system is specified as there are no joint boxes installed and all terminations are readily accessible at the switches and ceiling roses.

With a joint-box system normally only one cable is run to each wiring outlet. Where such joint boxes are installed beneath floors they should be accessible by leaving a screwed trap in the floorboard directly above the joint box. All conductors should be correctly colour identified.

On a composite cable installation where the conductors other than brown are used as a phase conductor they should be fitted with a brown sleeve at their terminations.

All conductors must be contained within a non-combustible enclosure at wiring outlets (i.e. the sheathing of the cable must be taken into the wiring accessory). Throughout the lighting installation a circuit-protective conductor (cpc) must be installed and terminated at a suitable earthing terminal in the accessory/box.

Where an earthing terminal may not be fitted in a PVC switch **pattress**, the cpc may be terminated in a connector. Where the sheathing is removed from a composite cable the cpc must be fitted with an insulating sleeve (green and yellow); this provides equivalent insulation to that provided by the insulation of a single-core non-sheathed cable of appropriate size complying with BS 6004 or BS 7211.

One-way switching

The most basic circuit possible is the one-way switch controlling one light, as shown in Figure 4.56. In this system, one terminal of the one-way switch receives the switch feed; the switch wire leaves from the other terminal and goes directly to the luminaire (a). Once operated, the switch contact is held in place mechanically and therefore the electricity is continually flowing through to the light (b).

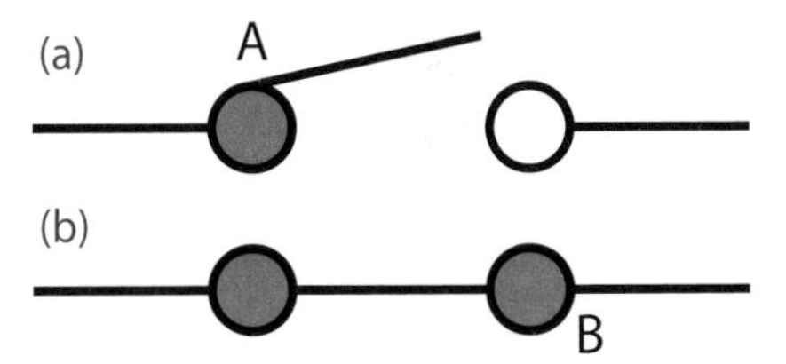

Figure 4.56 One-way switching

In other words, we supply the switch feed terminal, point (A).

Operate the switch and it comes out at the switch wire terminal, point (B).

Figures 4.57 to 4.59 show the full circuit when wired using single-core cables, which would be run in either conduit or trunking.

Using the new cable colours again, Figure 4.59 now shows a second light point fed from the same switch wire. This means that the second light is now wired in parallel with the first light.

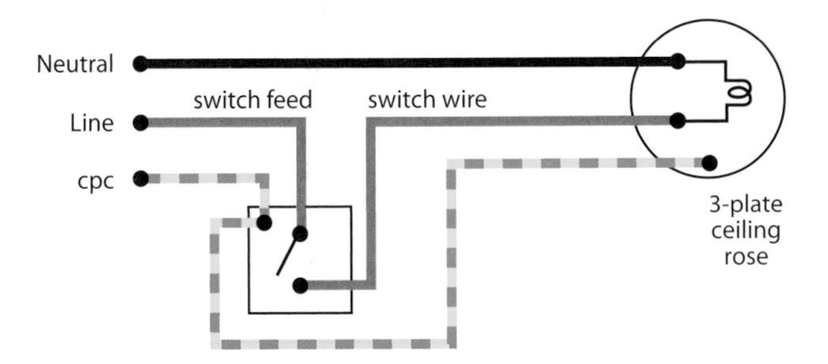

Figure 4.57 One-way switching for wiring with single-core cables (old cable colours)

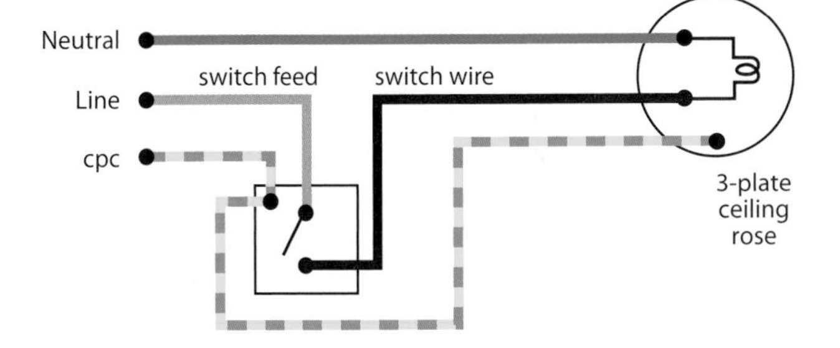

Figure 4.58 One-way switching for wiring with single-core cables (new cable colours)

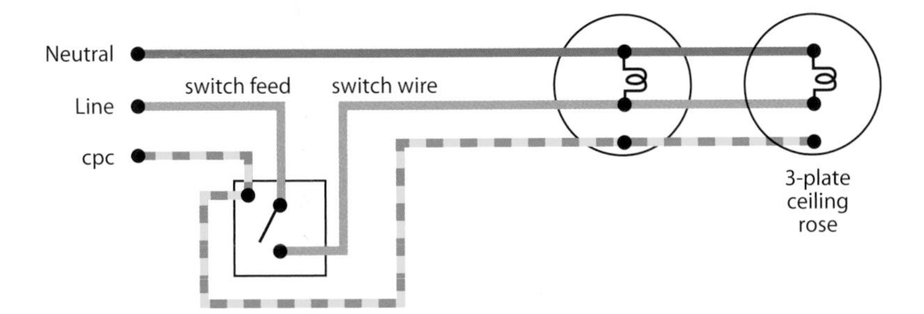

Figure 4.59 Extra lighting fed from the same switch, wired in parallel (new cable colours)

Two-way switching

Sometimes we need to switch a light on, or off, from more than one location, e.g. at opposite ends of a long corridor. When this is required, a different switching arrangement must be used, the most common being the two-way switch circuit. In this type of circuit, the switch feed is feeding one two-way switch, and the switch wire goes from the other two-way switch to the luminaire(s). Two wires known as 'strappers' then link the two switches together. In other words:

We supply the switch feed terminal, point (A).

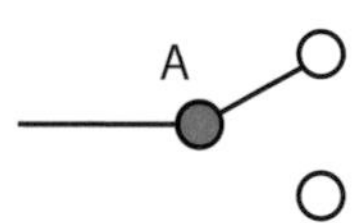

However, depending on the switch contact position, the electricity can come out on either terminal B or terminal C. In the following diagram it is shown energising terminal B.

If we now operated the switch, the contact would move across to energise terminal C.

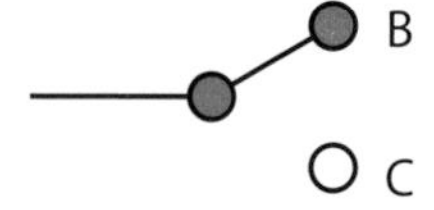

Please note that actual switch terminals are not marked in this way.

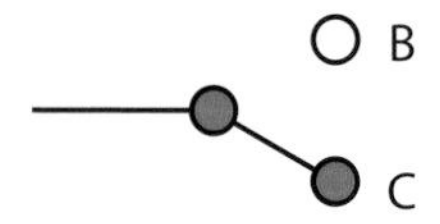

Switching of lighting circuits

By connecting the two two-way switches together we now have the ability at each switch to either energise the switch wire going to the light or to de-energise it (this is why this system is ideal for controlling lighting on corridors or staircases). In the first diagram below, the luminaire is off.

However, operate the second switch and we energise the common terminal (C) and the luminaire will now come on.

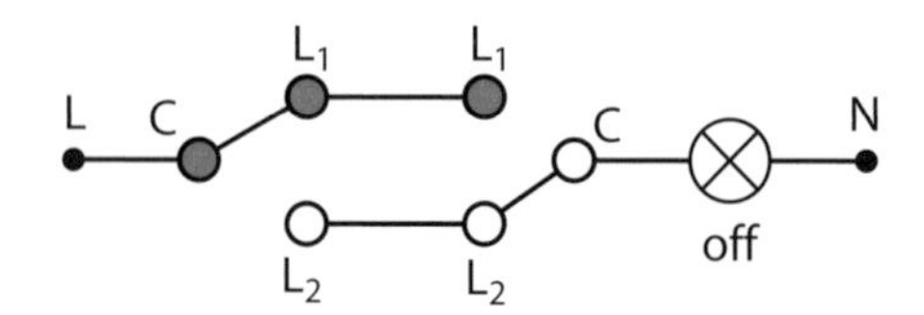

Figure 4.60 shows the full circuit when wired using single-core cable in conduit or trunking.

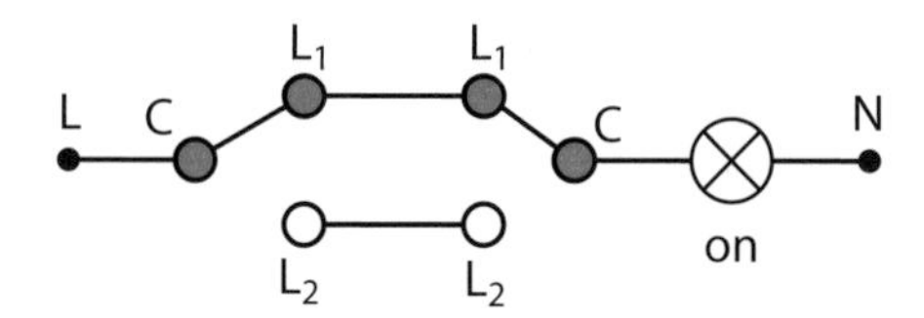

We can also use two-way switches for other purposes. Figure 4.61 shows one two-way switch to control two indicator lamps. This sort of system is frequently used as an

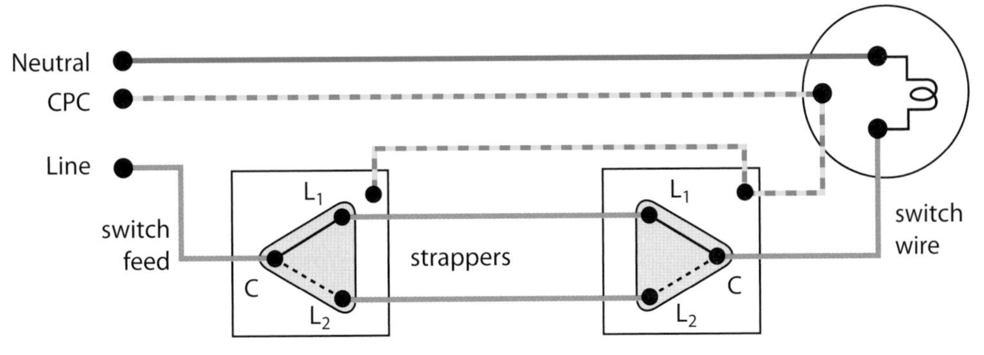

(a) New cable colours

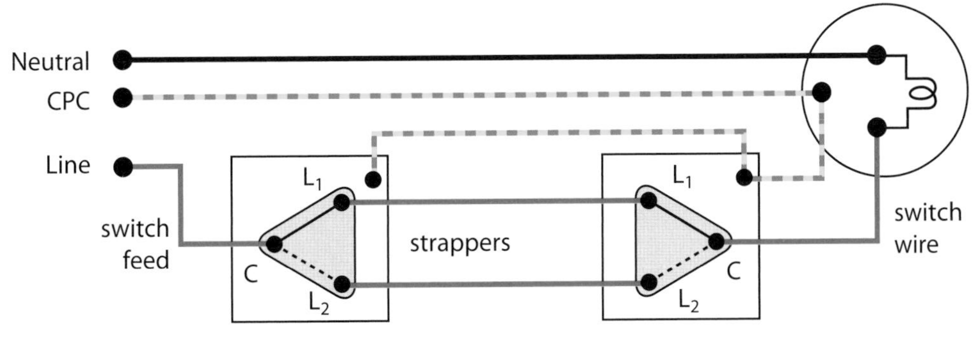

(b) Old cable colours

Figure 4.60 Full circuit wired with single-core cable. Old and new colours

entry system outside offices or dark rooms, where the two lamps can be marked, for example as 'available' and 'busy'.

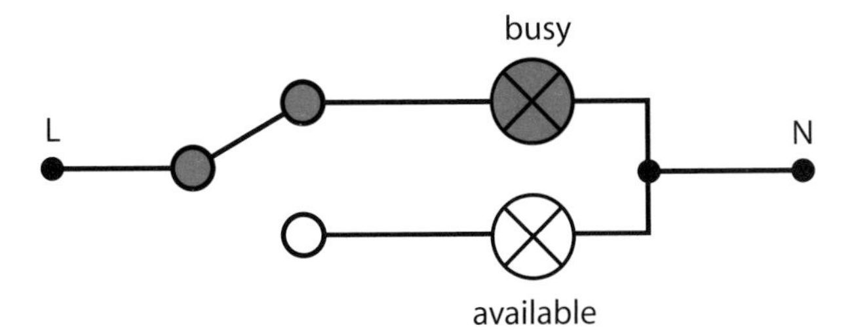

Figure 4.61 Two-way switch controlling two lamps

Intermediate switching

If more than two switch locations are required, e.g. in a long corridor with other corridors coming off it, then intermediate switches must be used – see (d). The intermediate switches are wired in the 'strappers' between the two-way switches.

The action of the intermediate switch is to cross-connect the 'strapping' wires. This gives us the ability to route a supply to any terminal depending upon the switch contact positions.

When we operate the switch into position two, the switch contacts cross over.

(a)

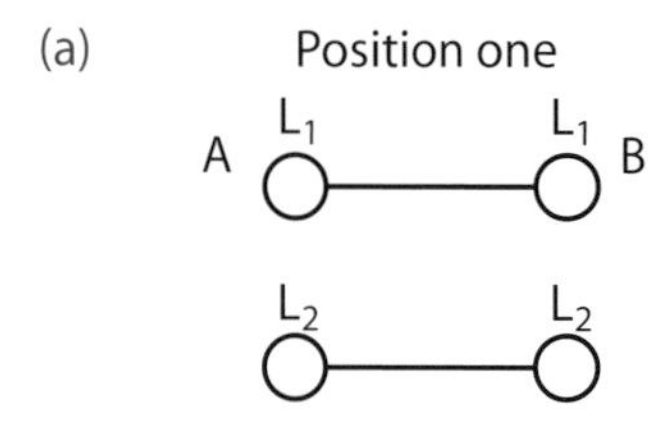

(b)

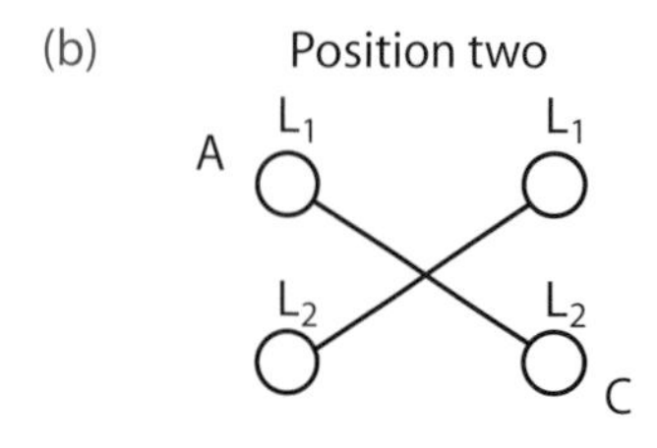

This means that a signal sent into terminal A can always be directed onto either terminal B or terminal C as required.

Ignoring terminal markings, (c) shows an arrangement where the switch wire is de-energised and therefore the luminaire is off.

However, by operating the intermediate switch, we can route the switch feed

(c)

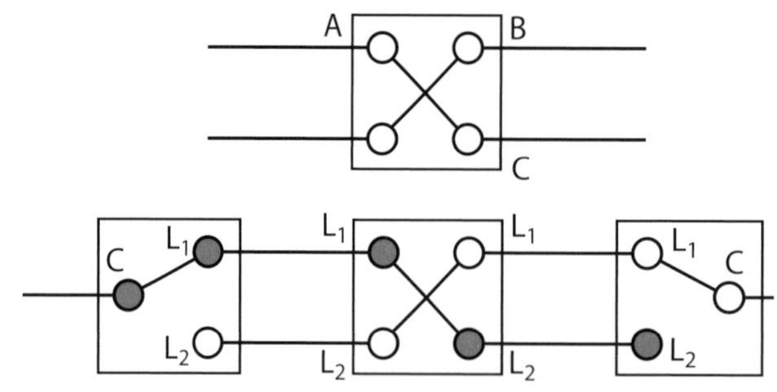

along another section of the 'strappers' and energise the switch wire terminal, and therefore the luminaire will come on, as shown in (d).

To use the example of a long hotel main corridor with other minor corridors coming off it, if we have an intermediate switch at each junction with the main corridor,

(d)

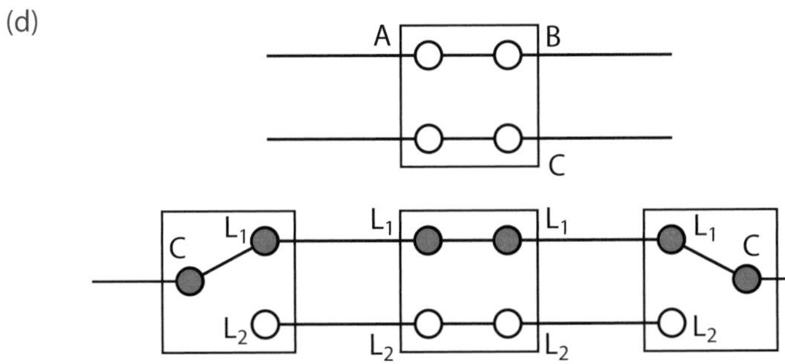

Figure 4.62 Intermediate switching

anyone joining or leaving the main corridor now has the ability to switch luminaires on or off.

Figure 4.63 shows the full circuit in the new colours, when wired using single-core cable in conduit or trunking.

Note: in this diagram the light will be off.

Remember

Any number of intermediate switches may be used between two-way switches and they are all wired into the 'strappers'. In other words, an intermediate switch has two positions

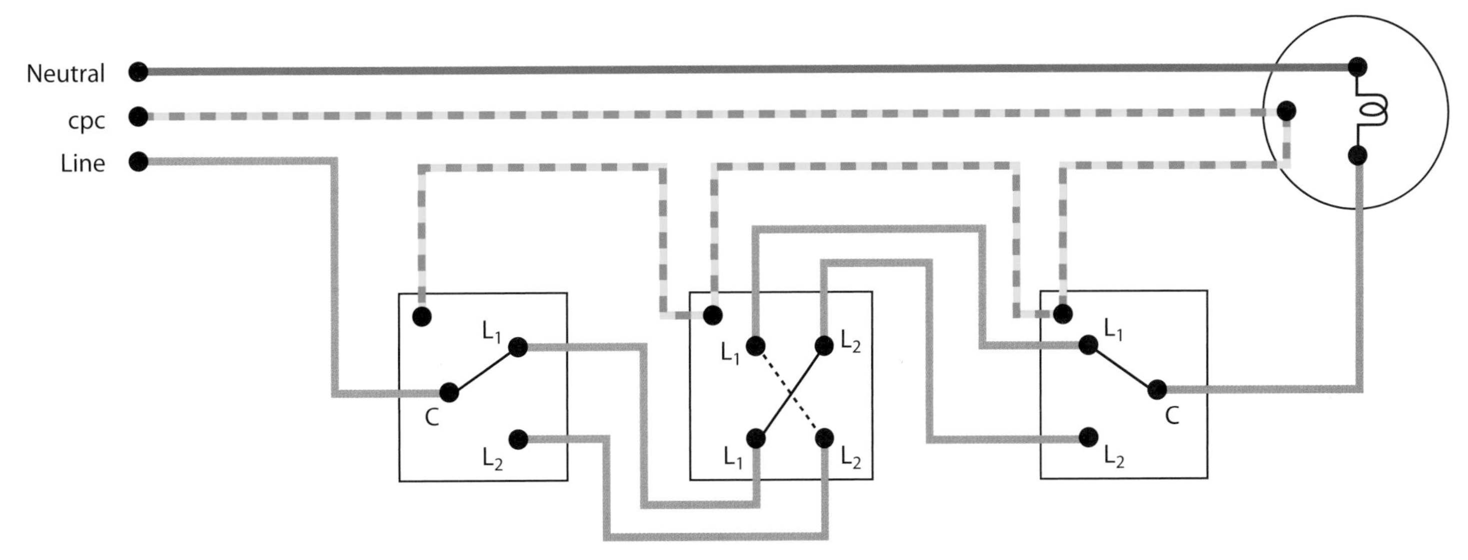

Figure 4.63 New cable colours

Wiring with multicore cables

Multicore cables (commonly referred to as twin and earth) are basically a three-core cable consisting of a line, neutral and earth conductor. Both the line and neutral conductors consist of copper conductors insulated with coloured insulation (see Figures 4.64 and 4.65), the earth being a non-insulated bare copper conductor sandwiched in between the line and neutral. In the old colours, the correct use of this cable type would be to use two red conductors (switch feed and switch wire) from switch locations up to luminaires. However, for ease, it was common practice for many to use a cable containing red and black conductors.

This means that the black conductor should be sleeved at both ends to indicate that it is a 'live' conductor. (Please be aware that many did not sleeve the conductors and therefore a luminaire connected across two black conductors is actually connected to a switch wire and a neutral.)

The outer white or grey covering is the sheath, and its main purpose is to prevent light mechanical damage of the insulation of the conductors.

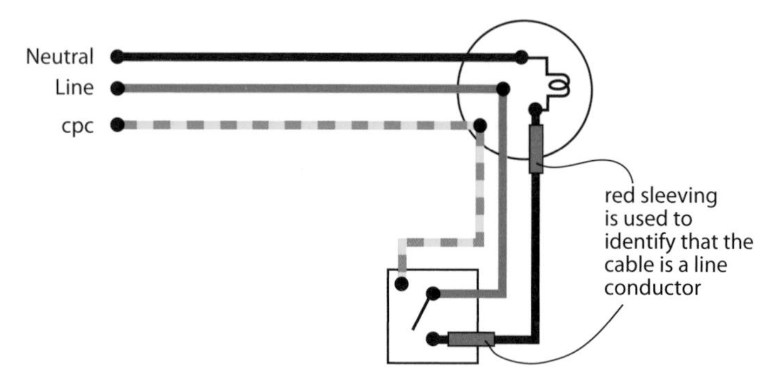

Figure 4.64 Old cable colours

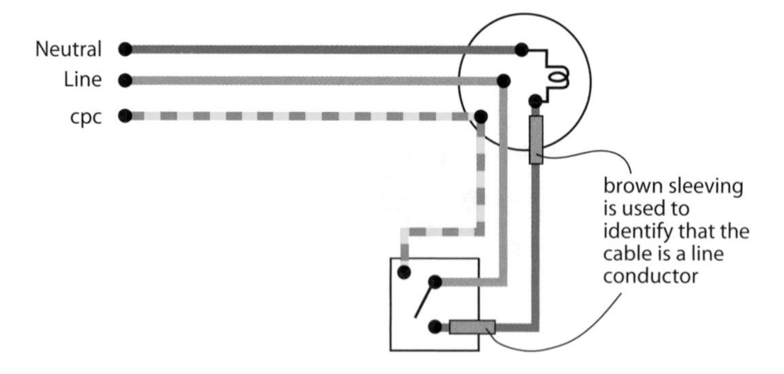

Figure 4.65 New cable colours

This cable is often used for wiring domestic and commercial lighting circuits, and you would normally use 1.5 mm^2 cable. Using this type of cable instead of singles is slightly more complicated because you are restricted as to where you plan your runs.

Two-way switching and conversion circuit

This method of switching using multicore cables requires the use of four-core cable. This is a cable that has three coloured and insulated conductors and a bare earth conductor. The three coloured conductors are in the old colours, coloured red, yellow, and blue. This type of cable is normally only stocked in 1.5 mm² and used in a lighting circuit where its application is for switching that requires more than one switch position, i.e. two-way and intermediate switching.

It is also used for converting an existing one-way switching arrangement into a two-way switching arrangement. The first step in this process is to replace the existing one-way switch with a two-way switch. This would then be connected as shown in Figure 4.66.

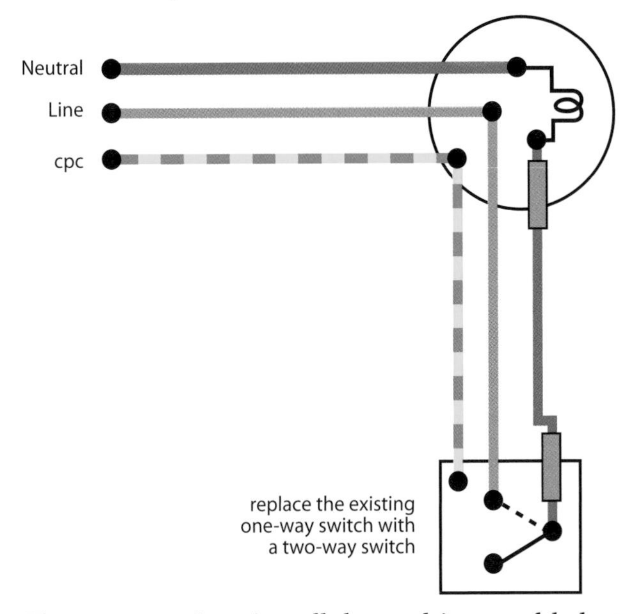

Figure 4.66 Conversion of one-way lighting into two-way switching

The next step is to install the multicore cable between this location and the new two-way switch location and complete the circuit connections as in the diagram below.

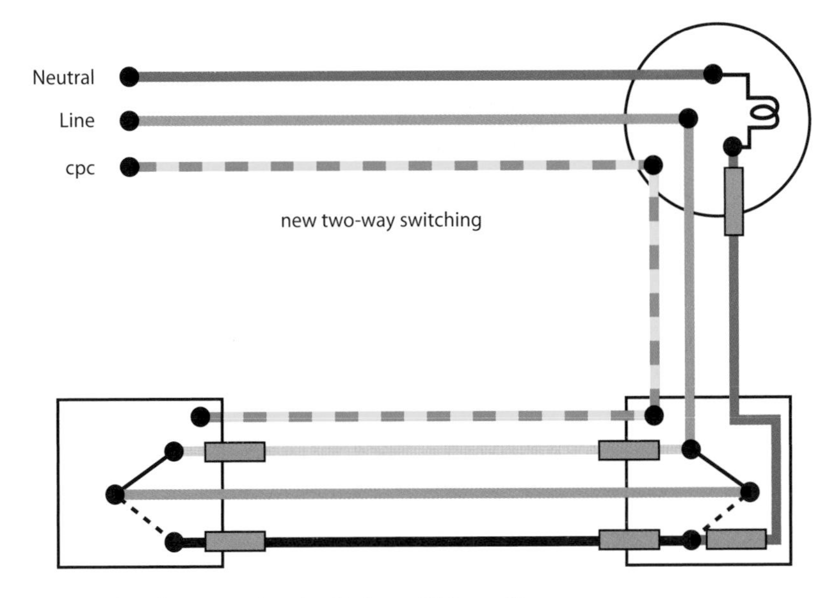

Figure 4.67 Conversion of one-way lighting into two-way switching

Note: In Figure 4.67 the light will be off.

Intermediate switching

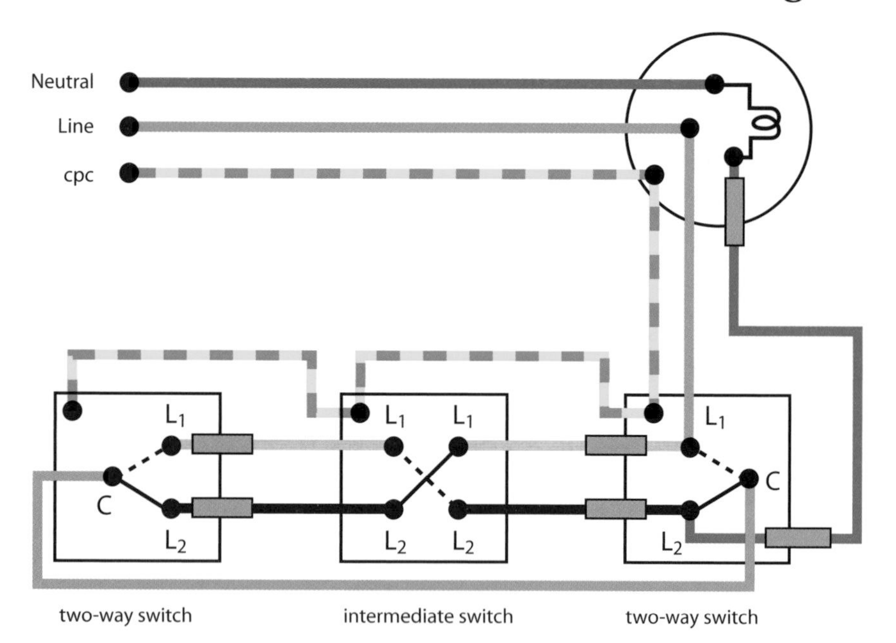

The diagram illustrates an intermediate lighting circuit using multicore cables.

Note: In Figure 4.68 the light will be on.

Figure 4.68 Intermediate switching

Wiring using a junction/joint box

This method of wiring is considered somewhat old-fashioned, as it has now been superseded by the loop-in method. However, that is not to say that you will not come across this method in the millions of households in the country that have a junction-box method installed. Care should be taken when wiring junction boxes, and the following precautions should be observed:

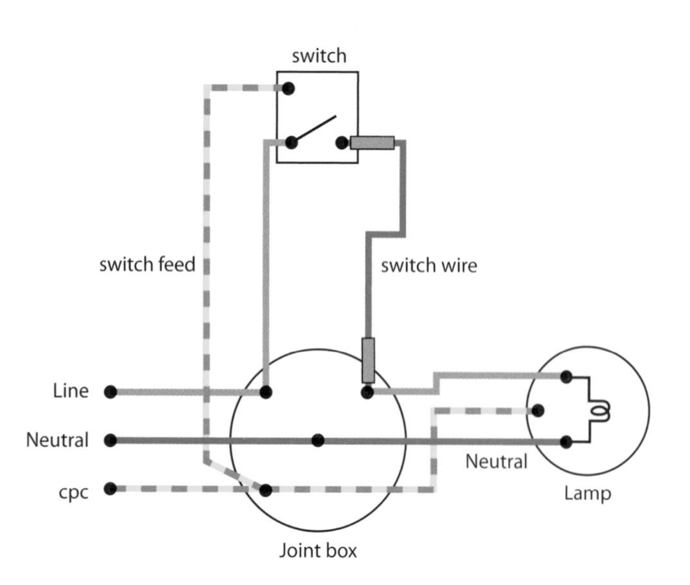

Figure 4.69 One-way switch using a joint box

- the protective outer sheath should be taken inside the junction-box entries to a minimum of 10 mm
- where terminations are made into a connector, only sufficient insulation should be removed
- sufficient slack should be left inside the joint box so as to prevent excess tension on conductors
- cables should be inserted so that they are not crossing. They should be neat and fitted so that the lid fits without causing damage
- correct size joint boxes should be used
- joint boxes should be secured to a platform fitted between the floor and ceiling joists.

Figure 4.69 illustrates the control of one light via a one-way switch using this method.

Rings, radials and spurs

This section sets out the different design options for ring and radial circuits inside households or similar premises in accordance with Regulation 433.1. It also looks at the use of socket outlets and fused connection units. There are also other design aspects that need to be remembered. These are:

- BS 7671 – Chapter 41 – Protection against electric shock
- BS 7671 – Chapter 42 – Protection against thermal effects
- BS 7671 – Chapter 43 – Protection against overcurrent
- BS 7671 – Part 5 – Selection & erection of equipment

The ring final circuit

In this type of circuit, the line, neutral and c.p.c. start at the distribution board, passing through the respective terminals of each socket outlet in the circuit and then return to the same terminals in the distribution board where it is connected to a 30 A or 32 A overcurrent protective device, thus forming a ring as shown in Figure 4.70.

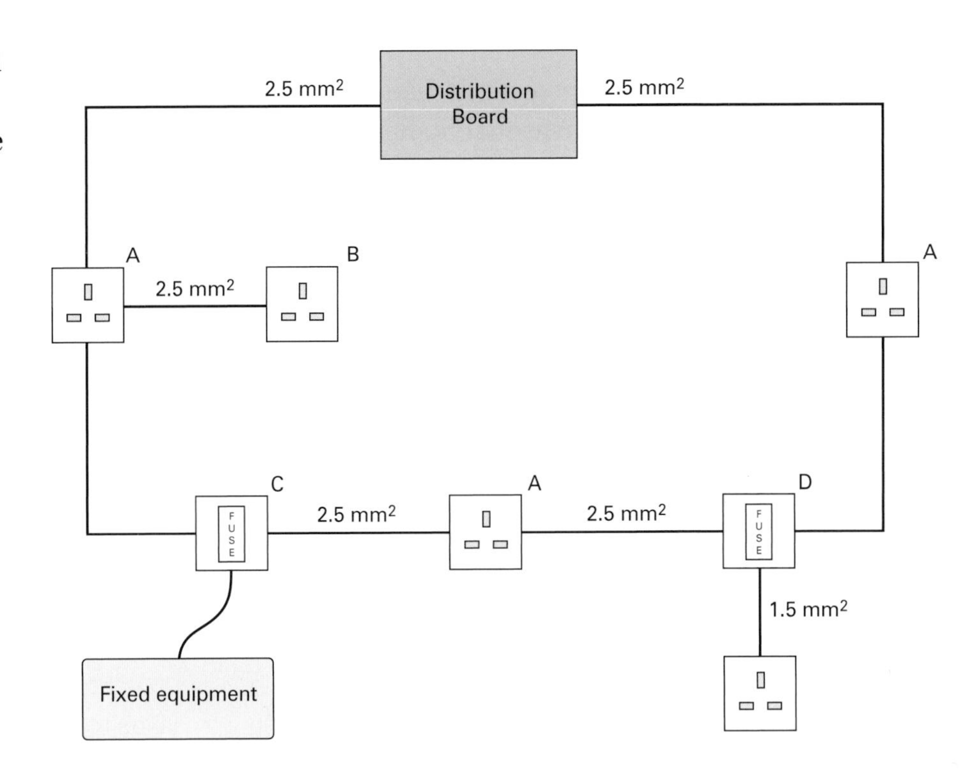

Figure 4.70 Ring final circuit

Please note the following relative to Figure 4.70:

- **A** – BS 1363 single or twin 13 A socket outlet. Each outlet of twin or multiple sockets is regarded as one socket outlet.
- **B** – BS 1363 socket fed via an unfused spur. An unfused spur should only feed one single or twin socket outlet. The unfused spur may be also connected to the origin of the circuit in the distribution board.
- **C** – Fused Connection Unit (FCU) supplying fixed equipment. The fuse in the FCU must not exceed 13 A.
- **D** – BS 1363 socket outlet fed via a fused connection unit. The number of sockets supplied from the FCU depends upon the load characteristics having taken diversity into account.

Conductor sizes shown relate to the live conductors only within flat 'twin & earth' cable. Reduced circuit protective conductor size is permitted (e.g. 2.5 mm^2 Live with 1.5 mm^2 c.p.c.). Although an unlimited number of socket outlets can be installed, there are some key issues that need to be taken into account.

The load current in any part of the circuit should be unlikely to exceed, for long periods, the current carrying capacity of the cable (Regulation 433.1.5). This can generally be achieved by:

- not supplying immersion heaters, electric space heaters or similar from the ring
- positioning the sockets so that the load is reasonably shared around the ring
- connecting cookers, ovens and hobs with a rated power greater than 2 kW on their own dedicated radial circuit
- taking account of the floor area being served by the ring. As a rule of thumb, a limit of 100 m^2 has been adopted.

Where more than one ring main is installed in the same premises, it is good practice to reasonably share the load over the ring main circuits so that the assessed load is balanced. Care must be taken to ensure that the requirements of Regulations 411.3.3 and 522.6.6 to 522.6.8 have been met in terms of additional protection by an RCD if required.

Wherever possible, cables should be fixed in such a position as not to be covered by thermal insulation. Should a cable be partially or completely covered by thermal insulation then reference should be made to Regulation 523.7.

Standard circuits

The following table illustrates a selection of standard arrangements for final circuits providing power to domestic socket outlets:

Type of circuit	Overcurrent protective device		Minimum conductor size (mm^2)		Maximum floor area served (m^2)
	Rating (A)	Type	Copper conductor PVC/XLPE insulation	Copper conductor MI installation	
Ring	30 or 32	Any	2.5	1.5	100
Radial	30 or 32	Cartridge fuse or circuit breaker	4.0	2.5	75
Radial	20	Any	2.5	1.5	50

Radial circuits

In this type of circuit, the line, neutral and c.p.c. start at the distribution board connected via an appropriate overcurrent protective device, then pass through the respective terminals of each socket outlet in the circuit and finish at the last outlet.

There are two types of radial circuit, one protected by a 20 A overcurrent protective device and the other by a 30 A/32 A overcurrent protective device, as shown in Figure 4.71.

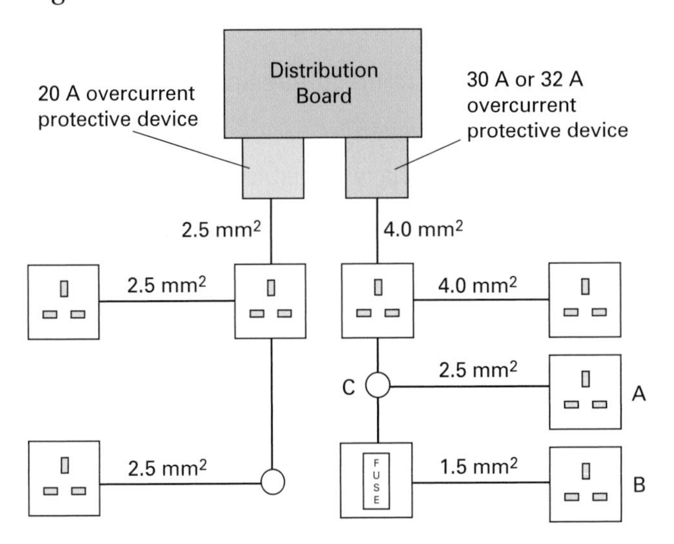

Figure 4.71 Radial circuits with overcurrent protective device

Please note the following relative to Figure 4.71:

- **A** – BS 1363 socket fed via an unfused spur. An unfused spur run in 2.5 mm^2 cable should only feed one single or twin socket outlet. Although shown here as being connected from a junction box, the unfused spur may be also connected to the origin of the circuit in the distribution board.
- **B** – BS 1363 socket outlet fed via a fused connection unit. The number of sockets supplied from the FCU depends upon the load characteristics having taken diversity into account.
- **C** – represents a junction box. Junction boxes with screw terminals must be accessible for inspection, testing and maintenance. Alternatively, use maintenance free terminals/connections (Regulation 526.3).

Conductor sizes shown relate to the live conductors only within flat 'twin and earth' cable. Reduced circuit protective conductor size is permitted (e.g. 4.0 mm^2 live with 1.5 mm^2 c.p.c.). Where the radial circuit is protected by a 20 A overcurrent protective device, the floor area being served by the ring is generally limited to 50 m^2. Where the radial circuit is protected by a 30 A/32 A overcurrent protective device the floor area being served by the ring is generally limited to 75 m^2.

Care must be taken to ensure that the requirements of Regulations 411.3.3 and 522.6.6 to 522.6.8 have been met in terms of additional protection by an RCD if required.

Wherever possible, cables should be fixed in such a position as not to be covered by thermal insulation. Should a cable be partially or completely covered by thermal insulation then reference should be made to Regulation 523.7.

Knowledge check

1. List all the tools you would expect to find in an electrician's toolbox.
2. Describe precautions you would need to take when using cutting and sawing tools.
3. List the different types of cable tray that are commercially available.
4. Describe in words how to construct a 90° inside bend using a crimping tool.
5. Which type of diagram uses symbols to represent all circuit components and shows how these are connected?
6. List the different type of screws that can be found in electrical training workshops.
7. Describe what magnesium oxide is and where it can be found within the electrical industry.
8. What is the maximum voltage drop that Regulations 525.1 to 525.4 state must not be exceeded?
9. To prevent electrolytic action when an aluminium cable is terminated onto copper busbars, what action should you take?
10. Describe the method for removing tongue-and-groove floorboards.
11. List different methods by which you can connect to terminals.
12. What is the purpose of a MICC seal?
13. Draw a simple diagram of a bending machine and label the main components.
14. List six different fixing devices for steel conduit and describe their use.
15. What is the maximum space factor allowed in trunking?
16. When terminating the sheathing of a conductor at an accessory, what must be done?
17. What size cable is normally used on a lighting circuit?
18. Approximately how much protective sheath should be taken into a junction box at the entry?
19. What are blue cores of cables covered with to identify them as being live?
20. How many socket outlets can be installed in a standard domestic ring circuit?
21. What is the maximum floor area standard domestic radial circuits can serve?
22. What is the maximum floor area that a domestic ring circuit can serve when protected by a 30/32 A protective device?

Chapter 5

Electrical Regulations and special locations

Overview

Electrical Regulations exist to protect both you as the electrical contractor, and the user of any systems you install or work that you do. This chapter will help you to interpret and use the Regulation book and explain what effect Regulations will have on the design of electrical circuits. It will also look at installations in potentially explosive atmospheres and petrol filling stations. This chapter will cover:

- **Electrical Regulations**
  - plan and style of the Regulations
  - the Parts of the Regulations
  - Appendices of the Regulations
  - the Electricity At Work Regulations 1989
- **Special installations or locations**
  - Potentially explosive atmospheres

Although every effort has been made to ensure the accuracy of information given in terms of compliance with BS 7671:2008, at the time of printing certain key documents, such as the IEE On-Site Guide and the IEE Guidance Note 3 were unavailable. Therefore please verify information given with these documents as they become available.

Electrical Regulations

This section covers:

- plan and style of the Regulations
- the Parts of the Regulations
- Appendices of the Regulations
- the Electricity At Work Regulations 1989.

When the first electrical regulations were laid down virtually no one had electricity in their homes or workplaces, now everyone has access to electricity. The Regulations have moved on since 1882 and we are now up to the *17th Edition 2008*. Over the years the Regulations have taken account of the new types of electrical equipment available and its usage, whether that be computers, lighting, overcurrent protection etc. and affect all aspects of the installation process.

The Regulations are designed to protect people, property, and livestock from electric shock, fire and burns and injury from mechanical movement of electrically operated equipment. They are not designed to instruct untrained persons, take the place of a detailed specification or to provide information for every circumstance.

The use of other British Standards is needed to supplement the information contained in these regulations, such as BS 5266 which deals with emergency lighting.

In 1992 the 16th Edition of the *IEE Wiring Regulations* became a British Standard, BS 7671. These regulations are not a statutory document (not legislated by an Act of Parliament). However, compliance with the Regulations within this book will allow compliance with such legislation as *Electricity at Work Regulations 1989, Health and Safety at Work Act 1974* and *Electricity Supply Regulations 1988*, which are all statutory documents and are enforceable by law.

Many other British Standards are referred to throughout these Regulations and in some cases these other standards have a BS EN number which refers to European harmonisation standards whereby all such standards will become applicable throughout Europe. These harmonised standards are co-ordinated by representatives from all the countries in the European Union via an organisation known as CENELEC.

In addition to BS 7671 the IEE has previously published a series of 'Guidance Notes', which are books that seek to clarify or simplify the requirements in BS 7671.

Did you know?

The first regulations covering the use of electricity came into existence in 1882 and were called 'Rules and regulations for the prevention of fire risks arising from Electric Lighting'. That is over 120 years ago!

Find out

Throughout Chapter 5 there will be many references to the Regulations; it would be advisable to look up the Regulations referred to when they are mentioned

Remember

A useful book is the *IEE On Site Guide*, which is an abbreviated version of BS 7671 which also gives practical explanations, and in some cases drawings, of the more commonly used regulations as well as useful cable tables etc. Another useful reference is produced by the union Unite

Plan and style of the Regulations

There are seven parts to the Regulations and appendices. These are based on an internationally agreed arrangement that follows the pattern laid out within IEC 60364. In this numbering system the first digit signifies a part of the Regulations, the second digit combines with the first to signify a chapter, the third digit combines with the first two to signify a section and any further digits signify sub-sections and the specific regulation number.

With a copy of the Regulations in front of you, we can show this by looking at the following regulation: 412.1.3.

412.1.3	First digit	Part 4 of the Regulations (Protection for Safety)
412.1.3	Second digit	First chapter within Part 4 (Chapter 41: Protection against electric shock)
412.1.3	Third digit	Second section of Chapter 41 (Protective measure: Double or reinforced insulation)
412.1.3	Fourth digit	First sub section of section 2 (General)
412.1.3	Fifth digit	Specific regulation within sub-section 1 ('Where this measure is used as the sole…')

The parts of the Regulations

Part 1 Scope, object and fundamental principles

There are three chapters within this part and numerous sections within each chapter, the first chapter (Chapter 11) dealing with what the Regulations actually cover and what they do not.

The Regulations apply to the design, erection and verification of electrical installations such as:

- residential premises
- commercial premises
- public premises
- industrial premises
- agricultural and horticultural premises
- prefabricated buildings
- caravans, caravan parks and similar sites
- construction sites, exhibitions, shows, fairgrounds and other installations for temporary purposes including professional stage and broadcast applications
- marinas
- external lighting and similar installations
- mobile or transportable units
- photovoltaic systems
- low voltage generating sets
- highway equipment and street furniture.

However, they do not cover the following:

- systems for the distribution of electricity to the public
- railway traction equipment, rolling stock and signalling equipment
- equipment of motor vehicles
- equipment on board ships covered by BS 8450
- equipment for mobile and fixed offshore installations
- equipment of aircraft
- those aspects of mines and quarries covered by statutory regulations
- radio interference suppresion equipment (unless it affects installation safety)
- lightning protection systems covered by BS EN 62305
- aspects of lift installations covered by BS 5655 and BS EN 81-1
- electrical equipment of machines covered by BS EN 60204.

The Regulations also include requirements for:

- circuits supplied at nominal voltages up to and including 1000 V a.c. or 1500 V d.c.
- all consumer installations external to buildings
- fixed wiring for information and communication technology.

Chapter 12 (Objects and effects) explains that the Regulations contain the rules for the design and erection of electrical installations so as to provide for safe and proper functioning for the intended use.

Chapter 13 (Fundamental principles) (131.1 General), states that the requirements of this chapter "are to provide for the safety of persons, livestock and property against danger and damage that may arise in the reasonable use of electrical installations". In electrical installations, risk of injury may arise from:

- shock currents
- excessive temperatures likely to cause burns and fires
- ignition of potentially explosive atmospheres
- undervoltages, overvoltages and electromagnetic influences
- mechanical movement of electrically actuated equipment
- power supply interruption
- interruption of safety services
- arcing or burning.

Sections 131.2 through to 131.8 then clarify the above points, and it should be noted that the 17th edition Regulations see an important change of terminology within this section. Although generally the concepts remain the same, Regulation 131.2.1 introduces the phrase 'Basic Protection' to replace protection against direct contact and Regulation 131.2.2 introduces the phrase 'Fault Protection' to replace protection against indirect contact.

Part 2 Definitions

In this section of BS 7671 the terms used throughout the Regulations are given a specific meaning, so that when one person talks about a circuit breaker, for example, then everyone else knows what they mean by that word.

Look upon this section as a type of dictionary relating to electrical words.

Part 3 Assessment of general characteristics

Part 3 has six chapters and states that an assessment shall be made of the following characteristics of an installation, where each aspect then forms one of the six chapters:

- The purpose of the installation and its supplies (Chapter 31).
- The external influences it will be exposed to (Chapter 32).
- The compatibility of its equipment (Chapter 33).
- Its maintainability (Chapter 34).
- Recognised safety services (Chapter 35).
- Assessment for continuity of service (Chapter 36).

Part 4 Protection for safety

Introducing more changes in terminology, this part of the Regulations has four chapters.

Chapter 41 – Protection against electric shock

BS EN 61140 states the fundamental rule of protection against electric shock as being "that hazardous live parts shall not be accessible and that accessible conductive parts shall not be hazardous-live, either in normal use without a fault or under fault conditions".

According to BS EN 61140, protection under normal conditions is met by 'basic protection' provisions and protection under fault conditions is met by 'fault protection' provision.

These provisions are referred to as 'protective measures', where Chapter 41 states that a protective measure shall consist of:

- an appropriate combination of a provision for basic protection and an independent provision for fault protection OR
- an enhanced protective provision (e.g. reinforced insulation) that provides both basic and fault protection.

Sections within Chapter 41 then deal with the following protective measures:

- Section 411 – Automatic disconnection of supply (formerly EEBADS).
- Section 412 – Double or reinforced insulation.
- Section 413 – Electrical segregation.
- Section 414 – Extra low voltage (SELV and PELV).

Chapter 42 – Protection against thermal effects

This chapter deals with protection against the effects of heat or thermal radiation, the ignition, combustion of materials, flames and smoke from an electrical installation and against fire services being cut off by the failure of electrical equipment.

Chapter 43 – Protection against overcurrent

This chapter deals with protection of live conductors from the effects of overcurrent.

Chapter 44 – Protection against voltage disturbances and electromagnetic disturbances

This chapter deals with protection against the likes of temporary overvoltages due to earth faults in the H.V. system, overvoltages of atmospheric origin and where a reduction in voltage could cause danger.

Part 5 Selection and erection of equipment

Revised under this edition of the Regulations, this part provides common rules for compliance with measures of protection for safety, requirements for proper functioning of intended use and requirements pertinent to the external influences. This is given over the following six chapters:

- Common rules (Chapter 51).
- Selection and erection of wiring systems (Chapter 52).
- Protection, isolation, switching, control and monitoring (Chapter 53).
- Earthing arrangements and protective conductors (Chapter 54).
- Other equipment (Chapter 55).
- Safety services (Chapter 56).

Part 6 Inspection and testing

Every installation shall during erection and upon completion, before being put into service, be inspected and tested to verify that the requirements of the Regulations have been met.

During this process, precautions shall be taken to prevent danger to persons and to protect property and installed equipment from damage.

Part 6 has three chapters dealing with the subject as follows:

- Initial verification (Chapter 61).
- Periodic inspection and testing (Chapter 62).
- Certification and reporting (Chapter 63).

This part of the Regulations does not give any detail on how to carry out any of the tests required, this information normally being found in the IEE *On Site Guide* or the *Unite Guide to Good Electrical Practice*. However, we must refer to the Regulations to ensure that any results obtained are within acceptable limits.

In this respect it should be noted that the minimum values of insulation resistance have changed to the following.

Circuit normal voltage	d.c. test voltage (V)	Minimum insulation resistance (MΩ)
SELV and PELV	250	0.5
Up to and including 500 V with the exception of the above	500	1.0
Above 500 V	1000	1.0

Part 7 Special installations or locations

This part of the Regulations has been modified and now comprises the 15 sections listed below, each of which either supplements or modifies the general requirements of other parts of the Regulations:

- Locations containing a bath or shower (701).
- Swimming pools and other basins (702).
- Rooms and cabins containing hot air saunas (703).
- Construction and demolition site installations (704).
- Agricultural and horticultural premises (705).
- Conducting locations with restricted movement (706).
- Caravan/camping parks and similar locations (708).
- Marinas and similar locations (709).
- Medical locations – reserved for future use (710).
- Exhibitions, shows and stands (711).
- Solar photovoltaic (pv) power systems (712).
- Mobile or transportable units (717).
- Installations in caravans and motor caravans (721).
- Temporary installations for structures, amusement devices and booths at fairgrounds, amusement parks and circuses (740).
- Floor and ceiling heating systems (753).

Some of the more relevant sections are covered opposite.

Section 701: Locations containing a bath or shower

It should be noted that this section in particular sees radical change, namely:

- Zones 0, 1 and 2 are retained, but Zone 3 has been removed.
- All circuits supplying the bathroom (irrespective of the points they are serving) have to be protected by 30 mA RCDs.
- Supplementary equipotential bonding is not required when all of the following are met:
  - All circuits for the bathroom meet the requirements for ADS.
  - All final circuits have 30 mA RCD protection.
  - All extraneous conductive parts of the bathroom are connected to the protective equipotential bonding of the installation (411.3.1.2).
- SELV socket outlets and shaver sockets are permitted outside Zone 1. 230 V socket outlets are permitted provided they are more than 3 m horizontally from Zone 1.

Figure 5.1(a)–(i) illustrate the new zone arrangements and dimensions.

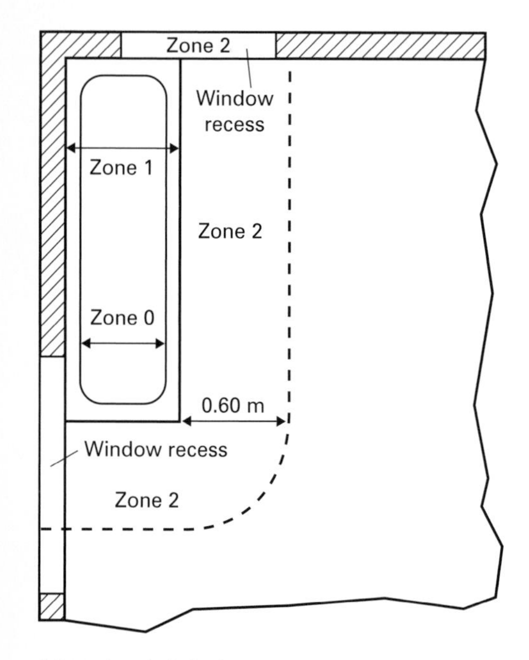

(a) Bath tub (plan)

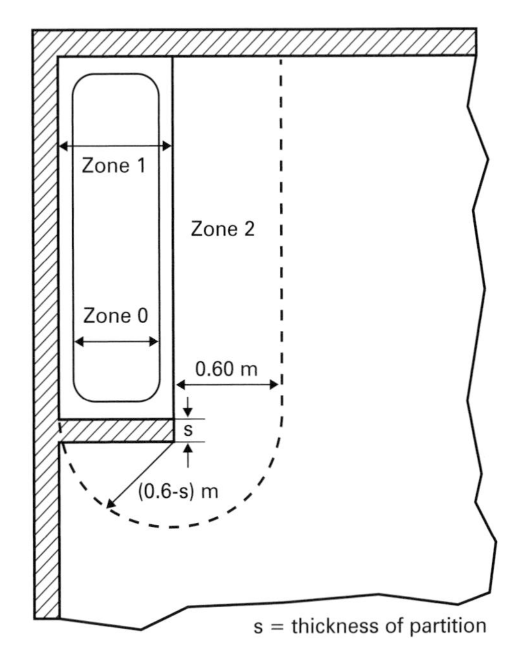

(b) Bath tub, with permanent fixed partition (plan)

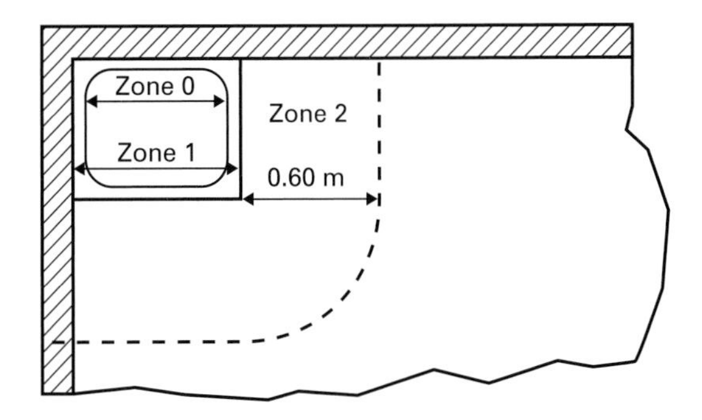

(c) Shower basin (plan)

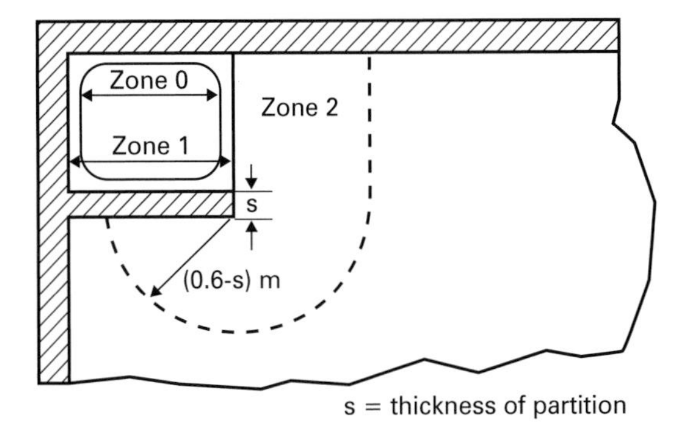

(d) Shower basin with permanent fixed partition (plan)

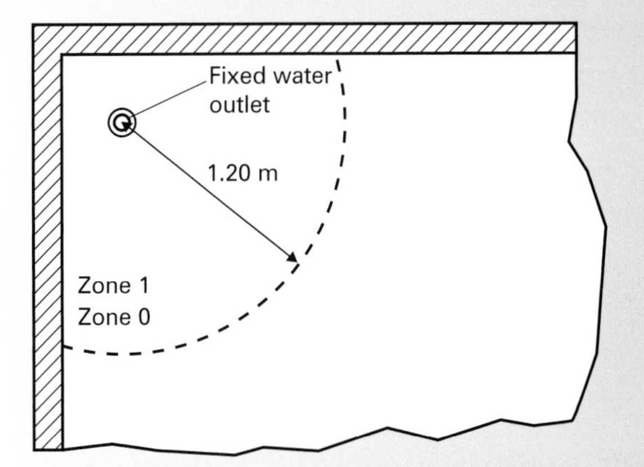

(e) Shower, without basin (plan)

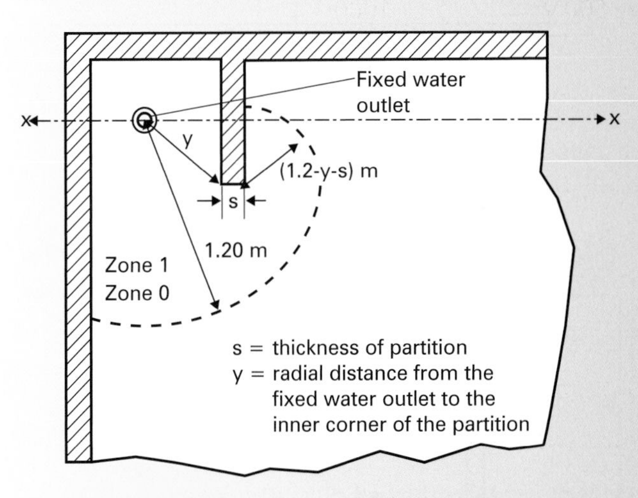

(f) Shower, without basin, but with permanent fixed partition (plan)

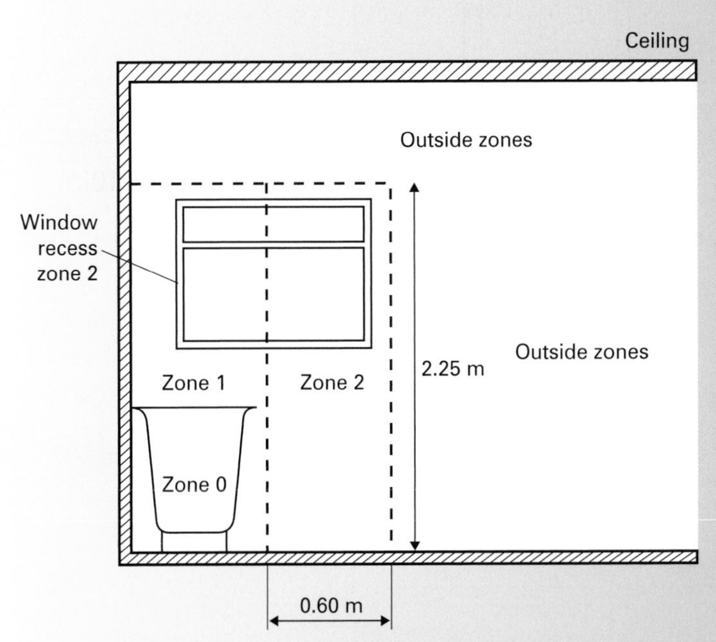

(g) Zone dimensions for a bath tub (elevation)

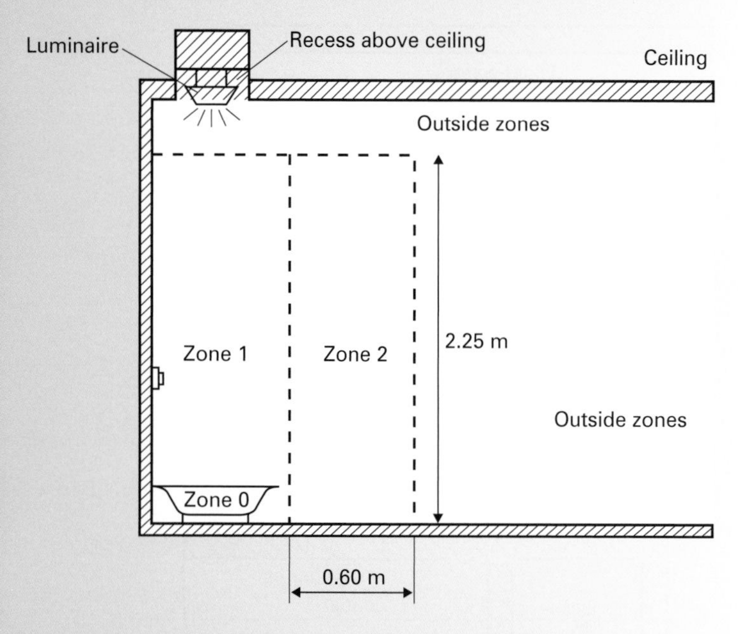

(g) Zone dimensions for a shower basin (elevation)

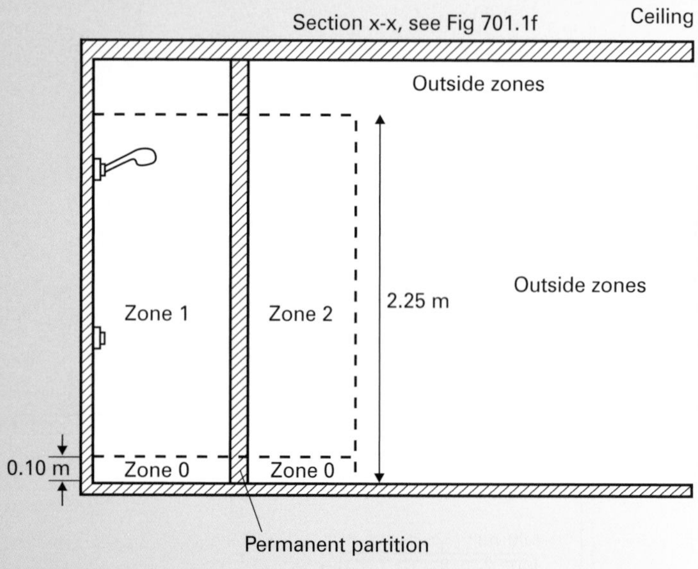

(g) Zone dimensions for a shower basin with no permanent fixed partition

Figure 5.1 Zone arrangements and dimensions (elevation)

Section 702: Swimming pools and other basins

This section sees the introduction of the words 'other basins' into the title. This reflects the fact that this section refers to the basins of swimming pools, paddling pools and fountains and, where designated as swimming pools, natural waters such as the sea and lakes.

The zone arrangements are slightly modified from the previous regulations and are shown in Figures 5.2 to 5.5.

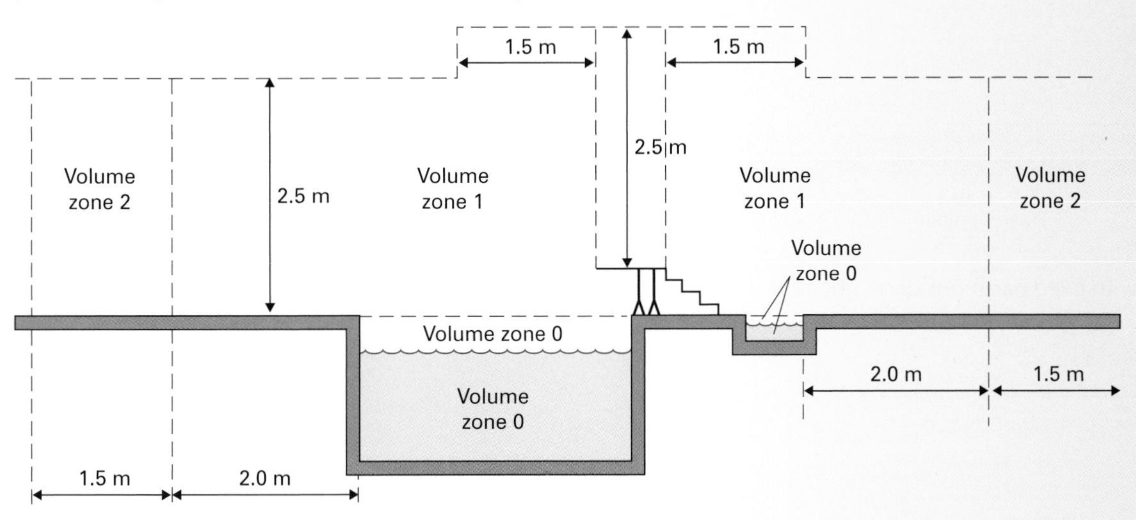

Figure 5.2 Zone dimensions for swimming pools and paddling pools

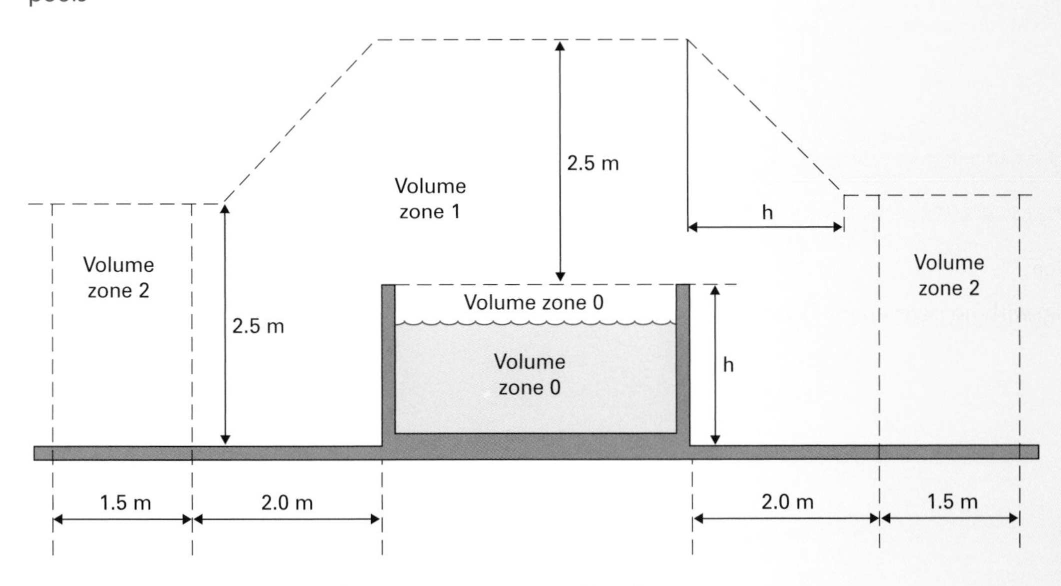

Figure 5.3 Zone dimensions for basin above ground level

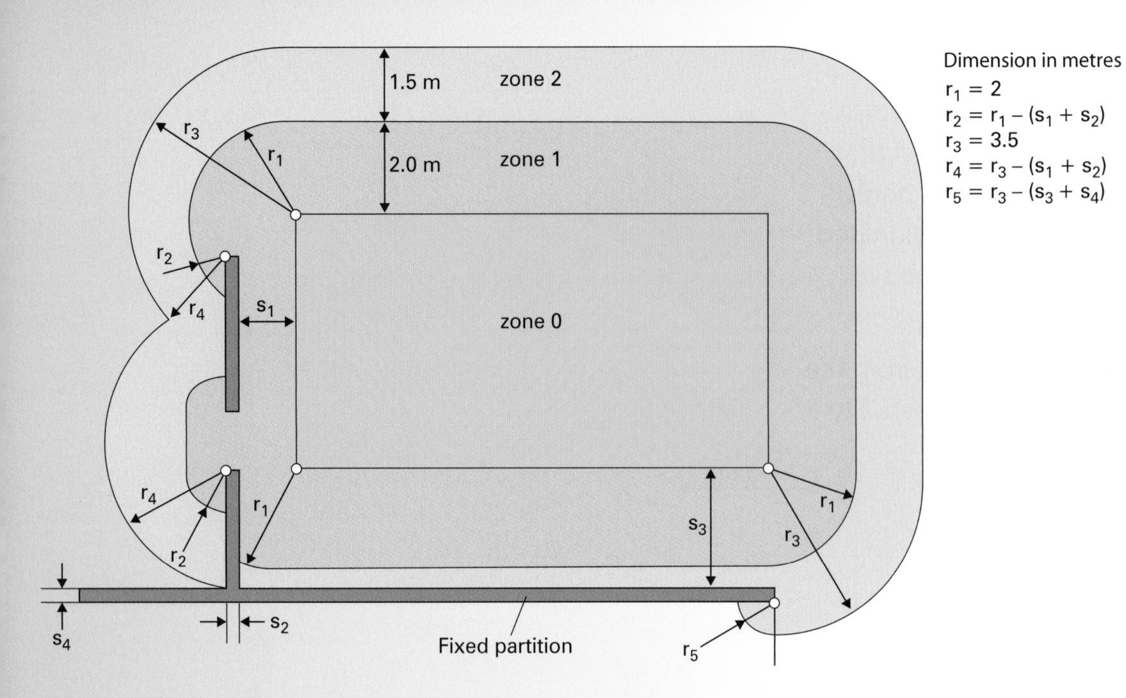

Figure 5.4 Zone dimensions with fixed partitions of height at least 2.5 m

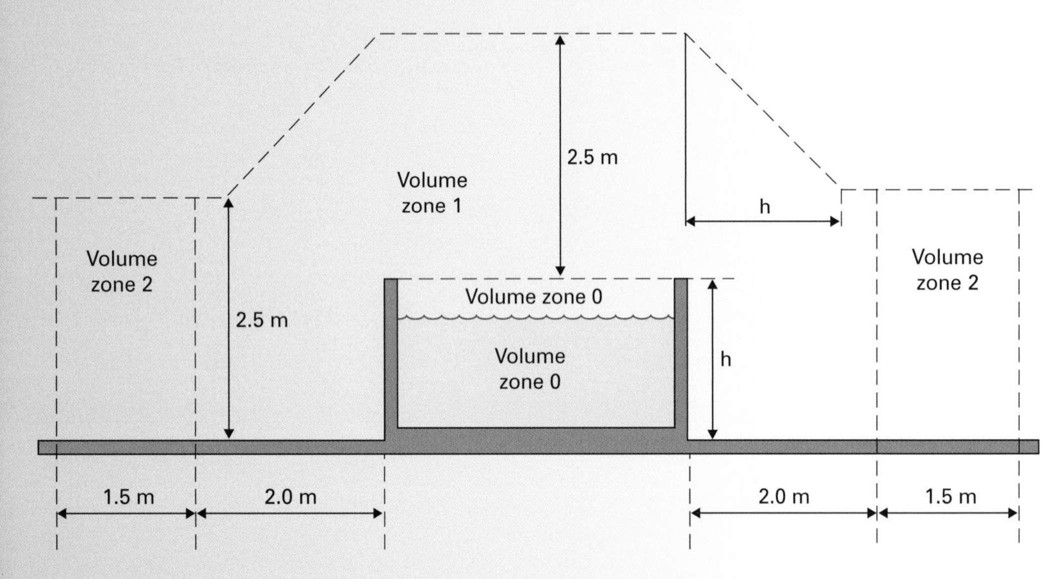

Figure 5.5 Zone dimensions for basin above ground level

Section 703: Rooms containing saunas

In this section, Regulation 703.411.3.3 now specifies that additional protection shall be provided for all circuits of the sauna by the use of appropriate RCDs.

However, RCD protection is not required for the sauna heater unless recommended by the manufacturer. The zone arrangements are also slightly modified from the previous regulations and shown in Figure 5.6.

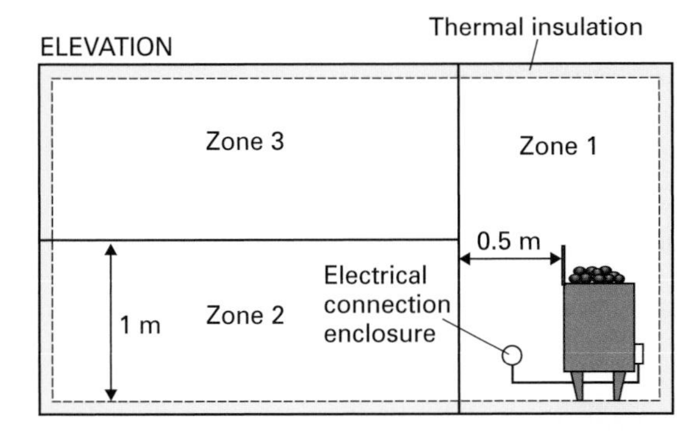

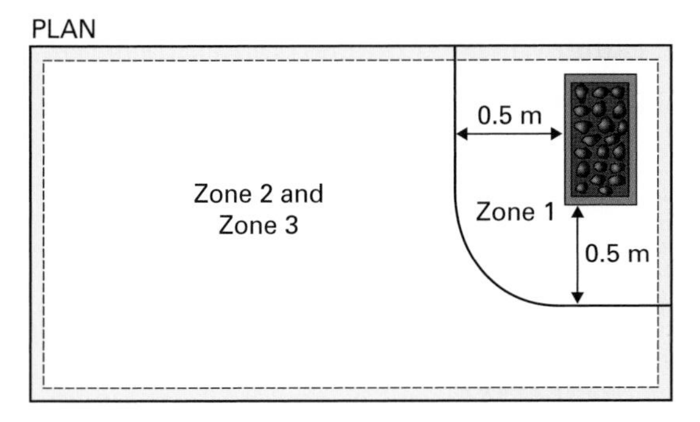

Figure 5.6 Zone dimensions for a sauna

Section 705: Agricultural and horticultural premises

Regulation 705.411.1 now specifies that irrespective of the earthing arrangements, the following disconnection device shall be provided:

- For final circuits supplying socket outlets not exceeding 32 A, an RCD not exceeding 30 mA.
- For final circuits supplying socket outlets exceeding 32 A, an RCD not exceeding 100 mA.
- In all other circuits, RCDs not exceeding 300 mA.

In livestock locations and where accessible to the livestock, all exposed conductive parts and extraneous conductive parts must be connected together with supplementary equipotential bonding. An example of this is shown in Figure 5.7.

Section 708 (Caravan and camping parks)

The note to regulation 708.521 states the preferred method of supply is by underground distribution cables buried to a depth of at least 0.6 m and away from caravan pitches or areas where tent pegs or similar could be present.

Overhead supplies are permitted, but must be no less than 6 m above ground where there is vehicular movement.

Relevant to the actual caravan pitch:

- The socket outlet and its enclosure that forms part of the caravan pitch shall be to IP44 and mounted at a height of between 0.5 and 1.5 m.
- At least one socket outlet shall be provided for each pitch.
- Socket outlet rating shall be no less than 16 A.
- Each socket outlet will be protected individually by overcurrent protection and a 30 mA RCD.

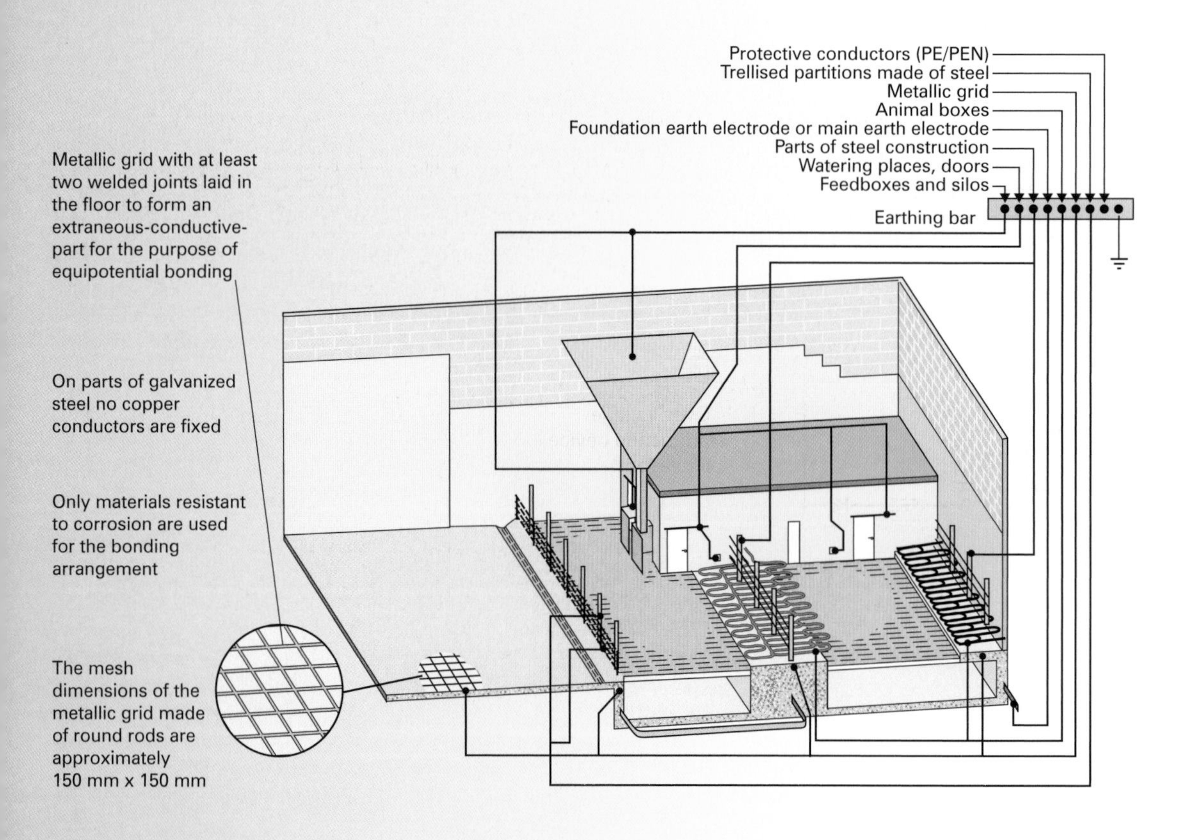

Figure 5.7 Supplementary equipotential bonding within a cattle shed

Section 711 (Exhibitions, shows and stands)

This section only applies to temporary exhibitions and shows. It does not apply to permanent shows or to the fixed supplies and installations of exhibition halls that house temporary exhibitions.

Care must be taken over the assessment of the event, in that it could range from an art exhibition through to an agricultural show. As a result:

- All supply cables shall be protected by a 30 mA RCD.
- The protective measures of obstacles and placing out of reach are not permitted.
- Cable size shall be a minimum of 1.5 mm^2.
- The installation shall be inspected and tested after each assembly.

Section 712 (Solar photovoltaic [PV] power systems)

Another new section to the Regulations, this system is based upon inter-connected PV cells that turn daylight into electrical energy. As these systems are normally installed outside on a roof, the electrical installation must be suitable for the environment and therefore have a suitable IP rating. The equipment must conform to the relevant BS for this type of system.

Equally, to allow maintenance of the PV converter, means of isolating it from both the a.c. and d.c. sides shall be provided.

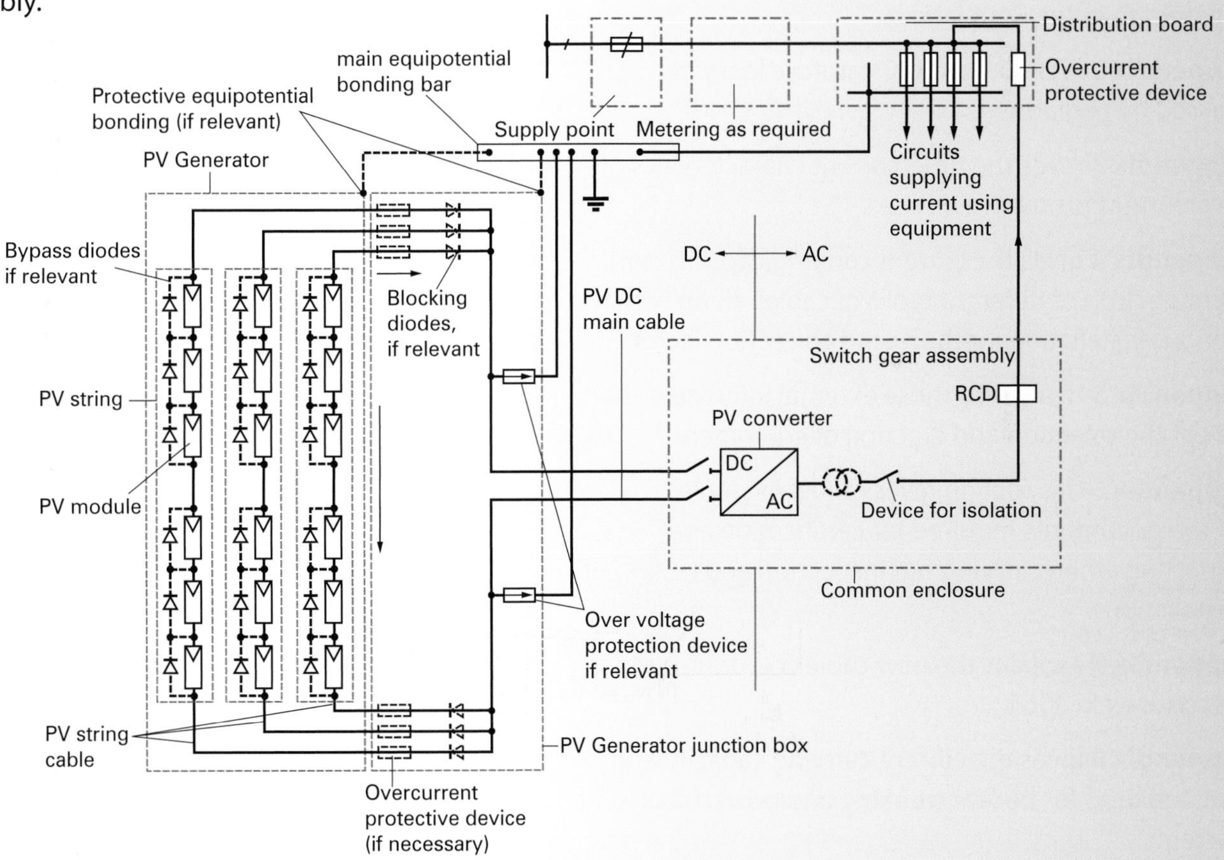

Figure 5.8 PV installation for one array

Section 753 (Floor and ceiling heating systems)

This section is also new to the 17th edition of the Regulations, with such systems most commonly installed in domestic premises. Consequently for protective measure ADS, 30 mA RCDs shall be used.

To avoid overheating of the system, then within the zone of its installation, the unit shall be limited to a temperature of 80°C. Additionally, to protect against burns to the skin, the temperature of the floor shall be limited to 35°C.

Appendices

BS 7671 2008 moves from 7 appendices to 15, namely:

Appendix 1 lists those British and European Standards referred to in the Regulations.

Appendix 2 lists relevant UK statutory legislation passed by parliament.

Appendix 3 gives the time/current characteristics of overcurrent protective devices.

Appendix 4 gives the current carrying capacity and voltage drop of different types of cables along with the various installation methods and correction factors.

Appendix 5 deals with those external influences that effect the selection and erection of equipment.

Appendix 6 this section gives model examples of the documents required for certification and reporting when carrying out inspection and testing of installations.

Appendix 7 explains the new cable colouring system introduced in 2004.

Appendix 8 gives the current carrying capacity and voltage drop for busbar trunking and powertrack systems.

Appendix 9 gives definitions of earthing arrangements relative to multiple source, d.c. and other systems.

Appendix 10 relates to the protection of conductors in parallel against overcurrent.

Appendix 11 relates to the effect of harmonic currents on balanced 3-phase systems.

Appendix 12 gives the maximum values of voltage drop in consumers installations.

Appendix 13 gives method for measuring the insulation resistance value of floors/walls to earth or to the protective conductor system.

Appendix 14 gives consideration to the increase of resistance in conductors with increase in temperature, relative to the measurement of earth fault loop impedance.

Appendix 15 gives details of ring and radial final circuit arrangements.

The Electricity at Work (EAW) Regulations 1989

The Electricity at Work Regulations were passed through Parliament in 1989 and came into force on 1 April 1990. Their purpose is to require precautions to be taken against the risk of death or personal injury from electricity in work activities.

The Regulations were made under the Health and Safety at Work Act 1974, which imposes duties on employers, employees and the self-employed. These Regulations (EAW) are more specific and concentrate on work activities at or near electrical equipment and make one person primarily responsible to ensure compliance in respect of systems, electrical equipment and conductors; this person is referred to as the 'duty holder'.

They were also designed to include all the systems etc. that BS 7671 does not cover, such as voltages above 1000 volts a.c.

There are 33 Regulations and 3 appendices within EAW 1989 but not all of them apply to all situations. This chapter contains an overall look at the relevant Regulations and what they mean to give you an appreciation of what they are about. For detailed information read *The memorandum of guidance on the Electricity at Work Regulations 1989.*

Regulation 1 only states that these Regulations came into force on 1 April 1990.

Regulation 2 contains definitions of what is meant by certain words or phrases.

Regulation 3 deals with duty holders and the requirements imposed on them by these Regulations. There are three categories of duty holder, namely:

- employers
- employees
- self-employed persons.

The duty holder is the person who has a duty to comply with these Regulations because they are relevant to circumstances within their control. Such a person must be competent.

Whenever a Regulation does not use the phrase 'so far as is reasonably practicable' this means it is an absolute duty; **it must be done regardless of cost or any other consideration**.

Whenever a Regulation does use the phrase **'as far as is reasonably practicable'** this means that the duty holder must assess the magnitude of the risks against the costs in terms of physical difficulty, time, trouble and expense involved in minimising that risk(s). The onus is on the duty holder to prove in a court of law, that he or she took all steps, as far as is reasonably practicable.

Regulation 4 has four parts. The first part deals with the construction of the electrical systems (all parts) in that the equipment should be suitable for its intended use so that it does not give rise to danger 'as far as is reasonably practicable'.

The second part deals with the maintenance of systems, whereby all systems should be maintained to prevent danger (this includes portable appliances) 'as far as is reasonably practicable'. Records of maintenance including test results should be kept.

The third part deals with ensuring safe work activities near a system including operation, use and maintenance 'as far as is reasonably practicable'. This could include non-electrical activities such as excavation near underground cables, and erecting scaffolding near overhead lines. Safe work activities include things such as:

- company health and safety policy
- permit to work systems
- clear communication
- use of competent people
- personnel attitudes.

The fourth part deals with provision of protective equipment such as insulated tools, test probes, insulating gloves, rubber mats, etc. which must be suitable for use, maintained in that condition and properly used. This is an absolute duty.

Regulation 5 has four parts. This Regulation states that 'no electrical equipment must be used where its strength and capability may be exceeded and give rise to danger', for example switchgear should be capable of handling fault currents as well as normal load currents, correct size cable etc.

An example is the installation of a socket near a back door of a house. There may be nothing plugged into it at present but it is reasonable to assume that it could be used in the future for plugging in an appliance to be used outside, such as a lawnmower, which would therefore necessitate protection by an RCD. This is an absolute duty.

Regulation 6 deals with the siting and/or selection of electrical equipment and whether it would be exposed to, or you could foreseeably predict it being exposed to, adverse or hazardous environments.

The following list provides what needs to be considered when siting and/or selecting electrical equipment, not just for the present situation but for what you could reasonably expect could be the situation in the future 'as far as is reasonably practicable':

- protection against mechanical damage
- effects of weather, natural hazards, temperature or pressure
- effects of wet, dirty, dusty or corrosive conditions
- any flammable or explosive substances, including dusts, vapours or gases.

Regulation 7 is concerned with conductors in a system and whether they present a danger to persons. All conductors must either be suitably covered with insulating material and protected or, if not insulated (such as overhead power lines), placed out of reach 'as far as is reasonably practicable'. The definition of 'placed out of reach' is in Part 2 of BS 7671.

Regulation 8 deals with the requirements for earthing or other such suitable precautions that are needed to reduce the risk of electric shock when a conductor (other than a circuit conductor) becomes live under fault conditions. This consists of such things as earthing the outer conductive parts of electrical equipment that can be touched and other conductive metalwork in the vicinity such as water and gas pipes. Other methods could be reduced voltage systems, double insulated equipment and RCDs. This is an **absolute** Regulation.

Regulation 9 is about maintaining the integrity of referenced conductors, which in simple terms means that the neutral conductor must not have a fuse or switch placed in it. The only exception is that a switch may be placed in the neutral conductor if that switch is interlocked to break the phase conductor(s) at the same time. This is an **absolute** Regulation.

Regulation 10 requires that all joints and connections in a system must be mechanically and electrically suitable for its use. For example things like taped joints on extension leads are not allowed. This is an **absolute** Regulation.

Regulation 11 states that every part of a system must be protected from excess current that may give rise to danger. This means that suitably rated fuses, circuit breakers etc. must be installed so that in a fault situation they will interrupt the supply and prevent a dangerous situation happening. This is an **absolute** Regulation.

Regulation 12 deals with the need for switching off and isolating electrical equipment and, where appropriate, identifying circuits. Isolation means cutting off from every source of electrical energy in such a way that it cannot be switched back on accidentally, in other words a means of 'locking off' the switch securely. This is an **absolute** Regulation.

Regulation 13 refers to the precautions required when work is taking place at or near electrical equipment which is to be worked on, whether it be electrical or non-electrical work that is taking place. The electrical equipment must be isolated, locked off and tested for absence of live parts before the work takes place and must remain so until all persons have completed their work. A safe isolation procedure or written Permit to Work scheme should be used. This is an **absolute** Regulation.

Regulation 14 refers to the precautions needed when working on or near live conductors. An absolute duty is imposed that the conductors must be isolated unless certain conditions are met. Precautions considered appropriate are:

- properly trained and competent staff
- provision of adequate information regarding nature of the work and system
- use of appropriate insulated tools, equipment, instruments, test probes and protective clothing
- use of insulated barriers
- accompaniment of another person
- effective control of the work area.

This is an **absolute** Regulation.

Regulation 15 concerns itself with matters relating to work being carried out at or near electrical equipment whereby to prevent danger, adequate working space, adequate means of access and adequate lighting must be provided. This is an **absolute** Regulation.

Regulation 16 deals with the competency of persons working on electrical equipment to prevent danger or injury. To comply with this regulation a person should conform to the following or be under such a degree of supervision as appropriate given the type of work to be carried out:

- adequate understanding and practical experience of the system to be worked on
- an understanding of the hazards that may arise
- the ability to recognise at all times whether it is safe to continue.

The Regulation tries to ensure that no-one places themselves or anyone else at risk due to their lack of technical knowledge or practical experience.

This is an **absolute** Regulation.

Regulations 17 to 28 apply to mines and quarries only.

Regulation 29 is what is known as the 'Defence' Regulation. If an offence is committed by the duty holder under these Regulations (the absolute ones) and criminal proceedings are brought by the HSE (Health and Safety Executive), then if the duty holder can prove that they took all reasonable steps and exercised due diligence to avoid committing that offence, they will not be found guilty.

Regulation 30. A duty holder can apply to the HSE for exemption from these Regulations for the items listed:

- any person
- any premises
- any electrical equipment
- any electrical system
- any electrical process
- any activity.

Exemptions will only be granted by the HSE provided they do not prejudice the health and safety of any persons.

Regulation 31 deals with work activities and premises outside of Great Britain. If the activity or premises is covered by sections 1 to 59 and sections 80 to 82 of the Health and Safety at Work Act 1974 then these regulations apply.

Regulation 32 details what these regulations do not apply to:

- sea going ships
- aircraft or hovercraft moving under their own power.

Regulation 33 deals with changes and modifications to these regulations since they were brought in.

Appendix 1 lists the HSE guidance publications available for help in understanding and applying the regulations.

Appendix 2 lists various other codes of practice and British Standards that could help in the understanding and application of these regulations.

Appendix 3 deals with the legislation concerning working space and access regulations.

Installations in potentially explosive areas

Within hazardous areas there exists the risk of explosions and/or fires occurring due to electrical equipment 'igniting' the gas, dust or flammable liquid.

These areas are not included in BS 7671 but are instead covered by IEC Standard BS EN 60079 as follows:

- BS EN 60079 Part 10 – Classification of hazardous areas
- BS EN 60079 Part 14 – Electrical apparatus for explosive gas atmospheres
- BS EN 60079 Part 17 – Inspection/maintenance of electrical installations in hazardous areas.

The BS EN 60079 has been in place since 1988, replacing the old BS 5345. However, many installations obviously still exist that were completed in accordance with BS 5345 and new European Directives (ATEX) address safety where there is a danger from potentially explosive atmospheres.

Other statutory regulations such as the Petroleum Regulation Acts 1928 and 1936 and local licensing laws govern storage of petroleum.

Definition of hazardous areas

A hazardous area can be defined as: 'an area in which explosive gas/air mixtures are, or may be expected to be, present in quantities such as to require special precautions for the construction and use of electrical apparatus'.

Zoning

Hazardous areas are defined in the Dangerous Substances and Explosive Atmospheres Regulations 2002 (DSEAR) as 'any place in which an explosive atmosphere may occur in quantities such as to require special precautions to protect the safety of workers'. In this context, 'special precautions' is best taken as relating to the construction, installation and use of apparatus, as given in BS EN 60079 -10.

Area classification is a method of analysing and classifying the environment where explosive gas atmospheres may occur. The main purpose is to facilitate the proper selection and installation of apparatus to be used safely in that environment, taking into account the properties of the flammable materials that will be present. DSEAR specifically extends the original scope of this analysis, to take into account non-electrical sources of ignition, and mobile equipment that creates an ignition risk.

Hazardous areas are classified into zones based on an assessment of the frequency of the occurrence and duration of an explosive gas atmosphere, as follows:

- Zone 0: An area in which an explosive gas atmosphere is present continuously or for long periods
- Zone 1: An area in which an explosive gas atmosphere is likely to occur in normal operation
- Zone 2: An area in which an explosive gas atmosphere is not likely to occur in normal operation and, if it occurs, will only exist for a short time.

Various sources have tried to place time limits on to these zones, but none have been officially adopted. The most common values used are:

- Zone 0: Explosive atmosphere for more than 1000 h/yr
- Zone 1: Explosive atmosphere for more than 10, but less than 1000 h/yr
- Zone 2: Explosive atmosphere for less than 10 h/yr, but still sufficiently likely as to require controls over ignition sources.

Where people wish to quantify the zone definitions, these values are the most appropriate, but for the majority of situations a purely qualitative approach is adequate.

When the hazardous areas of a plant have been classified, the remainder will be defined as non-hazardous, sometimes referred to as 'safe areas'.

Definitions

- **Explosive limits** are the upper and lower percentages of a gas in a given volume of gas/air mixture at normal atmospheric temperature and pressure that will burn if ignited.
- **Lower explosive limit (LEL)** is the concentration below which the gas atmosphere is not explosive.
- **Upper explosive limit (UEL)** is the concentration of gas above which the gas atmosphere is not explosive.
- **Ignition energy** is the spark energy that will ignite the most easily ignited gas/air mixture of the test gas at atmospheric pressure. Hydrogen ignites very

easily, whereas Butane or Methane require about 10 times the energy.

- **Flash point** is the minimum temperature at which a material gives off sufficient vapour to form an explosive atmosphere.
- **Ignition temperature or auto ignition temperature** of a material is the minimum temperature at which the material will ignite and sustain combustion when mixed with air at normal pressure, without the ignition being caused by any spark or flame. This is not the same as flash point, so don't confuse them!

Selection of equipment

DSEAR sets out the link between a zone and the equipment that may be installed in that zone. This applies to new or newly modified installations. The equipment categories are defined by the ATEX equipment directive, set out in UK law as the Equipment and Protective Systems for Use in Potentially Explosive Atmospheres Regulations 1996.

Standards set out different protection concepts, with further subdivisions for some types of equipment according to gas group and temperature classification. Most of the electrical standards have been developed over many years and are now set at international level, while standards for non-electrical equipment are only just becoming available from CEN.

The DSEAR ACOP describes the provisions concerning existing equipment.

There are different technical means (protection concepts) of building equipment to the different categories. These, the standard current and the letter giving the type of protection are listed below.

Correct selection of electrical equipment for hazardous areas requires the following information:

- temperature class or ignition temperature of the gas or vapour involved according to Table 5.1 overleaf.
- classification of the hazardous area, as in zones shown in Table 5.2 overleaf.

If several different flammable materials may be present within a particular area, the material that gives the highest classification dictates the overall area classification. The IP code considers specifically the issue of hydrogen containing process streams as commonly found on refinery plants.

Consideration should be shown for flammable material that may be generated due to interaction between chemical species.

Ignition sources – identification and control

Ignition sources may be:

- flames
- direct fired space and process heating
- use of cigarettes/matches etc.
- cutting and welding flames
- hot surfaces
- heated process vessels such as dryers and furnaces
- hot process vessels
- space heating equipment
- mechanical machinery
- electrical equipment and lights
- spontaneous heating
- friction heating or sparks
- impact sparks
- sparks from electrical equipment
- stray currents from electrical equipment
- electrostatic discharge sparks
- lightning strikes
- electromagnetic radiation of different wavelengths
- vehicles (unless specially designed or modified are likely to contain a range of potential ignition sources).

Temperature classification	Maximum surface temperature, °C	Ignition temperature of gas or vapour °C
T1	450	>450
T2	300	>300
T3	200	>200
T4	135	>135
T5	100	>100
T6	85	>85

Table 5.1 Ignition temperatures

Zone 0	Zone 1	Zone 2
Category 1	Category 2	Category 3
'ia' intrinsically safe EN 50020, 2002	'd' flameproof enclosure EN 50018 2000	Electrical Type 'n' EN 50021 1999 Non electrical EN 13463-1, 2001
Ex s – Special protection if specially certified for Zone 0	'p' pressurised EN 50016 2002	
	'q' powder filling EN 50017, 1998	
	'o' oil immersion EN 50017, 1998-	
	'e' increased safety EN 50019, 2000 'ib' Intrinsic safety EN 50020, 2002	
	'm' encapsulation EN 50028, 1987	
	's' special protection	

Table 5.2 Classification of hazardous areas

Sources of ignition should be effectively controlled in all hazardous areas by a combination of design measures, and systems of work:

- using electrical equipment and instrumentation classified for the zone in which it is located. New mechanical equipment will need to be selected in the same way (see above)
- earthing of all plant/equipment (see Technical Measures Document on Earthing)
- elimination of surfaces above auto-ignition temperatures of flammable materials being handled/stored (see above)
- provision of lightning protection
- correct selection of vehicles/internal combustion engines that have to work in the zoned areas (see Technical Measures Document on Permit to Work Systems)
- correct selection of equipment to avoid high intensity electromagnetic radiation sources, e.g. limitations on the power input to fibre optic systems, avoidance of high intensity lasers or sources of infrared radiation
- prohibition of smoking/use of matches/lighters
- controls over the use of normal vehicles
- controls over activities that create intermittent hazardous areas, e.g. tanker loading/unloading
- control of maintenance activities that may cause sparks/hot surfaces/naked flames through a Permit to Work System.

Petrol filling stations

The primary legislation controlling the storage and use of petrol is the Petroleum (Consolidation) Act 1928 (PCA). This requires anyone who keeps petrol to obtain a licence from the local Petroleum Licensing Authority (PLA). The licence may be, and usually is, issued subject to a number of licence conditions. The PLA set the licence conditions, but they must be related to the safe keeping of petrol.

The Local Authority Co-ordinating Body on Food and Trading Standards (LACOTS) has issued a set of standard licence conditions which most, if not all, PLAs apply to their sites.

Installations within petrol filling stations are effectively also covered by BS EN 60079 Parts 10, 14 and 17.

Additionally there is industry-developed guidance for this sector in the form of the electrical section of IP/APEA's Guidance for the Design, Construction, Modification and Maintenance of Petrol Filling Stations (Institute of Petroleum and the Association for Petroleum and Explosives Administration). This guidance replaced most of HS(G)41 and was published in 1999 by IP/APEA with input from HSE.

FAQ

Q Why do I need BS 7671 to make lights and sockets work? All you need is common sense!

A You are perfectly correct, however evidence shows that common sense and the need to be commercially competitive do not always go hand in hand. BS 7671 is a collection of common-sense rules to give guidance on how to make installations work under fault conditions. Anybody can install additional lights and sockets, they do so every weekend buying their materials at DIY outlets, but because they have no knowledge of the Regulations they believe that just because the lights work when they turn them on that they have done a good job. What they may not know is that the circuit can function even under fault conditions. It's under such fault conditions that people's lives and property are at risk.

Q Why do we need additional Regulations for 'special locations'?

A This section of the Regulations recognises that in certain locations there are additional risks, and more stringent safeguards need to be applied to ensure that people, property and livestock are not put at risk. An example of one such special location would be the bathroom; here the presence of water/moisture on the skin has the effect of reducing the body's natural electrical resistance, increasing the risk of a fatality should an electric shock situation arise. The additional Regulations required for such areas are intended to eliminate this risk.

Knowledge check

1. Describe the limitations found in Zone 1 in a room containing a bath or shower.
2. What do the abbreviations PELV and SELV represent?
3. List the locations which require supplementary equipotential bonding.
4. List the rules which apply to all zones in relation to the wiring systems used.
5. List the types of fixed current using equipment that may be installed in Zone 2 if it is suitable for Zone 2.
6. List the current using equipment which can be installed in Zone 1 if it is suitable for Zone 1.
7. In Zone 0 what type of switchgear or accessories may be fitted?
8. In Zone 1 what is the external classification of equipment that can be found in that area?
9. Define the terms, explosive limit, ignition energy and flash point.

Example

A small business premises with 12 offices is supplied with a 400/230 volt, three-phase four-wire supply. It has the following installed loads:

- 15 × fluorescent lighting points, each rated at 60 W
- 4 × standard ring final circuits supplying 13 A socket outlets
- 2 × 20 A radial circuits supplying 13 A socket outlets
- 1 × 4 kW shower (instantaneous type)
- 1 × 3 kW immersion heater with thermostatic control
- 1 × 4 kW cooker
- 1 × 3 kW cooker

Apply diversity as required by the *On-site Guide* Table 1B and 'spread' the loads over three-phase supply to produce the most effective load balanced situation?

$$\frac{15 \times 60 \times 1.8230}{230}$$

Answer

Lighting

$$\frac{15 \times 60 \times 1.8}{230} = \frac{1620\,\text{W}}{230\,\text{V}} = 7.04\,\text{A}$$

90% = 6.34 A (after diversity)

Ring circuits

= 30 A + 3 circuits at 15 A = 75 A (after diversity)

Radial circuit

= 20 A + 15 A = 35 A

4kW shower

$$\frac{4000\,\text{W}}{230\,\text{V}} = 17.39\,\text{A}$$

3kW immersion heater

$$\frac{3000\,\text{W}}{230\,\text{V}} = 13.04\text{A}$$

Cookers

$$100\% \text{ of } \frac{4000\,\text{W}}{230\,\text{V}} = 17.39\,\text{A}$$

$$\text{and } 80\% \text{ of } \frac{3000\,\text{W}}{230\,\text{V}} = 10.43\text{A}$$

	Red phase	Yellow phase	Blue phase
Ring circuits	30 A	15 A	45 A
	15 A		
Radial circuits		20 A	15 A
Shower	17.39 A		
Heater (immersion)			13.04 A
4 kW cooker		17.39 A	
3 kW cooker			10.43 A
Lighting			6.34 A
Totals	62.39 A	52.39 A	59.81 A

Suggested load balancing (other combinations could be acceptable).

The method used in this calculation is slightly different to the method used in the example question on page 210. There is no hard and fast rule as to calculating diversity for calculating maximum demand. The recommendation is that qualified electrical engineers may use other methods of determining maximum demand.

Glossary

a.c. alternating current.

adjacent in trigonometry, the side of a ttriangle that is adjacent to that angle under consideration and to the right angle.

airway breathing passage which needs to be kept clear in an unconscious casualty to prevent suffocation or choking.

bill of quantities a list of all the materials required, their specification and the quantities needed.

breach of contract when one party does not fulfil the contract terms.

catenary wire steel support wire, used to span the gap between two buildings, and from which electrical cables are then suspended.

chest compressions used in cardiopulmonary resuscitation (CPR) when trying to resuscitate a casualty whose heart has stopped beating.

combustion rapid chemical combination of two or more substances accompanied by production of heat and light; commonly known as burning.

completion order record that all material on a purchase order has been delivered.

conductor any material allowing electrical charges to flow easily: includes metal pipework, metal structures of buildings, salt water or ionised gases.

conduit mild steel or PVC tubing through which electrical cabling can be run.

contract legally binding agreement between two or more parties.

control panel programmable 'brains' of any system to which all parts of the system are connected.

cosine in trigonometry, function that in a right-angled triangle is the ratio of the length of the adjacent side to that of the hypotenuse.

coulomb SI unit of electrical charge equal to the quantity of electricity transported in one second by a current of one ampere; symbol: C. Named after Charles-Augustin de Coulomb, a French military engineer.

critical path analysis (CPA) mathematical network analysis technique of planning complex working procedures with reference to the critical path of each alternative system.

critical path network (CPN) diagram that represents tasks, how long they take and how they relate to one another, used to calculate the time a project will take.

current flow of free electrons in a conductor; measured in amperes.

customer anyone to whom a service is given, either directly or indirectly.

d.c. direct current.

delta connection triangular arrangement of electrical three-phase windings.

dermatitis inflammation of the skin normally caused by the hands making contact with irritating substances.

directly proportional increase in one property which results in an increase in another: e.g. the harder a coin is flicked along a smooth flat surface, the further it will travel. (See also **indirectly proportional**.)

disability a person has a disability if they have a physical or mental impairment that has a substantial and long-term affect on their ability to carry out normal everyday activities.

discharge consent written authorisation that must be obtained from the Environment Regulator before discharging any sewage, effluent or contaminated surface water to surface waters or groundwater.

efficiency ratio of a system's output against the input, normally expressed as a percentage. In machines, degree to which friction and other factors reduce actual work output from the theoretical maximum of 100%.

electromagnetic induction production of an e.m.f. in a conductor, by moving it through a magnetic field across the lines of flux which will, in a closed circuit, cause an electric current to flow in the conductor.

electromotive force (e.m.f.) total force from a source such as a battery that causes the motion of electrons due to potential difference between two points.

electrons particles in an atom that circle in orbit round the nucleus, said to possess a negative charge.

electronic variable speed drive device which provides variable voltage and frequency supply enabling motor speed and torque (current) to be precisely controlled. Used throughout industry for process automation, air handling, variable speed conveyor systems.

explosive atmosphere area in which explosive dust/gas/air mixtures are, or may be expected to be, present in such quantities as to require special precautions for construction, installation and use of electrical apparatus.

Fleming's left-hand (motor) rule used to explain relationship between direction of motion, magnetic field and current flow in electromagnetism. The hand is held with the thumb, first and second fingers at right angles to each other, indicating direction of motion, field and electric current respectively. Named after Sir John Ambrose Fleming, who devised the rule.

flux density the strength of a magnetic field at any point, calculated by counting the lines of magnetic flux at that point.

friction force that opposes motion, occurring when two substances rub together.

fused spur spur connected to a ring final circuit through a fused connection unit, the fuse incorporated being related to the current-carrying capacity of cable used for the spur, but not exceeding 13 A. (See also **spur**.)

Genuine Occupational Requirements where an employer can demonstrate that there is a genuine identified need for somene of specific race, gender, etc., to the exclusion of other races, genders, etc.

Gantt chart type of bar chart showing activities against time taken, used to monitor progress of a contract. Named after its inventor, Robert Gantt.

half saddle see strap saddle

hazard anything that can cause harm.

hygroscopic the ability to absorb water.

hypotenuse side in a right-angled triangle opposite the right angle.

improvement notice notice issued by a Health and Safety Executive (HSE) inspector detailing any health and safety problems, together with remedial action to be taken within a given time period.

indirectly proportional increase in one property which results in decrease of another: if a coin is flicked along a smooth angled surface, the greater the angle of the surface, the less distance the coin will travel.

ions electrically charged atoms formed by the loss or gain of one or more electrons, leaving them with a net positive or negative charge.

isolator mechanical switching device which can cut off supply from all, or a section of, an installation by separating it from every source of electrical energy.

kinetic energy energy due to movement.

Kirchhoff's Law the sum of the voltage drops around any closed loop in the network must equal zero.

mass amount of matter contained in an object regardless of volume or any forces acting on it. Not to be confused with weight.

mechanical advantage relationship between effort needed to lift something (input) and load (output). A machine which puts out more force than is put into it, is said to give good mechanical advantage.

method statement document detailing good practices that could be established in order to avoid or minimise statutory nuisance, particularly to neighbours.

mouth-to-mouth ventilation part of the approved resuscitation procedure whereby the lungs of an unconscious person are inflated by another person breathing into the casualty's mouth.

moveable pulley pulley (a rope, chain or belt wrapped round a wheel to lift a heavy object) that moves with the load, using less effort than the weight of the load.

mutagens agents, such as radiation or chemical substances that cause genetic mutation.

networking interconnection of computers, using structured cable installation, enabling people to communicate via email, access data directly from other computers and the Internet, and manage many aspects of an organisation from their desks.

neutrons sub-atomic particles within the nucleus of an atom, which have no charge and are said to be electrically neutral.

newton metric unit of force which, if applied for one second, will cause a one-kilogram object starting from rest to reach a speed of one m/s. Named after the English mathematician and physicist Sir Isaac Newton.

noggin piece of wood fixed between wall or ceiling frame sections to allow back boxes or ceiling roses and suchlike to be attached.

non-fused spur spur usually directly connected to a circuit at the terminal of socket outlets. (See also **spur**.)

nucleus centre of an atom, made up of protons and neutrons.

Ohm's Law principle establishing that current flowing in a circuit is directly proportional to voltage and indirectly proportional to resistance in a constant temperature. Named after the German physicist G. S. Ohm.

olive metal ring used in MI cable glands that is tightened under a threaded nut to form a seal on to the cable's metal sheath.

opposite in trigonometry, the side of a triangle opposite the angle being considered.

oxidation chemical changes brought about by exposure to oxygen. Burning: rapid oxidation; rusting: slow oxidation.

P.A.T. (portable appliance test) test detecting any adverse condition making equipment electrically unsafe, such as damaged flexes or exposed live parts.

pattress plate fixed through a structure; normally a plastic box fixed to the surface of a wall or ceiling to receive an electric light switch, socket outlet or ceiling rose.

peak value maximum value of a.c. or d.c. waveform. In a.c. this can refer to the negative half cycle.

Permit to Work official document giving authorisation to work within defined circumstances.

phase voltage voltage measured across a single component in a three-phase source or load.

potential difference the amount of energy used up by one coulomb in its passage between any two points in a circuit.

potential energy the stored energy of an object.

power factor number less than 1.0, used to represent relationship between the apparent and true power of a circuit.

PPE (personal protective equipment) all equipment designed to be worn or held to protect against risk to health or safety. Includes most types of eye, hand, foot and head protection.

prohibition notice notice issued by a Health and Safety Executive (HSE) inspector, immediately prohibiting any activity likely to cause serious personal injury and not allowing it to be resumed until remedial action has been taken.

protons particles within the nucleus of an atom said to possess a positive charge.

pulley See **moveable pulley**, and **single fixed pulley**.

RCD (residual current device) device which monitors the electrical current flowing both in the phase and neutral conductors.

resistance opposition to the flow of electrons (current).

resistivity nature of a circuit containing pure resistance in which all power is dissipated by resistor(s). Voltage and current are in phase with each other.

risk the chance of being harmed by a hazard.

Safety Data Sheet sheet giving information on how chemicals should be handled, stored and disposed of. Must accompany any potentially hazardous material supplied.

scaled drawings drawings used to show the size of everything with a fixed ratio (scale) to the size of the actual object.

schematic diagram symbolic and simplified diagram or other representation.

sine in trigonometry, function that in a right-angled triangle is the ratio of the length of the opposite side to that of the hypotenuse.

sine wave waveform representation of how alternating current (a.c.) varies with time (current or voltage varies with the sine of the elapsed time).

single fixed pulley pulley (a rope, chain or belt wrapped round a wheel to lift a heavy object) using more effort than the weight of the load to lift the load from the ground.

solenoid long, hollow cylinder round which is wound a uniform coil of wire. When a current is sent through the wire, a magnetic field is created inside the cylinder.

special waste waste material that has hazardous properties, or that may contain residues of hazardous or dangerous substances.

spur radial branch taken from a ring final circuit, perhaps to feed a new socket outlet.

star connection Y-shaped arrangement of three-phase electrical windings.

star–delta system of connection sometimes used in induction motors, whereby a star connection switches to a delta connection.

statutory requirement that is binding by law.

strap saddle fixing device sometimes used for securing conduit to cable tray or steel framework; also known as a half saddle.

tangent in trigonometry, function that in a right-angled triangle is the ratio of the length of the opposite side to that of the adjacent side.

thesaurus type of dictionary that gives word synonyms.

Trade Effluent Agreement agreement that must be entered into with the Statutory Sewerage Undertaker before discharging any effluents into a public sewer.

transferable skills skills that you gain which can be applied to different working areas and thus make you an attractive proposition to employers.

transposition method using principles of mathematics to rearrange a formula or equation in order to find an unknown quantity.

trigonometry mathematical study of how sides and angles of a triangle are related to each other.

variation order document raised when work done varies from the original agreed in the contract and listed in the job sheet.

voltage difference in electrical charge between two points in a circuit, expressed in volts; force available to push current round a circuit.

waste transfer notes signed records that must be kept of all waste material received or transferred, according to the Controlled Waste Regulations 1998.

weight force on a body due to attraction of gravity.

work plan strategy applied to a whole or part of a project to ensure installation is carried out safely and efficiently.

Index